ADVANCES IN PROTEIN CHEMISTRY

Volume 62

Unfolded Proteins

John T. Edsall
(1902-2002)

ADVANCES IN PROTEIN CHEMISTRY

EDITED BY

FREDERIC M. RICHARDS
Department of Molecular Biophysics
and Biochemistry
Yale University
New Haven, Connecticut

DAVID S. EISENBERG
Department of Chemistry and Biochemistry
University of California, Los Angeles
Los Angeles, California

JOHN KURIYAN
Department of Molecular Biophysics
Howard Hughes Medical Institute
Rockefeller University
1230 York Avenue
New York, NY 10021

VOLUME 62

Unfolded Proteins

EDITED BY

GEORGE D. ROSE
Department of Biophysics
Johns Hopkins University
Baltimore, Maryland

ACADEMIC PRESS

An imprint of Elsevier Science

Amsterdam   Boston   London   New York   Oxford   Paris
San Diego   San Francisco   Singapore   Sydney   Tokyo

Academic Press
An imprint of Elsevier Science.
525 B Street, Suite 1900, San Diego, California 92101-4495, USA
http://www.academicpress.com

Academic Press
84 Theoblad's Road, London WC1X 8RR, UK
http://www.academicpress.com

International Standard Book Number: 0-12-034262-6

PRINTED IN THE UNITED STATES OF AMERICA
02 03 04 05 06 07   MM   9 8 7 6 5 4 3 2 1

CONTENTS

The Expanded Denatured State: An Ensemble of Conformations Trapped in a Locally Encoded Topological Space

David Shortle

Identification and Functions of Usefully Disordered Proteins

A. Keith Dunker, Celeste J. Brown, and Zoran Obradovic

Unfolded Proteins Studied by Raman Optical Activity

L. D. BARRON, E. W. BLANCH, AND L. HECHT

What Fluorescence Correlation Spectroscopy Can Tell Us about Unfolded Proteins

CARL FRIEDEN, KRISHNANANDA CHATTOPADHYAY, AND ELLIOT L. ELSON

Unfolded Peptides and Proteins Studied with Infrared Absorption and Vibrational Circular Dichroism Spectra

TIMOTHY A. KEIDERLING AND QI XU

Is Polyproline II a Major Backbone Conformation in Unfolded Proteins?

ZHENGSHUANG SHI, ROBERT W. WOODY, AND NEVILLE R. KALLENBACH

Toward a Taxonomy of the Denatured State: Small Angle Scattering Studies of Unfolded Proteins

IAN S. MILLETT, SEBASTIAN DONIACH, AND KEVIN W. PLAXCO

Determinants of the Polyproline II Helix from Modeling Studies

TREVOR P. CREAMER AND MARGARET N. CAMPBELL

Hydration Theory for Molecular Biophysics

MICHAEL E. PAULAITIS AND LAWRENCE R. PRATT

Insights into the Structure and Dynamics of Unfolded Proteins from Nuclear Magnetic Resonance

H. JANE DYSON AND PETER E. WRIGHT

Unfolded State of Peptides

XAVIER DAURA, ALICE GLÄTTLI, PETER GEE, CHRISTINE PETER, AND WILFRED F. VAN GUNSTEREN

A New Perspective on Unfolded Proteins

ROBERT L. BALDWIN

JOHN T. EDSALL AND ADVANCES IN PROTEIN CHEMISTRY

Today the advances in almost any aspect of science are proceeding at an ever increasing rate. There is a strong tendency for each individual to feel that his/her area is indeed the most exciting. However, these same individuals do not read and think any faster now than their forebears did, with the result that science on a daily basis is breaking into smaller and smaller segments of the overall pie. Each small segment usually develops its own vocabulary which then inhibits easy communication between segments, even in very similar areas. This can easily be seen in *Nature Magazine* which, in just a few years, has fragmented itself into seven or so different journals for research articles (and five separate volumes for reviews). The number seems likely to increase markedly in future years.

Because of this behavior at the level of the primary literature, the so-called secondary literature (i.e., reviews) has become increasingly important. There are two limiting major types of reviews: (1) A list of what has been published in several different related segments over some usually short time interval with minimal attention to tying it all together. Such reviews can be useful, but will still be uninterpretable to those at any distance from the general field. (2) Reviews, usually triggered by some recent publication, but based on fundamental points, drawing on work from diverse areas, selecting literature from studies that may go far back in time, and resulting in a synthesis that may change the direction and attitude of an entire field. Both types of reviews serve useful functions, but the most sought after, and the hardest to find, are the latter.

The rising importance of the whole area of reviews is seen in the output from commercial publishers and many scientific societies. This has generated a competition for the best formats for conveying the desired information. Developments in electronic communication have not yet solved the basic problems or eliminated the need for paper, but they certainly have altered the playing field. The younger members of the scientific community, say those under 70, tend to think of the current explosion in reviewing as a recent phenomenon, but this seems not to be so.

The first volume of *Advances in Protein Chemistry* appeared in 1944, near the end of the Second World War. The papers it contained were written during the height of the battles, a *tour de force* considering the difficulty of communications at that time. The following quotations are taken from the Preface of Volume 1:

Paragraph 1: "In the last generation, protein chemistry, which was once a relatively narrow branch of organic and biological chemistry, has spread out into the most varied fields of physics, chemistry, and biology. Enzymes, viruses, and many substances of immunological importance are now known to be proteins. The techniques now used for the study of proteins range from the most elaborate form of X-ray analysis to quantitative measurements of antibodies. Workers in the most diverse fields of science have not only contributed to the development of techniques, but have become interested themselves in applying the techniques they helped develop in the study of the problems of protein chemistry..."

Paragraph 2: "The rapid pace of the advances in protein chemistry, the varied character of the work being done, and its practical applications to industry and medicine have given rise to an increasing need for thoughtful and critical evaluation of the results achieved, and of their implications. We hope that this series of volumes will give the opportunity to workers in special subjects to present their views in more organized form than is possible in the regular journals, and also to express their personal judgment on problems which are still unsettled. We hope too that, as the reviews accumulate, they will provide a useful and comprehensive picture of the changing and growing field of protein chemistry, and a stimulus to its further development."

This statement accurately describes today the feelings and intents of the current editors. The lack of any nonnegotiable length of a manuscript we feel has contributed to the construction of some important and first class reviews.

This review series and all of its attitudes, approaches, and coverage was started by John T. Edsall and M. L. Anson (known to everyone as "Tim"). Interestingly enough, much of the pressure on, and help for, these two laboratory scientists was provided by Kurt Jacoby, who was then Vice President of the new version of Academic Press set up in

America by German immigrants displaced by the war. Later members of the senior staff, most recently Shirley Light, have maintained their strong support for this series for at least 60 years. The Series Editors have been a self-reproducing group. In chronological order, by addition or replacement, they have been John T. Edsall, Vols. 1–48 (emeritus); Tim Anson Vols. 1–23; Kenneth Bailey Vols. 5–17; Christian B. Anfinsen Vols. 12–47; Frederic M. Richards Vols. 19– ; David S. Eisenberg Vols. 39– ; Peter Kim Vols. 48–56; John Kuriyan Vols. 59– .

As these numbers show, John Edsall's tenure as an Editor of *Advances in Protein Chemistry* dwarfs the work of the rest of us. For most of this period, his eminence as one of the pioneering researchers of protein chemistry was crucial to attracting high-quality manuscripts to the series. Edsall, always a stickler for proper notation and clear writing, edited every submitted review in detail, often in successive drafts. He also worked diligently to encourage authors to finish important reviews that they had begun, but somehow had difficulty in finishing. One author of a review long past the deadline reported that he was surprised to see Dr. Edsall, then late in his ninth decade, at the door of his office, and in a city far from his home university. Edsall greeted the author by saying that the long-delayed review was so important that he had come in person to request that it be completed so that he, Edsall, could live to read it in print. The astonished author promised to finish the review quickly.

Advances in Protein Chemistry and the references contained therein provide a documented history of the field of protein chemistry. Not all applications of the theories and techniques are covered in this series, but the foundations are laid. There are occasional articles in which a single publication changes the thinking and approaches of an entire field. For example: Fred Sanger in Volume 7 in 1952 described the very start of the sequencing era, giving both methods and results on a pure protein, insulin. This work solved a long-standing argument on the covalent structure and biological significance of linear polypeptide chains. (The following year, the famous paper of Watson and Crick on DNA set Sanger off on what would be his next great sequencing success.) W. Kauzmann in Volume 14 pointed out the importance of nonpolar interactions, the hydrophobic effect, in both structure and function, drawing data from many fields in chemistry. J. Porath in Volume 17 introduced Sephadex as a resin for column separations based solely on the Stokes radius of the particle and not specific binding properties. Gel filtration is now a very widely used procedure for all manner of macromolecular materials. G. N. Ramachandran and V. Sasisekharan in Volume 23, with the primitive computers available at that time, produced the "Ramachandran Map," the starting point to this day for the analysis of the structure of polypeptide chains. Jane Richardson in Volume 34 provided a classification of the structures of the then known proteins, and developed by hand (no computers) the basic drawing procedures for macromolecular structures which have been further extended in the currently available computer-produced modeling programs. Charles Tanford produced the enormous review on denaturation (a total of 255 pages in three parts appearing in Volumes 23 and 24) which now serves as a starting summary for all the later reviews on disordered structures, a current area of great interest in many biological systems, and particularly as background for this current volume.

While protein chemistry could easily be divided into a variety of subfields, each with its own review journal, the editors decided a few years ago on a different approach. They now pick a subfield and, perhaps more importantly, an individual to serve as editor for that particular volume. This editor is free to choose the authors asked to write the chapters for the whole volume focused on the selected subject area. So far this has worked quite well. The volumes are small enough so that we hope that all members of the field will read all of the volumes both in and outside of their particular area. Such an action would be a small step in trying to keep a broad overview alive. The mix of techniques, biological systems, and results will, of course, vary in each volume. There may be some overlap, but in our view that would be good. George Rose has done an excellent job of collecting a very interesting series of papers for this volume, and "by good luck" (a euphemism for "careful planning"), they all happen to be in a part of protein chemistry that has long been one of John Edsall's favorites.

David S. Eisenberg
John Kuriyan
Frederic M. Richards

GETTING TO KNOW U

Review articles fall somewhere between two extremes: the he said this, she did that category and the what it all means category—summaries vs. syntheses. The reviews in this volume tend toward the latter type. Indeed, the whole volume can be regarded as one long synthesis describing the unfolded state of proteins as depicted in multiple, complementary perspectives.

For years, the reigning paradigm for the unfolded state has been the random coil, whose properties are given by statistical descriptors appropriate to a freely jointed chain. Is this the most useful description of the unfolded population for polypeptide length scales of biological interest? The answer given by this volume is clear: there is more to learn. But first a word about the occasion that prompted this volume.

John T. Edsall will be 100 in November 2002, and this volume is our birthday present to him. Fred Richards has written the dedicatory Preface on behalf of the three Series Editors: David Eisenberg, John Kuriyan, and himself. John Edsall was a founding editor of *Advances,* an editor of legendary prowess as described by Richards. I cannot resist recounting a personal story about his effectiveness in this role. About 20 years ago, Lila Gierasch, John Smith, and I agreed to write a review on peptide and protein chain turns for this series (Rose *et al.,* 1985). Several years came and went, with the promised review still "in aspiration." I confess to being the rate-limiting author. As it happens, I was under consideration for promotion about this time, and the dean solicited John's opinion. Apparently, John neglected to mention my failings, and promotion was forthcoming. Soon after, the phone rang. The voice was unmistakable:

"George, John Edsall here. How are you?"
"Very well," I stammered.

"I want to congratulate you on your promotion. I trust my letter was helpful."

More stammering: "Yes, it was, John, thank you very much."

"Very good", he responded. And then, without noticeable pause, "turning now to the matter of your overdue review..."

It was ready for the next issue.

In contrast to my own delinquency, contributors to this volume have been quite punctual. The occasion of John's 100th birthday was such an effective forcing function for all of us that no editorial nagging was needed, and perhaps it would have been unnecessary in any case. But, there are no transferable hints for future editors. An event like this occurs only once!

BACKGROUND

Arguably, the discipline called *protein folding* was established early in the last century in publications of Wu (Wu, 1931; Edsall, 1995) and Mirsky and Pauling (1936). Both historic papers sought to provide a theory of protein denaturation. At a time when proteins were widely regarded as colloids, these prescient articles had already recognized that many disparate properties of proteins are abolished coordinately on heating. Was this mere coincidence? To Wu and Mirsky and Pauling it seemed more plausible that such properties are all a consequence of some root-level cause—hypothesized to be the protein's structure. When that structure is melted out, the properties are abolished. Yesterday's far-sighted hypothesis is today's fact, as confirmed by Kauzmann (Simpson and Kauzmann, 1953).

These early investigations of protein denaturation set the stage for the contemporary era. Less than a decade after Mirsky and Pauling (1936), Anson summarized earlier evidence that denaturation is a reversible process (Anson, 1945). Any lingering doubt was dispelled by the work of Anfinsen and co-workers showing that urea-denatured ribonuclease could be renatured spontaneously and reversibly (Anfinsen and Scheraga, 1975). This demonstration placed the folding problem squarely in the realm of equilibrium thermodynamics. Instructions for folding are encoded in the amino acid sequence and, therefore, in the genetic code, and no cellular components are needed, i.e., the problem can be studied *in vitro*. These instructions establish the link between the one-dimensional world of replication, transcription, and translation and the three-dimensional world of folded, functioning proteins. What process could be more elegant or more profound. All of this is implicit in

Anfinsen's thermodynamic hypothesis, which states that the native state for a protein in its physiological milieu is the one of lowest Gibbs free energy (Anfinsen, 1973).

The three overall questions arising from this understanding of folding are how to characterize the folded state, the unfolded state, and the transition between these two populations. The characterization of folded proteins is on familiar ground, with more than 15,000 structures currently in the Protein Data Bank (Berman *et al.*, 2000). Given that folding is spontaneous and reversible, the transition can be well characterized by equilibrium thermodynamics as described by Tanford (Tanford, 1968; Tanford, 1970). There is no better introduction to this topic. For small, biophysical proteins, the transition is effectively two-state under experimental conditions of interest, meaning it can be described by only two rate constants, $k_{forward}$ and $k_{reverse}$, with a single activated state that can be obtained from these rates with respect to temperature, denaturant concentration, etc. As shown by Ginsburg and Carroll (1965) for temperature-denatured ribonuclease, diverse probes yield identical folding curves on normalization. Consequently, we study the unfolding transition for a protein, not a family of such curves, each corresponding to a different probe.

The apparent two-state folding process is disappointing news for the chemist seeking intermediates, but welcome news for the protein thermodynamicist, who can now treat the transition as $U \rightleftharpoons N$, i.e., the native state population, N, is in equilibrium with the unfolded population, U. Almost all contemporary work has been along these lines (Schellman, 2002). On closer inspection, the definition of the denaturation process has been somewhat fuzzy from the beginning. Experimentalists have come to accept an operational definition such as a change in observable properties, e.g., Ginsburg and Carroll (1965), while theoreticians tend to opt for an abstract definition, such as the random chain model. In either case, the actual situation is more complex than simple two-state behavior: partially folded intermediates are present within the equilibrium population (Mayne and Englander, 2000).

We turn now to the remaining question—how to characterize the unfolded population, the topic of this volume.

The equilibrium population is said to have a *structure* when a substantial fraction of the molecules adopts similar conformations. But the phrase *lacking structure* does not imply that individual molecules comprising the ensemble lack a conformation; rather, the population is too heterogeneous to be readily characterized using a coherent, structure-based descriptor. The unfolded state resists ready characterization because it is so diverse. Typical biophysical methods report ensemble-averaged properties in which important components of the population may be concealed beneath the background or the intrinsic distributions may be lumped into a misleading average. It is here that our thinking depends most critically on underlying models.

The current model, which originated with Flory (Flory, 1953; Flory, 1969) and Tanford (Tanford, 1968; Tanford, 1970), treats unfolded proteins as random chains as summarized in Baldwin's concluding chapter. Specifically, under conditions that favor unfolding, the chain is free to adopt all sterically allowed (Ramachandran and Sasisekharan, 1968) values of ϕ,ψ-angles, and when it does so, the population is found to be Gaussian-distributed around the radius of gyration expected for random, freely jointed chains of the same length in good solvent, but with excluded volume constraints.

It is important to realize that the random-chain model need not imply an absence of residual structure in the unfolded population. Formative articles—many of them appearing on the pages of *Advances in Protein Chemistry*—recognized this fact. Kauzmann's famous review raised the central question about structure in the unfolded state (Kauzmann, 1959):

> "For instance, one would like to know the types of structures actually present in the native and denatured proteins. ... The denatured protein in a good solvent such as urea is probably somewhat like a randomly coiled polymer, though the large optical rotation of denatured proteins in urea indicates that much local rigidity must be present in the chain" (pg. 4).

Tanford cautioned that a random chain and an α-helix are expected to have similar values of the radius of gyration at chain lengths approximating ribonuclease and lysozyme (Tanford, 1968, pg. 133). Indeed, the problem is one of scale. As Al Holtzer once remarked, "even a steel I-beam behaves like a random chain if it's long enough," but this realization does not figure heavily in designs on the length scale of my office at Hopkins.

The energy landscape for authentic random chains is maximally disordered and therefore temperature-independent, and this model prompts key questions about the unfolded population. Subjected to folding conditions, how do chains search out the native state in biological real-time while avoiding metastable traps along the way (Levinthal, 1969)? Such questions have engendered the 3Fs: frustration, funnels, and foldability, e.g., Sali *et al.* (1994); Bryngelson *et al.* (1995); Dill and Chan (1997); Tiana *et al.* (2000); Brooks *et al.* (2001).

Meanwhile, evidence continues to mount that the unfolded state is far from a random chain at length scales of interest, even under strongly

Fig. 1.   A 12-residue segment of polyalanine in a P_{II} helix shown for reference. Atoms are shown in conventional CPK coloring (carbon, black; nitrogen, blue; oxygen, red; hydrogen,white). Unlike the more familiar α-helix, a P_{II} helix is left-handed (ϕ,ψ = $-76°$, $+149°$).There are three residues per helical turn, such that the side chains form three parallel, collinear columns spaced uniformly around the long axis of the helix. A segment of P_{II} in solution would be expected to undergo significant fluctuations from the idealized structure shown here.

denaturing conditions. Critical examination of that evidence is the main purpose of this volume. In particular, the left-handed polyproline II helix (P_{II}) is found to play a central role in the unfolded population, as discussed in many chapters. It should be noted that the name of this helix is something of a misnomer in that the conformation often occurs in a proline-free sequence (Stapley and Creamer, 1999).

In P_{II} conformation, the backbone dihedral angles are near $\phi = -76°$, $\psi = 149°$, a favored location on the ϕ,ψ-map (Han et al., 1998; Poon and Samulski, 2000; Pappu and Rose, 2002), resulting in a left-handed helix with a perfect threefold repeat. Unlike the α-helix, a P_{II} helix lacks intrasegment hydrogen bonds (Fig. 1, see color insert); more-over, residues are free to fluctuate independently in this part of the ϕ,ψ-map (Pappu et al., 2000). Consequently, a high degree of fraying across the entire P_{II} segment is expected.

Although the P_{II} conformation is found in folded proteins (Adzhubei and Sternberg, 1993; Stapley and Creamer, 1999), it is not abundant. Approximately 10% of all residues have P_{II} ϕ,ψ-values. This too is expected because in P_{II}, the backbone cannot make either intrasegment H-bonds (like an α-helix) or intersegment H-bonds (like a β-sheet), only H-bonds to water. In sum, the P_{II} helix is a highly plausible unfolded conformation because hydrogen bonding promotes protein:solvent interactions at the expense of protein:protein interactions, and the structure is compatible with both preferred ϕ,ψ-angles and dynamic disorder (i.e., fraying).

The preceding issues underscore the crucial importance of models. Although $\sim$10% of all residues in proteins of known structure do have ϕ,ψ-values in the P_{II} region, the prevalence of P_{II} in the unfolded population could not have been deduced from this inventory.

Chapters in This Volume

Turning now to the chapters in this volume, a variety of complementary techniques and approaches have been used to characterize peptide and protein unfolding induced by temperature, pressure, and solvent. Our goal has been to assemble these complementary views within a single volume in order to develop a more complete picture of denatured peptides and proteins. The unifying observation in common to all chapters is the detection of preferred backbone conformations in experimentally accessible unfolded states.

Shortle has focused on the unfolded state for more than a decade, leading up to his recent demonstration using residual dipolar couplings that staphylococcal nuclease retains global structure in 8 M urea. His chapter on "The Expanded Denatured State" sets the stage. Dunker et al. then explore the complementary world of disordered regions within folded proteins, asking how disorder, rather than order, can be specifically encoded by the sequence in these cases.

Most of the experimental information about P_{II} in the unfolded state comes from optical spectroscopy. The chapter on "Unfolded Proteins Studied by Raman Optical Activity" by Barron *et al.* provides a definitive exposition of this technique that emphasizes structural signatures present in unfolded proteins, with particular attention to P_{II}. "What Fluorescence Correlation Spectroscopy Can Tell Us About Unfolded Proteins" by Chattopadhyay *et al.* describes how fluorescence can be used to follow changes in conformation by measuring changes in the diffusion coefficient over a broad range of time scales. "Unfolded Peptides and Proteins Studied with Infrared Absorption and Vibrational Circular Dichroism Spectra" by Keiderling and Xu emphasizes identification of residual structure in the unfolded state, including P_{II}. "Is Polyproline II a Major Backbone Conformation in Unfolded Proteins" by Shi *et al.* consolidates existing CD and other data on unfolded peptides and proteins, again emphasizing P_{II}.

"Toward a Taxonomy of the Denatured State: Small Angle Scattering Studies of Unfolded Proteins" by Millett *et al.* assesses denatured states induced by heat, cold, and solvent for evidence of residual structure, while "Insights into the Structure and Dynamics of Unfolded Proteins from NMR" by Dyson and Wright describes their extensive investigations of residual structure in the unfolded state.

Three theory papers are also included. "Determinants of the Polyproline II Helix from Modeling Studies" by Creamer and Campbell reexamines and extends an earlier hypothesis about P_{II} and its determinants. "Hydration Theory for Molecular Biophysics" by Paulaitis and Pratt discusses the crucial role of water in both folded and unfolded proteins. "Unfolded State of Peptides" by Daura *et al.* focuses on the unfolded state of peptides studied primarily by molecular dynamics.

A final summary is provided by Baldwin, who articulates the new direction implied by these chapters and points the way toward the future.

REFERENCES

Adzhubei, A. A., and Sternberg, M. J. (1993). *J. Mol. Biol.* **229**, 472–493.

Anfinsen, C., and Scheraga, H. (1975). *Adv. Prot. Chem.* **29**, 205–300.

Anfinsen, C. B. (1973). *Science* **181**, 223–230.

Anson, M. L. (1945). *Adv. Prot. Chem.* **2**, 361–386.

Berman, H. M., Westbrook, J., Feng, Z., Gilliland, G., Bhat, T. N., Weissig, H., Shindyalov, I. N., and Bourne, P. E. (2000). *Nucleic Acids Res.* **28**, 235–42.

Brooks, C. L., Onuchic, J. N., and Wales, D. J. (2001). *Science* **267**, 1619–1620.

Bryngelson, J. D., Onuchic, J. N., Socci, N. D., and Wolynes, P. G. (1995). *Proteins: Structure, Function, and Genetics* **21**, 167–195.

Dill, K. A., and Chan, H. S. (1997). *Nature Structural Biology* **4**, 10–19.

Edsall, J. T. (1995). *Adv. Prot. Chem.* **46**, 1–26.

Flory, P. J. (1953). Principles of Polymer Chemistry. Cornell University Press.

Flory, P. J. (1969). Statistical Mechanics of Chain Molecules. Wiley, New York.

Ginsburg, A., and Carroll, W. R. (1965). *Biochemistry* **4**, 2159–2174.

Han, W.-G., Jalkanen, K. J., Elstner, M., and Suhai, S. (1998). *J. Phys. Chem. B* **102**, 2587–2602.

Kauzmann, W. (1959). *Adv. Prot. Chem.* **14**, 1–63.

Levinthal, C. (1969). Mossbauer Spectroscopy in Biological Systems, Proceedings. *University of Illinois Bulletin* **41**, 22–24.

Mayne, L., and Englander, S. W. (2000). *Protein Science* **9**, 1873–1877.

Mirsky, A. E., and Pauling, L. (1936). *Proc. Natl. Acad. Sci. USA* **22**, 439–447.

Pappu, R. V., and Rose, G. D. (2002). *Proc. Nat. Acad. Sci. USA*, in press.

Pappu, R. V., Srinivasan, R., and Rose, G. D. (2000). *Proc. Natl. Acad. Sci. USA* **97**, 12565–70.

Poon, C.-D., and Samulski, E. T. (2000). *J. Am. Chem. Soc.* **122**, 5642–5643.

Ramachandran, G. N., and Sasisekharan, V. (1968). *Adv. Prot. Chem.* **23**, 283–438.

Rose, G. D., Gierasch, L. M., and Smith, J. A. (1985). *Adv. Prot. Chem.* **37**, 1–109.

Sali, A., Shakhnovich, E. I., and Karplus, M. (1994). *Nature* **477**, 248–251.

Schellman, J. A. (2002). *Biophysical Chem.*, in press.

Simpson, R. B., and Kauzmann, W. (1953). *J. Am. Chem. Soc.* **75**, 5139–5152.

Stapley, B. J., and Creamer, T. P. (1999). *Protein Science* **8**, 587–595.

Tanford, C. (1968). *Adv. Prot. Chem.* **23**, 121–282.

Tanford, C. (1970). *Adv. Prot. Chem.* **24**, 1–95.

Tiana, G., Broglia, R. A., and Shakhnovich, E. I. (2000). *Proteins* **39**, 244–251.

Wu, H. (1931). Studies on denaturation of proteins. *Chinese Journal of Physiology* **V**, 321–344.

George D. Rose

JOHN T. EDSALL
(November 3, 1902–June 12, 2002)

John Edsall passed away just as this volume was headed to press. It was hoped by the editors that this volume would be one of the 100th birthday presents that John would receive on November 3. This was not to be.

No brief biography can convey the enormous impact that John Edsall had on the students, scientists, and scholars who crossed his path. He seemed larger than life: someone who virtually personified the unselfish search for truth that science is supposed to be. As a researcher, he spent decades applying physical chemistry to proteins, showing, among other things, that proteins bristle with charges. As a person of action during World War II, he and his co-workers learned how to fractionate blood into proteins that can be used in medicine and surgery. As a community leader during the McCarthy era, he stood bravely in defense of the right of Linus Pauling and other scientists to take unpopular political stands [this episode is described in my brief biographical sketch on Edsall in *Protein Science*, **1**, 1399–1401 (1992)]. As a teacher, he inspired generations of undergraduate and graduate students to become biochemists, bolstered by his example as a gentle, humane scientist and his example that biochemists can contribute to society as a whole as well as to our own discipline. As a scholar, he worked to understand the history of ideas in protein science. As a writer of texts and monographs, he conveyed the most difficult concepts of energetics and function to new generations of scientists. More on all of this will be available in the November issue of the *Journal of Biophysical Chemistry* dedicated to John Edsall.

John Edsall's mind and body remained vigorous in years when most humans have surrendered to old age. When the Protein Society met in Boston in the hot summer days of 1997, John, nearing 95, went on foot and subway from his home in Cambridge to the Boston Convention

Center to attend scientific sessions. Nearly everyone watched with awe as the tall elderly gentleman in a blue suit entered the symposium room, still curious about developments in a field that he had helped to found 70 years before. A year later, my wife and I visited him in his Cambridge apartment. As we entered, we found John at his desk reading a book on philosophy in French. He was then, as ever, after the truth.

David S. Eisenberg

THE EXPANDED DENATURED STATE: AN ENSEMBLE OF CONFORMATIONS TRAPPED IN A LOCALLY ENCODED TOPOLOGICAL SPACE

By DAVID SHORTLE

Department of Biological Chemistry, The Johns Hopkins University School of Medicine, Baltimore, Maryland 21205

I. INTRODUCTION

A. Early Studies Suggested Denatured Proteins Are Essentially Random

When a protein reversibly unfolds in solution, it enters a large and diverse ensemble of conformations known as the denatured state. Studies of this reaction over the past 40 years have suggested it is a single,

ADVANCES IN
PROTEIN CHEMISTRY, Vol. 62

all-or-nothing transition that ends with the loss of a very substantial fraction of the native state's noncovalent structure. All tertiary structure and almost all helices and sheets appear to break down in this transition, leaving the polypeptide chain virtually devoid of features detectable by fluorescence, UV absorption, or near-UV circular dichroism (Dill and Shortle, 1991). Weak signals may be seen in the far-UV CD spectrum, suggesting the chain is not entirely random; however, uncertainty about the "random coil" spectrum makes these findings difficult to interpret.

Characterization of the size of denatured proteins by viscometry, gel filtration, and small-angle X-ray scattering has usually focused on the impact of denaturant concentration and indicated that, under highly denaturing conditions such as 6 M GuHCl, protein chains swell to approximately the size expected for a random coil (Tanford, 1968, 1970). Again, uncertainty as to the calculated size of a random coil from polymer physical chemistry and appropriate corrections for glycine and proline have made it difficult to settle the issue of whether or not denatured protein chains expand to the dimensions of a random coil (Miller and Goebel, 1968). The point is often overlooked that having the hydrodynamic radius predicted for a random coil does not prove a chain is a random coil.

In view of these studies and the lack of compelling evidence to the contrary, protein chemists have accepted the general premise that the denatured state is essentially random. Even though this may be a poor approximation under mildly denaturing conditions, so the argument goes, it is likely to improve under strongly denaturing conditions. A random coil or fully unfolded state has the added benefit of serving as a much more useful reference state for analysis of the thermodynamic issues of protein folding. If denatured proteins are random, the interactions of side chains with each other and the backbone can be ignored. Consequently, the effects of denaturants, temperature, and mutations on the native state can be more easily interpreted because their effects on the denatured state are defined as zero.

B. A More Complete but Different Picture from NMR-Based Experiments

With the development of NMR-based experiments for studying folded proteins, structural analysis of denatured proteins entered a new phase (Wuthrich, 1994; Dobson *et al.*, 1994; Shortle, 1996). Whereas other spectroscopies and hydrodynamic studies give data that correspond to a complex sum of the properties of all residues, NMR spectroscopy extracts information about individual residues. For the first time, local structural features could be directly inferred from the behavior of sets of residues along the sequence.

Nevertheless, it must be emphasized that the structural description obtained from NMR analysis is fundamentally different when the protein is not fully folded. The measured NMR parameters consist of signals that have been averaged over all conformations visited during a time interval from picoseconds to milliseconds, the latter being the most common. Consequently, NMR data, like other physical data collected for denatured proteins, are inherently ensemble-averaged and usually do not provide insight into the range of deviation from the mean structure. In many situations, the ensemble-average or "mean" structure itself may not correspond to a single physically realistic conformation. Thus the statistical nature both of the denatured state and of the NMR parameters used to characterize it must be kept at the forefront of all models used to interpret the data.

C. Reasons to Study the Denatured State

Because the denatured state has long been thought to be essentially random and because of the inherent difficulties in defining and interpreting structural data averaged in very complex ways, protein chemists have been slow to take up the structural characterization of denatured proteins. Yet a more complete description of the relatively small amounts of persistent structure they display is not simply of academic interest. There are compelling reasons to pursue structural studies of proteins that are not folded.

First, proteins refold from the denatured state, not from the hypothetical random coil state. It is the starting point of all refolding reactions, whether in a cell or in a test tube. To understand any chemical reaction, structural features of the reactant and the product must be compared to quantify the changes that occur, for these changes define the reaction.

Second, the denatured state is not only the physical starting point for refolding but also the conceptual starting point for defining the "problem of protein folding" and seeking a computational solution. To estimate the amount of entropy that must be lost as a protein folds, and hence the amount of bond enthalpy required to stabilize the folded state, the residual structure of the denatured state must be understood in some detail. The back-of-the-envelope calculation that Levinthal (1968) and others used to estimate the magnitude of the entropy loss is based on a very questionable estimate of the number of conformations likely to be populated in the denatured state. To the extent that the denatured state ensemble is reduced in size as a result of persistent structure, the magnitude of Levinthal's paradox will be diminished, perhaps disappearing altogether.

But the most compelling reason for studying denatured state structure is a practical one: the availability of improved methods based on
heteronuclear NMR experiments (Dyson and Wright, 2001). It is always
difficult to estimate the information content hidden in some unknown
realm of nature, but history has proved it is usually wise to take a look.
When one knows nothing about a phenomenon, the typical response is
to assume that it probably is not important. Yet surprises, often major
ones, lurk in places where scientific investigation has not yet shone its
bright light.

D.   Overview

In this review, NMR analysis of the denatured state of staphylococcal nuclease is briefly reviewed in nontechnical language. Most of the
work has come from the author's laboratory over the past eight years.
The initial experiments, which only measure local structural parameters,
reported small amounts of persisting helical structure, two turns, and indirect evidence for perhaps a three-strand beta meander. When applied
to the denatured state in 6 M urea, the same experiments indicated that
most of these features are lost.

However, recent application of two types of NMR methodologies that
provide long-range structural information have painted a very different
picture. From these experiments, the denatured state in 0 M urea and
8 M urea appear to be highly similar, both retaining the same overall
spatial positioning and orientation of the chain seen in the folded conformation.

Simple computer simulations to estimate the importance of side
chain–backbone interactions in encoding the long-range structure in
denatured proteins suggest that these simple steric effects are of overriding importance. If this conclusion is correct, it appears that much of
the estimated entropy available to a "random coil protein chain" is lost
during its synthesis. As it emerges from the ribosome, each protein finds
itself trapped in a topological space that preconfigures it to fold rapidly
and uniquely to the native state.

II.   Nuclease $\Delta 131 \Delta$: Local Structure

A.   The Protein System

Staphylococcal nuclease is a small $\alpha + \beta$ protein of 149 amino acids
that is stable to denaturation by only $+5.5$ kcal/mol (Anfinsen *et al.*,

1972). To make nuclease more unstable without grossly perturbing its denatured state, nine amino acids were removed from both ends of the protein (Alexandrescu *et al.*, 1994). Fourteen of these residues are unstructured; of the five residues integral to the native state, only one is hydrophobic (Leu7). The combined effect of these truncations is a 131-residue fragment termed Δ131Δ that has a stability of less than −3.0 under physiological conditions. On addition of calcium ion and either single-stranded DNA or the tight-binding ligand pdTp, this large fragment can refold to a fully catalytically active form. Under the conditions of most NMR experiments, namely pH 5.3 and 32°C, less than one molecule in a thousand populates the native conformation.

In the absence of salt, Δ131Δ gives sharp NMR lines at pH 5.0 to 5.3, even at concentrations of 5 mM protein (Zhang *et al.*, 1997), suggesting there is no appreciable aggregation at the high concentrations. Ultracentrifugation in the absence of added salt at a concentration of 1 mM also reveals no signs of aggregation (J. Sinclair and D. Shortle, unpublished data), and preliminary small-angle X-ray scattering analysis gave a radius of gyration of 27 Å, compared to 17 Å for folded nuclease (M. Kataoka, personal communication). However, titration of the protein with salt broadens the NMR lines, and visible precipitation begins to develop above 100 mM NaCl.

B. Initial NMR Characterization

For proteins that are not folded, the amide proton (^{1}H) and the amide nitrogen (^{15}N) provide the best separated two-dimensional set of resonances (Shortle, 1996). Side chain proton and ^{13}C chemical shifts are extensively averaged, so that resonances from different residues of the same amino acid type form a single sharp peak. Fortunately, there is a small amount of chemical shift dispersion for the ^{13}CA, ^{13}CB, and ^{13}CO resonances. For practical purposes, the ^{1}H–^{15}N two-dimensional correlation spectrum is the primary route to reading out most NMR parameters in proteins that are not folded. Using a large variety of pulse sequences, a surprisingly long list of chemical shifts, coupling constants, NOEs, and relaxation information can be transferred to these two spins for measurement.

The number of NMR parameters available for measurement is rather small, consisting of the chemical shift, relaxation rates (t_1 and t_2), scalar (J) couplings, dipolar (D) couplings, cross-relaxation rates (the NOE), and hydrogen exchange rates. All of these have been quantified for many of the amide protons of Δ131Δ, and most of the data suggest the presence of little persistent structure.

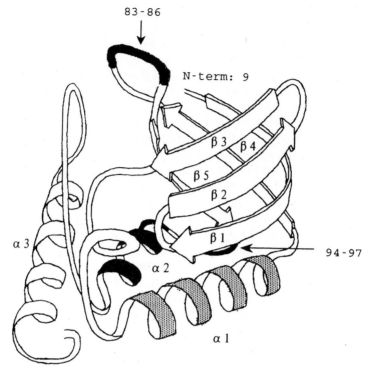

83-86

N-term: 9

β 3   β 4

β 5

β 2

α 3

β 1

94-97

α 2

α 1

FIG. 1. Schematic diagram of nuclease $\Delta 131\Delta$ in the folded conformation. The alpha helices and beta strands are labeled. NMR analysis suggests the two turns and one helix in black are modestly populated in the denatured state, whereas the shaded helix is slightly populated. Strands $\beta 1$–$\beta 2$–$\beta 3$ form an extended structure about which little is known. Reproduced from Barron, L. D., Hecht, L., Blanch, E. W., and Bell, A. F. (2000). *Prog. Biophys. Mol. Chem.* **73**, 1–49. © 2000, with permission from Elsevier Science.

For example, the temperature dependence of the H_N chemical shifts and hydrogen exchange rates measured by magnetization transfer and deuterium exchange suggest there are few if any populated hydrogen bonds (Alexandrescu *et al.*, 1994; Wang and Shortle, 1996; Mori *et al.*, 1997). Medium-range NOEs to H_N and H_A suggest two tight beta turns are significantly populated and the three helices may be slightly populated (Fig. 1). A very sophisticated NOESY experiment that can detect side chain NOEs by transferring the signal down the side chain to the H_N revealed a small number of new NOEs, most of which confirmed the presence of the two tight beta turns (Zhang *et al.*, 1997). Initial results with a very sensitive experiment for detecting NOEs between one type of protonated side chain and backbone amide protons in an otherwise fully deuterated protein indicate that any long-range NOEs are likely to be very weak (M. S. Ackerman and D. Shortle, unpublished data).

What conclusions can be drawn from the paucity of long-range NOEs is not clear. Whereas two protons must be separated by a short distance (<5 Å) to give rise to an NOE, they must also retain a relatively fixed spatial relationship for 1 ns or longer. Therefore, extensive contact between widely separated chain segments would go undetected if such structure were highly dynamic, with contacts lasting only 10–100 ps.

The most informative NMR parameter for analysis of $\Delta 131\Delta$ and other denatured proteins has been the secondary chemical shift of the ^{13}CA, ^{13}CB, and ^{13}CO spins. These chemical shifts depend sensitively on the dihedral angles phi and psi (Wishart and Sykes, 1994). When a residue does not have a random distribution of phi/psi angles, its chemical shift will be greater or less than the random coil value. From trends in secondary chemical shifts in ^{13}CA and ^{13}CO for contiguous sets of residues, the fractional population of helical structure can be reliably inferred (Reily et al., 1992). For $\Delta 131\Delta$ in buffer, these estimates are 0.1 for helix 1, 0.3 for helix 2, and 0.15 for helix 3. None of the three segments that form beta strands in the folded state and are detectable under physiological conditions ($\beta 4$, $\beta 5$, $\beta 6$) exhibits the secondary shifts expected for a fractional population of extended strand structure.

C. "Molten" Structure of the Beta Meander

The most obvious feature of the ^{15}N–^{1}H correlation spectrum of $\Delta 131\Delta$ is what is missing: All of the peaks corresponding to the first 30 residues are absent. In the native structure these residues form a three-strand beta meander ($\beta 1$–$\beta 2$–$\beta 3$), which combines with a two-strand beta hairpin ($\beta 4$, $\beta 5$) to form a tightly wound beta barrel. The most plausible explanation for their absence is severe line broadening by conformational exchange at a rate of ~1 ms.

This explanation is supported by tracking the ^{15}N–^{1}H correlation spectrum as a function of urea concentration (Wang and Shortle, 1995). At approximately 3–4 M urea, resonances from the first beta strand appear as a set. At approximately 5–6 M urea, the remaining peaks also appear in concert and grow stronger with increasing urea concentration. From the principle knows as "the power of positive thinking about a negative result," the tentative conclusion can be drawn that this segment of the chain forms a native-like array of beta strands.

While the CA and CO secondary chemical shifts cannot be determined using conventional triple-resonance experiments that employ the ^{1}H–^{15}N correlation spectrum, incorporation of individual

[13]C-carbonyl-labeled amino acids permitted some of the [13]CO resonances to be assigned (Wang and Shortle, 1996). In many cases, the resonances for the $\beta 1-\beta 2-\beta 3$ segment displayed the typical upfield shift that accompanies a beta strand structure, with convergence to more random coil values on titration of urea.

The kinetics of hydrogen exchange for the beta meander was measured by exchange in D_2O (Wang and Shortle, 1996). After a variable exchange time, the protein fragment was refolded with Ca^{2+} and pdTp, permitting quantification of exchange via peak intensities in the folded conformation. But no significant protection to exchange was detected, suggesting that, whatever the structure present in this "molten state," it does not consist of rigid, long-lasting hydrogen bonding between strands.

D. Restricted Motion as an Indicator of Structure

The concept of molecular structure implies a reduction in the freedom of motion for the involved atoms. Thus an indirect strategy for identifying structured segments is to search for restricted motion for contiguous sets of amino acid residues. Relaxation of the ^{15}N nucleus in the peptide bond provides a quantitative measure of the rates and angular range of motion experienced by individual amino acids under equilibrium conditions (Palmer, 2001).

Although the underlying physics and mathematics used to convert relaxation rates into molecular motions are rather complex (Lipari and Szabo, 1982), the most important parameter obtained from such analyses, the order parameter S^2, has a simple interpretation. In approximate terms, it corresponds to the fraction of motion experienced by a bond vector that arises from slow rotation as a rigid body of roughly the size of the macromolecule. Thus, in the interior of folded proteins, S^2 for H_N bonds is always close to 1.0. In very flexible loops, on the other hand, it may drop as low as 0.6 because subnanosecond motions partially randomize the bond vector before it rotates as a rigid body.

The first relaxation analysis of $\Delta131\Delta$ revealed motional behavior that is strikingly nonuniform, with some segments displaying relatively complex behavior indicative of slowly interconverting structures (Alexandrescu and Shortle, 1994). Nevertheless, the overall profile is clear. The middle one-third of the chain displays a significant restriction of motion, with a few residues displaying order parameters as large as 0.6 to 0.7. Residues at the ends have values as small as 0.1 to 0.2, and most residues fall between these extremes. A rough correlation was found between segment hydrophobicity and order parameters.

A second analysis of $\Delta 131\Delta$ confirmed this general pattern (Sinclair and Shortle, 1999). In addition, it was found that the "structure" in the central segment was little changed by deleting the first 30 residues (a fragment referred to as $\Delta 101\Delta$). In 6 M urea, however, most of the variation of S^2 along the chain has been removed, leaving the internal segments almost as unrestricted as the ends, all with order parameters of approximately 0.2. Thus, to a rough approximation, the relative restriction of the backbone of $\Delta 131\Delta$ in 6 M urea to motions for times shorter than 1 ns is nearly constant along the length of the chain.

III. Nuclease $\Delta 131\Delta$: Long-Range Structure

A. Solution Structure by Paramagnetic Relaxation Enhancement

Although the NMR studies presented above give limited indications of persistent local structure, the presence of long-range structure is not ruled out. To probe $\Delta 131\Delta$ for long-range structure, the little-used method of paramagnetic relaxation enhancement was employed (Kosen, 1989). This technique utilizes the large magnetic dipole of the free electron to measure distances between electron–proton pairs in a manner similar to NOEs between proton pairs. When a nitroxide free radical is attached at a unique site by chemical modification, the large magnetic field of the electron can enhance the relaxation of protons out to distances of 20 Å. The distance between the nitroxide label and an individual proton can be calculating by quantifying the changes in its relaxation and solving the Solomon–Bloembergen equation. This equation makes several fundamental assumptions, namely, a single structure with a fixed electron–proton distance whose only motion is rotation of the entire molecule. All of these assumptions are likely to be seriously violated by the residual structure of denatured proteins.

In the experiments reported by Gillespie and Shortle (1997a, 1997b), 14 unique cysteine mutants were introduced into $\Delta 131\Delta$, a PROXYL spin label was attached by alkylation, and the changes in t_1 and t_2 were measured at two magnetic fields. From these data sets, approximately 700 distances were calculated and used as restraints in structure calculations. To deal with the uncertainty in these distances due to violations of the simplifying assumptions, the largest range of allowed distances (± 5 Å) was used that produced conformations that were not entirely random.

Figure 2 (see color insert) shows five structures calculated using these distance restraints and an all-atom protein chain. Noteworthy is the relatively high degree to which their structures coincide. When these five conformations are compared to the structure of folded nuclease, it is obvious that chain segments of Δ131Δ are confined in approximately the same positions and orientations relative to the corresponding segments of the native structure. In other words, at a low level of resolution (i.e., the topology), this denatured protein is quite native-like. The only convincing exception to this conclusion involves strands β4 and β5, which are paired in the native state but drift apart in the denatured state. Except for the distance restraints obtained from the spin labels and a few medium-range NOEs, the only inputs into structure calculations were the standard atomic radii; electrostatics, hydrogen bonds, and dispersion forces were turned off. Thus these structures are based solely on the experimental data; in no manner was the native structure assumed or built in by the way the calculations were done.

B. Residual Dipolar Couplings and Alignment of Δ131Δ

To extend the structural analysis of Δ131Δ, residual dipolar couplings are being measured under a wide variety of conditions. This NMR parameter reflects a small change in the chemical shift of one nucleus that arises when a second nucleus assumes a fixed angular relationship to it and to the NMR tube (Prestegard *et al.*, 2000). Although the magnetic field contributed by a nearby ^1H, ^{15}N, or ^{13}C nucleus can change the resonance frequency by several kilohertz, the random tumbling of the molecule in free solution normally averages the net orientation and hence the coupling to zero. If the molecule is made to tumble in an oriented, asymmetric environment, a small net orientation of approximately 1 part per 1000 can be conferred on the molecule, giving rise to small residual dipolar couplings of a few hertz. These couplings provide information on the relative orientation of individual bond vectors relative to a unique set of axes (the alignment tensor) running through the molecule.

The most important feature of the information in dipolar couplings is that it is independent of distance. The data can be envisioned as reflecting the relative orientation of pairs of bond vectors, with the intervening distance having no effect (Meiler *et al.*, 2000). Thus dipolar couplings can potentially provide a method for characterizing the structure of denatured proteins, provided that the denatured state ensemble of conformations retains several levels of nonrandomness.

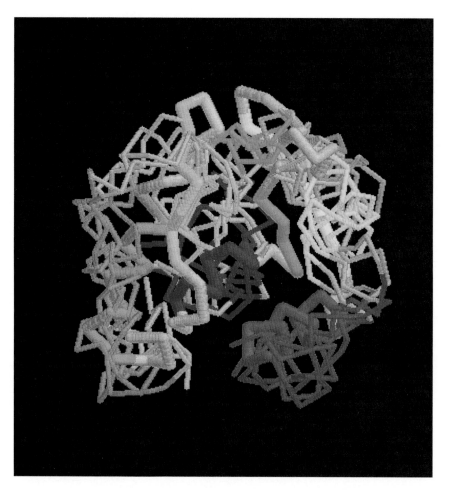

Fɪɢ. 2. Five structures for Δ131Δ, calculated from more than 600 weak distance restraints obtained by paramagnetic relaxation enhancement (Gillespie and Shortle, 1997a, 1997b), superimposed on the native conformation (broad line.) Helical segments are colored in shades of red and orange, strands in shades of yellow. Turn segments are gray. The orientation of the structures is the same as that in Figure 1.

To display dipolar couplings, a denatured protein must be alignable, and residues must retain a nonrandom orientation with respect to each other. In molecular terms, the ensemble-average shape cannot be spherically symmetric; it must have oblate or prolate ellipsoidal symmetry. The ellipsoidal axes must maintain a relatively fixed relationship to a unique set of axes running through the ensemble-average chain position. And residues that exhibit dipolar couplings must retain a fixed orientation with respect to these molecular axes, with the time-averaged projection of individual bond vector along this fixed orientation being nonzero. Thus the simple observation of dipolar couplings indicates the presence of significant amounts of ensemble-average structure.

$\Delta 131\Delta$ exhibits easily measurable residual dipolar couplings in spectra collected in strained polyacrylamide gels (Shortle and Ackerman, 2001), alkyl-PEG bicelles (Ackerman and Shortle, 2002), and cetylpyridinium bromide liquid crystals (Ackerman and Shortle, 2002). When plotted as a histogram, these data look remarkably like the histograms derived from folded proteins, having approximately the "powder pattern" shape expected for a more-or-less uniform distribution of orientations within the molecule. Except for the observation that significantly higher acrylamide concentrations and modestly higher concentrations of alkyl-PEG bicelles are required to achieve a workable range of alignment, $\Delta 131\Delta$ behaves like a folded protein.

C. Persistent Structure in 8 M Urea

To demonstrate that these dipolar couplings arise from long-range structure, attempts were made to eliminate them by addition of urea (Shortle and Ackerman, 2001). Surprisingly, urea at concentrations as high as 8 M has relatively little effect on the measured couplings. As seen in Figure 3 for data collected in strained polyacrylamide gels, a highly significant correlation persists between couplings in the absence and presence of urea. Although the magnitude of the couplings and the strength of the correlation gradually diminish with increasing urea concentration, the correlation remains highly significant.

To reduce the possibility that confinement in gel cavities is somehow responsible for the resistance of long-range structure to urea, the same experiments were carried out on $\Delta 131\Delta$ in alkyl-PEG bicelles (Ackerman and Shortle, 2002). The same results were obtained, with almost the same correlation coefficients as those seen in strained polyacrylamide gels.

The formal possibility that a compact, highly structured subset of conformations is responsible for this observation is highly unlikely for several

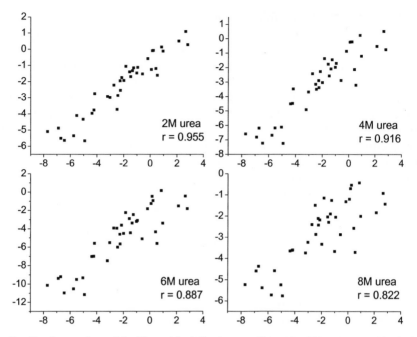

FIG. 3.   Scatterplots of the N_H residual dipolar couplings of $\Delta131\Delta$ measured in 2, 4, 6, or 8 M urea (y-axes) plotted against the same couplings measured in the absence of urea (x-axis); r is the Pearson correlation coefficient. Alignment was achieved with alkyl PEG bicelles (Ackerman and Shortle, 2002.)

reasons. (1) This subset of compact conformations would have to be extremely resistant to urea, since 8 M urea reduces the absolute magnitude of the couplings by just one-third. (2) To obtain the observed small averaged couplings and very sharp lines, this structured subset would have to be in rapid equilibrium with the much larger subset of random conformations, with interconversion times faster than one per millisecond, also unlikely in the presence of 8 M urea. (3) The residual dipolar couplings of this structured subset would have to be quite large to withstand a significant dilution by the large fraction of random conformations. Couplings larger than 30 Hz are typically not observed in folded proteins because the excessive orientation required causes severe line broadening.

Although the hydrodynamic radius of $\Delta131\Delta$ has not been determined in 8 M urea, WT protein in 5 M urea has a radius of gyration of 34 Å as measured by small-angle X-ray scattering (Flanagan *et al.*, 1992). Thus $\Delta131\Delta$ is likely to be as large if not larger. The simplest interpretation of these data is that Δ131Δ in 8 M urea retains approximately the same ensemble average shape and segment orientation that it has in 0 M urea. Because the structure in 0 M urea determined by paramagnetic relaxation enhancement (Figure 2B) is native-like, the conclusion seems unavoidable that Δ131Δ has a native-like topology in 8 M urea.

It is not yet clear how representative Δ131Δ will be of denatured proteins in general, but preliminary spectra of ubiquitin and eglin C denatured in 8 M urea at pH 3.0 give large ^1H–^{15}N dipolar couplings when aligned in alkyl-PEG bicelles (K. Briggman, M. Edgell, and D. Shortle, unpublished data). From the limited dispersion in ^1H and ^{15}N dimensions and the presence of the correct number of peaks, it appears that both of these proteins are probably in an expanded denatured state under these conditions. Further studies of other proteins will be needed to determine the generality of persistent native-like topology in expanded denatured states.

D. The Same Long-Range Structure in States Denatured by Urea, Acid, or Mutation

Conversion of dipolar coupling data into three-dimensional structures presents a major challenge for folded proteins; and for not-folded proteins, an additional set of obstacles must be confronted (Ackerman and Shortle, 2002). Nevertheless, the correlation of dipolar couplings measured under two sets of conditions establishes that the structures are likely to be highly similar.

Figure 4 shows such an analysis for two additional denatured states (Ackerman and Shortle, 2002): WT full-length nuclease in 4 M urea and WT in pH 3.0 buffer with no added salt. As seen, both sets of dipolar couplings correlate very well with those obtained for Δ131Δ in the absence of urea. This finding suggests that denaturation by at least three different agents—truncation, urea, and acid—gives rise to essentially the same persistent native-like topology.

In the case of the SDS denatured state of nuclease, only very small dipolar couplings have been observed (Ackerman and Shortle, 2002). Failure to observe dipolar couplings, however, does not argue for the absence of structure unless it can be established that the denatured protein has been aligned. Unfortunately, the only method for establishing alignment is the observation of dipolar couplings. (Prestegard et al., 2000). Although one possibility is that SDS truly effaces all structure, a more likely explanation is that SDS–protein micelles are nearly spherically symmetric, and thus are difficult to align.

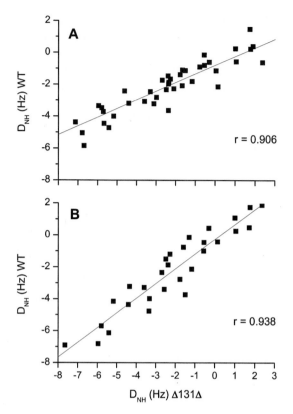

FIG. 4. Scatterplots of the N_H residual dipolar couplings from three different denatured states of staphylococcal nuclease plotted against each other. (A) Wild-type, full-length nuclease in 4 M urea (y-axis) versus $\Delta131\Delta$ in buffer (x-axis). (B) Wild-type, full-length nuclease denatured by acid (25 mM citrate, pH 3.0) (y-axis) versus $\Delta131\Delta$ in buffer (x-axis).

IV. PHYSICAL-CHEMICAL EXPLANATIONS OF LONG-RANGE STRUCTURE

A. The Paradox of Long-Range Structure in the Context of Little Local Structure

Even though $\Delta131\Delta$ exhibits little local structure in high concentrations of urea, it displays unmistakable long-range structure, a paradox that cannot be completely resolved. The structure reported by dipolar couplings represents the relative orientation of individual bond vectors after averaging over the entire ensemble of conformations. Yet the averaged scalar couplings and small or undetectable secondary chemical shifts suggest large fluctuations in the phi/psi angles of every residue. The ^{15}N relaxation data support the conclusion that, even with a time window of 1 ns, the N_H bond vector must wobble in a cone with a large angle. Nevertheless, the average relationship of residues to one another and the entire molecule remains nearly constant during the expansion that accompanies the change from 0 M to 8 M urea.

The absence of measurable secondary shifts for [13]CA and [13]CO resonances suggests that residues are frequently interconverting between the alpha and beta regions of the Ramachandran plot, a very substantial change in local structure. But this interpretation may be in error. Perhaps there is a mechanism to average backbone chemical shifts that does not require such large changes in backbone structure. If not, there must be very extensive coupling between bond rotations of neighboring residues (crankshaft motions) that keep the general direction of the polypeptide chain constant in the presence of large variations in phi and psi.

To explain the persistent structure in $\Delta131\Delta$ in 8 M urea, one must consider the major attractive interactions—dispersion forces, hydrogen bonds, electrostatic forces between opposite charges, and hydrophobic interactions. Because the first two are very short range and depend on the details of the local geometry, it seems unlikely they can play a major role. Electrostatic forces between unit charges, which are relatively long range, have never been ascribed a major role in stabilizing protein structure, so they also seem an unlikely candidate.

Hydrophobic interactions, on the other hand, are strong, indifferent to local details, and are relatively long range. If transient direct or water-separated contacts occur between nonpolar side chains, the net effect could be local organization and an overall compaction of the polypeptide chain. Whereas the strengths of hydrophobic interactions must be considerably reduced in 8 M urea, they clearly are not eliminated, as evidenced by the persistence of lipid bicelles. Thus hydrophobic interactions probably play some role in persistent global structure, the importance of which can be tested by replacing multiple hydrophobic side chains with similarly shaped polar ones.

A less obvious explanation is that the observed residual structure is not due to attractive interactions, but rather to repulsive ones. The steric repulsion between atoms forced to partially overlap is a dominant, if not the dominant, force in all of chemistry. These highly local interactions are known to be important in polymer conformations (Flory, 1969; Ramachandran and Sasisekharan, 1968). For homopolymers or simple alternating polymers, they can often be safely neglected by assuming they confer no net directionality to the chain. Polypeptide chains, however, are chiral and support specific sequences of 20 differently shaped and charged side chain groups. Consequently, the assumption that conformational averaging will confer no net directionality to the backbone has no secure basis.

B.   The Case for Local Side Chain Backbone Interactions

Textbook discussions of the steric effects of side chains on backbone conformation convey the impression that, except for glycine and proline, all side chains restrict the allowed range of phi and psi to approximately the same extent. If this were true, encoding of long-range structure by local steric repulsion would seem unlikely.

A detailed quantitative analysis of the preferences of amino acids in folded proteins for different regions of the Ramachandran plot reveals that the 18 nonglycine, nonproline residues exhibit different preferences (Shortle, 2002). Figure 5 shows the range of relative propensities displayed by these 18 amino acids for a somewhat arbitrary subdivision

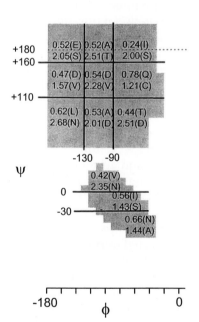

FIG. 5.   Subdivisions of the phi/psi or Ramachandran map labeled with the range of propensities for the 18 amino acids (glycine and proline are excluded) as they map to each subdivision calculated from a large collection of folded protein structures. The propensity is defined by the probability that amino acid x will be found in a subdivision divided by the probability that an "average" amino acid will be found in a subregion. Data are taken from Table 1 of Shortle (2002).

of the beta strand region into nine subdivisions and the alpha helical region into three subdivisions.

Although these differences in propensities are not large, their energetic importance can be estimated by converting them into equivalent free energy differences using the equation $-RT$ ln K, where K is equated to the propensity. [The propensity is approximately equivalent to the equilibrium constant for an imaginary exchange reaction, in which an "average" side chain is replaced with a specific one (Shortle, 2002).] Using this "Boltzmann hypothesis" (Finkelstein *et al.*, 1995), the preferences of amino acids for different subdivisions of the Ramachandran plot often translate into approximately -0.5 to $+0.5$ kcal/mol, a modest but significant term.

A reasonable test of the importance of these residue-specific phi/psi references is to use the calculated propensities in a pseudo-energy function for evaluating different conformations (Shortle, 2002). The simplest such function is to sum the logarithms of the propensities from Figure 5 for each residue after placing a protein sequence in a unique conformation. When a set of 121 proteins was broken into arbitrary overlapping fragments of length 40 and then "threaded" through approximately 1700 protein structures, the correct conformation had the best score versus 300,000 incorrect conformations approximately 20% of the time and a score among the best 0.1% of conformations more than 60% of the time. Using scoring functions based on composite propensities for several structural features leads to even higher rates of identifying the wild-type structure as the one with the optimal set of propensities. These data strongly suggest that very local side chain–backbone interactions represent a major determinant of the final conformation of folded proteins.

C. Prediction of Backbone Conformation from Phi/Psi Propensities

Another simple test of the structural information in these propensities is to use them to predict secondary structure. Figure 6 shows the results of one such test, in which the sequence of wild-type nuclease was broken into overlapping fragments of length 12. Beginning with the amino terminus and shifting the starting point of the fragment by three residues, sequence fragments were generated and then threaded through 2500 high-resolution crystal structures. The 25 structural fragments with the best scores are saved, along with their secondary structure. For each position in the sequence of nuclease, the fractions of saved fragments that were helical, strand, or turn (i.e., not helix or strand) are plotted as a function of position. Although the results

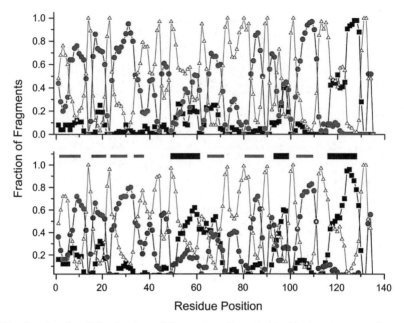

FIG. 6. Graphs of the fraction of fragments of length 12 with the corresponding sequence from staphylococcal nuclease found to have helical (■), strand (●), or turn (△) secondary structure. The locations of the three helices (▬) and seven strands (▬) are shown between the two graphs. Overlapping sequence fragments of length 12 beginning at position 1 and then shifted by 3 residues (1, 4, 7, etc.) were generated. For each sequence fragment, the 20 structural fragments with the best scores were retrieved from a set of 2500 high-resolution proteins, and the secondary structure at each position was averaged over all fragments. Top panel: The scoring function was the sum of the logarithms of propensities for phi/psi angles of I and $I \pm 1$, with 16 subdivisions for I, and 9 for $I \pm 1$. Bottom panel: The scoring function includes that used in the top panel plus the logarithm of rotamer probabilities (Shortle, 2002).

are not perfect, they demonstrate that the positions of all turns are correctly identified and each of the six beta strands is found. Only one of the three helices is well defined.

The analysis can be extended to include another important steric term, the side chain rotamer preference. If this log of the backbone-dependent rotamer probability is simply added to the phi/psi scoring function (Shortle, 2002), the analysis shows a higher level of success in predicting secondary structure (Fig. 6), with all three helices generating clear signals. Addition of a rotamer term to the scoring function in the fragment recognition test described above also significantly improves the results.

For a number of reasons, the statistical nature of the above estimate of the energetic importance of side chain–backbone interactions means it will almost certainly be an underestimate. (1) The binning of data in arbitrarily defined subdivisions of the Ramachandran plot involves loss of detail. (2) Each residue has been treated as an isolated entity, with all interactions beyond a side chain interacting with the two peptide bonds that flank it having been ignored. Thus the coupling of steric clashes involving flanking side chains or residues more than one removed have been ignored (Pappu *et al.*, 2000). (3) Steric effects of the side chain that involve more than simple description by the single dihedral angle chil have been ignored. (4) The frequencies of occurrence of phi/psi values for a residue type will be distorted by long-range interactions in the folded state, overriding local preferences to maximize packing and burial of nonpolar atoms.

Consequently, the impact of local steric interactions in the denatured state is likely to be significantly greater than that estimated above. Although these simple arguments in no way prove that a native-like topology is fully encoded in every polypeptide chain by local interactions, they do support the idea that long-range structure, which cannot be removed by strongly denaturing conditions, could arise predominantly from local steric hindrance.

V. Conclusions

A. *Implications of Locally Encoded Long-Range Structure*

Scientific thinking, like all human though, makes extensive use of analogies and metaphors to represent objects that cannot be seen. Long flexible cylinders such as a piece of cooked spaghetti or a garden hose have served as reasonable metaphors in classical polymer physical chemistry for representing polymer chains that are chemically simple and achiral. For polypeptide chains, however, the connotation of isotropic deformability implicit in these metaphors is clearly misleading, probably grossly so. A few minutes of manipulation of a plastic, space-filling CPK model of an octapeptide will convince most protein chemists of the severe restrictions to bending imposed by local interactions. Perhaps no common object from daily life can serve as a suitable metaphor for the exceedingly complex local restrictions to flexion of polypeptide chains.

In view of the large variations in phi/psi preferences for the different amino acids observed in folded protein (see Fig. 5), it follows that the task of encoding the general features of long-range structure does not fall entirely to specific side chain/side chain interactions. The role of these chemical interactions may be less to encode the topology of the chain and more to facilitate stabilization of one specific conformation out of the many trapped within the topological space encoded by local interactions.

Because simple lattice models take no account of local directional preferences, they fail to model these important local restraints on protein structure. Instead, they rely almost entirely on long-range interactions to encode the most stable conformation(s) (Dill *et al.*, 1995). Thus the ability of lattice models to reproduce protein-like behavior must be called into question. And though their simplicity makes them intellectually attractive, their use in teaching and modeling protein-like behavior must be qualified with a caveat that local directional preferences have been ignored.

B.   Three Physical States: Expanded Denatured, Compact Denatured, and Native

The changes in structure of denatured nuclease as a function of urea concentration (Fig. 3) suggest that, as hydrophobic interactions are weakened and the backbone becomes more highly solvated, the chain expands gradually. The data presented by Millet *et al.* in this volume suggest that this expansion does not continue asymptotically as predicted by simple polymer physical chemistry. This is the behavior expected for a polypeptide chain trapped in a small region of conformation space. Most, perhaps all, of the conformations accessible in the expanded denatured state may have a native-like topology.

If so, the role of hydrophobic interactions between nonpolar side chain atoms may be to drive the expanded denatured state to a more compact, but not quite native, state during folding. This collapse is likely to be a continuous transition rather than a cooperative, "first-order" transition, since hydrophobic interactions are indifferent to the details of distance and angular relationships (buried is buried). Because of the smaller entropy loss from closure of short loops, local hydrophobic interactions are likely to form before long-range contacts. The end product of this collapse would be a compact denatured state, the so-called "molten-globule state," which displays the characteristic property of severe broadening of NMR lines due to multiple conformations with different chemical shift environment interconverting on the millisecond time scale.

Finally, the native state is achieved from the compact denatured state in the major cooperative reaction of folding, which is accompanied by

Three Physical States of Proteins

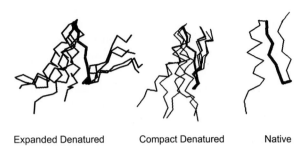

Expanded Denatured   Compact Denatured   Native

FIG. 7.   Simple schematic diagram representing three conformational states of proteins: The expanded denatured state whose structure is determined by steric interactions; the compact denatured state, with structure determined by hydrophobic buril; and the native state, with structure determined by dispersion forces, hydrogen bonds, and electrostatics.

a large release of heat under most conditions. Presumably, in this reaction tight packing between buried side chain and backbone atoms is driven by dispersion forces, and to a lesser extent by hydrogen bond formation/polarization.

This simple three-state model of protein folding, shown schematically in Figure 7, ascribes a separate force to shaping the structure of each state. Local steric interactions trap the protein chain in a large ensemble of conformations with the correct topology; hydrophobic interactions drive the chain to a smaller, more compact subset of conformations; then dispersion forces supply the enthalpy loss required to achieve a relatively fixed and rigid ensemble of native conformations.

Although additional experiments and simulations are needed to determine how much of reality is captured in this model, it does explain one important property of proteins: their rapid rate of refolding, which is independent of denaturation conditions. If the polypeptide chain is unable to escape from this steric trap and access conformations with the wrong topology, it could never wander far from the folded conformation and thereby avoid incorrect side chain/side chain interactions.

C.   The Unrecognized Role of the Ribosome in the Energetics of Protein Conformations

If the expanded denatured state is trapped in small region of conformation space where all chains have the correct topology, how did it get there? To this question there is but one answer. A fraction of the free energy of ATP hydrolysis consumed during protein translation is used to limit the entropy of the polypeptide chain. The peptide bond can form between two amino acids only when their side chains have moved into positions that no longer interfere with attack of one residue's amino group on the other residue's carboxyl group. Therefore, as each peptide linkage is formed on the ribosome, a subset of phi/psi/chi angles is effectively lost from the space of possible conformations. Not only does the ribosome determine the primary structure of each protein it makes, it also establishes the topological space in which that protein chain will be confined for the rest of its existence.

REFERENCES

Ackerman, M. S., and Shortle, D. (2002). *Biochemistry* **41,** 3089–3095.
Alexandrescu, A. T., and Shortle, D. (1994). *J. Mol. Biol.* **242,** 527–546.
Alexandrescu, A. T., Abeygunawardana, C., and Shortle, D. (1994). *Biochemistry* **33,** 1063–1072.
Anfinsen, C. B., Schecter, A. N., and Taniuchi, H. (1972). *Cold Spring Harbor Symp. Quant. Biol.* **36,** 249–255.
Dill, K. A., and Shortle, D. (1991). *Ann. Rev. Biochem.* **60,** 795–825.
Dill, K. A., Bromberg, S., Yue, K., Fiebig, K. M., Yee, D. P., Thomas, P. D., and Chan, H. S. (1995). *Protein Sci.* **4,** 561–602.
Dobson, C. M., Evans, P. A., and Radford, S. E. (1994). *Trends Biochem. Sci.* **19,** 31–37.
Dyson, H. J., and Wright, P. E. (2001). *Methods Enzymol.* **339,** 258–270.
Finkelstein, A. V., Badretdinov, A. Y., and Gutin, A. M. (1995). *Proteins: Struct. Funct. Genet.* **23,** 142–150.
Flanagan, J. M., Kataoka, M., Shortle, D., and Engelman, D. M. (1992). *Proc. Natl. Acad. Sci. USA* **89,** 748–752.
Flory, P. J. (1969). *Statistical Mechanics of Chain Molecules.* Wiley, New York.
Gillespie, J. R., and Shortle, D. (1997a). *J. Mol. Biol.* **268,** 158–169.
Gillespie, J. R., and Shortle, D. (1997b). *J. Mol. Biol.* **268,** 170–184.
Kosen, P. A. (1989). *Methods Enzymol.* **177,** 86–121.
Levinthal, C. (1968). *J. Chim. Phys.* **65,** 44–45.
Lipari, G., and Szabo, A. (1982). *J. Amer. Chem. Soc.* **104,** 4546–4559.
Meiler, J., Blomberg, N., Nilges, M., and Griesinger, C. (2000). *J. Biomol. NMR* **16,** 245–252.
Miller, W. G., and Goebel, C. V. (1968). *Biochemistry* **7,** 3925–3935.
Mori, S., van Zijl, P. C. M., and Shortle, D. (1997). *Proteins: Struct. Funct. Genet.* **28,** 325–332.
Palmer, A. G. (2001). *Ann. Rev. Biophys. Biomol. Struct.* **30,** 129–155.
Pappu, R. V., Srinivasan, R., and Rose, G. D. (2000). *Proc. Natl. Acad. Sci. USA* **97,** 12565–12570.
Prestegard, J. H., Al-Hashimi, H. M., and Tolman, J. R. (2000). *Quart. Rev. Biophys.* **33,** 371–424.
Ramachandran, G. N., and Sasisekharan, V. (1968). *Adv. Protein Chem.* **23,** 283–438.
Reily, M. D., Thanabal, V., and Omecinsky, D. O. (1992). *J. Amer. Chem. Soc.* **114,** 6251–6252.
Sinclair, J. F., and Shortle, D. (1999). *Protein Sci.* **8,** 991–1000.
Shortle, D. (1996). *Curr. Opin. Struct. Biol.* **6,** 24–30.
Shortle, D., and Ackerman, M. S. (2001). *Science* **293,** 487–489.

Shortle, D. (2002). *Protein Sci.* **11,** 18–26.
Tanford, C. (1968). *Adv. Protein Chem.* **23,** 121–282.
Tanford, C. (1970). *Adv. Protein Chem.* **24,** 1–95.
Wang, Y., and Shortle, D. (1995). *Biochemistry* **34,** 15895–15905.
Wang, Y., and Shortle, D. (1996). *Protein Sci.* **5,** 1898–1906.
Wishart, D. S., and Sykes, B. D. (1994). *Methods Enzymol.* **239,** 363–392.
Wuthrich, K. (1994). *Curr. Opin. Struct. Biol.* **4,** 93–99.
Zhang, O., Kay, L. E., Shortle, D., and Forman-Kay, J. D. (1997). *J. Mol. Biol.* **272,** 9–20.

IDENTIFICATION AND FUNCTIONS OF USEFULLY DISORDERED PROTEINS

By A. KEITH DUNKER,* CELESTE J. BROWN,* and ZORAN OBRADOVIC†

*School of Molecular Biosciences, Washington State University, Pullman, Washington 99164, †Center for Information Science and Technology, Temple University, Philadelphia, Pennsylvania 19122

Numerous apparently native proteins have disordered regions and some are wholly disordered, yet both types of protein often utilize their unordered amino acids to carry out function. How do such disordered proteins fit into the view that amino acid sequence codes for protein structure? Our hypothesis was that disorder, like order, is encoded by the amino acid sequence. To test this hypothesis, intrinsically disordered protein was compared with ordered protein. Initially, small sets of ordered and disordered sequences were compared by prediction of order and disorder. Success rates much greater than expected by chance indicated that disorder is encoded by the sequence. Once larger datasets of ordered and disordered proteins were collected, direct sequence comparisons could be made. The amino acid compositions, sequence attributes, and evolutionary characteristics of disordered sequences differed from the corresponding features of ordered sequences in ways commensurate with our hypothesis that disorder is encoded by the sequence. The differences between ordered and disordered sequences allowed the development of predictors of natural disordered regions (PONDRs). Sequence codes for structure and sequence codes for disorder; thus, it follows that disorder ought to be considered a category of native protein structure. Two questions arise from this categorization: How common is native protein disorder? What functions are

ADVANCES IN
PROTEIN CHEMISTRY, Vol. 62

carried out by this category of structure? Application of a particular PONDR to genomic sets of sequences indicated that disordered regions are extremely common, especially in eukaryotic cells. Because function depends on structure and disorder is a type of structure, disorder–function relationships were examined. A wide variety of functions were found to be associated with disordered structure, the common use of disorder in signaling and information networks being especially interesting.

I. Testing Whether Intrinsic Disorder Is Encoded by the Amino Acid Sequence

In our usage, an ordered protein contains a single canonical set of Ramachandran angles, whereas a disordered protein or region contains an ensemble of divergent angles at any instant and these angles interconvert over time. Intrinsically disordered protein can be extended (random coil–like) or collapsed (molten globule–like). The latter type of disorder typically includes regions of fluctuating secondary structure, so disorder does not mean absence of helix or sheet. Both types of disorder have been observed in apparently native proteins (Williams, 1989; Wright and Dyson, 1999; Demchenko, 2001; Dunker and Obradovic, 2001; Uversky, 2002).

Intrinsic disorder might not be encoded by the sequence, but rather might be the result of the absence of suitable tertiary interactions. If this were the general cause of intrinsic disorder, any subset of ordered sequences and any subset of disordered sequences would likely be the same within the statistical uncertainty of the sampling. On the other hand, if intrinsic disorder were encoded by the amino acid sequence, any subset of disordered sequences would likely differ significantly from samples of ordered protein sequences. Thus, to test the hypothesis that disorder is encoded by the sequence, we collected examples of intrinsically ordered and intrinsically disordered proteins, then determined whether and how their sequences were distinguishable.

A. Using Prediction

In our first attempts to determine whether sequence codes for disorder, lack of resources limited us to the collection of a small number of disorder examples, which contained only about 1200 residues in total. For such small numbers, differences between ordered and disordered sequences could not be discerned with statistical reliability. Thus, we turned to prediction as a way to estimate whether ordered and disordered sequences are the same or different (Romero *et al.*, 1997b).

In these initial studies, disordered regions were sorted according to length: short = 7–21, medium = 22–39, and long = 40 or more residues. As detailed previously (Romero et al., 1997b) and described in Section II,A, predictors were developed for each length class and for the three length classes merged together. For predictor training, an initialization is required: Five independent initializations were used for each predictor. Also, 5-cross validation on disjoint training and testing sets was used, so each result is based on $5 \times 5 = 25$ sets of predictions.

Since equal numbers of disordered and ordered residues were used for training and testing, prediction success would be about 50% if disordered and ordered sequences were the same. In contrast to this 50% value, prediction success rates for the short, medium, long, and merged datasets were 69% ± 3%, 74% ± 2%, 73% ± 2%, and 60% ± 3%, respectively (Romero et al., 1997b), where the standard errors were determined over about 2200, 2600, 2000, and 6800 individual predictions, respectively.

The success rate of every prediction set was greater than the value of 50% expected by chance. Specifically, the various sets of predictions differed from the 50% value by about 3 standard deviations (for the lowest success rate, which was for the merged data) to about 12 standard deviations (for the highest success rates, which were for the medium and long regions of disorder). Overall, these data provided very strong support for our hypothesis that disorder is encoded by the amino acid sequence (Romero et al., 1997b).

B. Database Comparisons

To obtain statistically significant comparisons of ordered and disordered sequences, much larger datasets were needed. To this end, disordered regions of proteins or wholly disordered proteins were identified by literature searches to find examples with structural characterizations that employed one or more of the following methods: (1) X-ray crystallography, where absence of coordinates indicates a region of disorder; (2) nuclear magnetic resonance (NMR), where several different features of the NMR spectra have been used to identify disorder; and (3) circular dichroism (CD) spectroscopy, where whole-protein disorder is identified by a random coil–type CD spectrum.

Once sufficient numbers of intrinsically disordered and ordered protein sequences were collected, it became possible to compare them directly. The sequences in these databases were examined for differences in amino acid composition, sequence attributes, and evolutionary characteristics.

TABLE I

Number of Proteins and Residues in Databases of Intrinsically
Disordered Protein Characterized by Various Methods

Detection method	Number of proteins	Number of residues
X-ray	59	3907
NMR	43	4108
CD	55	10,818
Merged	157	18,833

1. Databases of Ordered and Disordered Proteins

Three groups of disordered proteins have been assembled, with the groups defined by the experimental method used to characterize the lack of ordered structure. Because the focus has been on long regions of disorder, an identified disordered protein or region was not included in these groups if it failed to contain 40 or more consecutive residues. Disordered regions from otherwise ordered proteins as well as wholly disordered proteins were identified. Table I summarizes the collection of sequences in this database.

Three groups of ordered sequences have been developed from the Protein Data Bank (PDB) for various purposes. The first group, called Globular 3-D, was formed from NRL 3-D (Pattabiraman *et al.*, 1990) by deletion of the nonglobular proteins. NRL 3-D contains essentially all of the residues with backbone coordinates in PDB. Globular 3-D has the advantage of containing the largest number of ordered chains and residues, but also contains many proteins with high sequence similarity or even identity. The second group was constructed by deleting the unobserved residues and keeping the observed residues from a nonredundant subset of proteins called PDB_Select_25 (Hobohm and Sander, 1994), yielding the collection called O_PDBS25. Because PDB_Select_25 was formed by grouping PDB proteins into sets having 25% or more sequence identity, the sequence identity between any two proteins in O_PDBS25 is less than 25%. The third group was assembled from the proteins in PDB_Select_25 having no unobserved residues, giving the completely ordered subset or CO_PDBS25. These three groups of ordered proteins are described in Table II.

2. Comparing Amino Acid Compositions

Different protein folding classes can be identified by differences in their amino acid compositions (Nakashima *et al.*, 1986); thus, we reasoned that, if disorder were encoded by the sequence, then regions of disorder would be analogous to a new folding class and hence should

TABLE II

Number of Proteins and Residues in Databases of Intrinsically Ordered Protein

Name	Number of proteins	Number of residues
Globular 3-D	14,540	2,610,197
O_PDBS25	1021	222,116
CO_PDBS25	130	32,509

be distinguishable by amino acid compositional differences compared to ordered protein.

Figure 1 shows the amino acid compositions and compositional differences of the various protein groups versus amino acid type, where the amino acids are arranged using the "flexibility" scale of Vihinen and co-workers. In this arrangement, the tendency to be buried increases to the left and the tendency to be exposed increases to the right (Vihinen *et al.*, 1994). The compositions of the three disordered sets are very similar to one another (Fig. 1, top), and the compositions of the three ordered sets are even more similar to one another (Fig. 1, middle). The compositional differences show systematic distinctions between ordered and disordered protein (Fig. 1, bottom). Because the differences are calculated as (disorder − order)/(order), positive peaks represent fractional enrichments and negative peaks represent fractional depletions of amino acids in disordered as compared to ordered protein.

Although the disordered proteins were characterized by three different methods, all three datasets of disordered amino acids showed semiquantitatively similar changes for 16 of the 20 amino acids (Fig. 1, top). Because the three methods rely on completely different underlying biophysical principles for determining disorder, substantial compositional differences among the datasets were expected; surprisingly, however, such differences were not observed. Thus, the compositions of the disordered proteins (Fig. 1, top) likely indicate inherent tendencies of this type of protein.

A few proteins are extremely overrepresented in PDB, and CO_PDBS25 does not contain very many proteins; hence Globular 3-D, O_PDBS25, and CO_PDBS25 might exhibit compositional differences. The three groups, however, have nearly the same composition for every amino acid (Fig. 1, middle), so overrepresentation of some proteins in Globular 3-D and the smaller number of sequences in CO_PDBS25 did not lead to significant amino acid biases.

The compositional differences (Fig. 1, bottom) show that, compared to ordered protein, the three disordered datasets exhibit large and

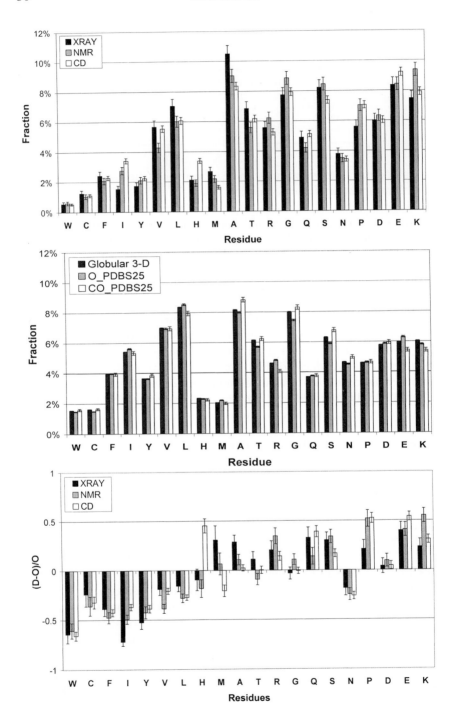

significant depletions of eight amino acids (W, C, F, I, Y, V, L, and N) enrichments in seven (K, E, P, S, Q, R and A), and inconsistent changes for five (H, M, T, G, and D).

All the disorder-specific depletions except one are, from the leftmost, typically buried amino acids. The one exception, N, is out of place in being both a surface-preferring and an order-promoting residue. Perhaps the short side chain, with its propensity to hydrogen-bond to the backbone (Presta and Rose, 1988; Richardson and Richardson, 1988), tends to induce local structure. This order-inducing tendency might explain the out-of-place behavior of N.

The disorder-specific enrichments are mainly, from the rightmost, typically exposed amino acids, with the exceptions of N, T, G, and D. Like N, T and D also have groups with hydrogen-bonding potential attached to the β-carbon and so can readily form hydrogen bonds with the backbone. Thus, the tendencies of T and D to be less disorder-promoting than their neighbors might also be due to the ordering effects of such hydrogen bonding.

Here G is classified as order–disorder neutral, whereas in a previous study G was classified as disorder-promoting (Williams et al., 2001). This change in classification arises from very small differences between the disordered data in the two studies. The data then and now are not statistically different from each other; neither are the data significantly different between order and disorder for this residue.

During the development of the flexibility scale, a dual behavior was noticed for G. Specifically, when flanked by residues with the high flexibility indices, G exhibited a high average flexibility index; but when flanked by residues with low flexibility indices, G exhibited a low average flexibility. The context dependence of the flexibility index of G was much larger than the context dependence of any other residue. To explain these data, it was suggested that when it is in a flexible region, G enhances the flexibility by being able to adopt many conformations; but when it is in a rigid, buried region, G enhances the rigidity by facilitating tight packing (Vihinen et al., 1994). This dual structural role for G may account for its neutrality with respect to the promotion of order or disorder.

FIG. 1. Comparisons of amino acid compositions of ordered protein and disordered protein. (Top) Amino acid compositions of three disordered datasets. (Middle) Amino acid compositions of three ordered datasets. (Bottom) Compositions of disordered datasets relative to the Globular 3-D dataset (from Romero et al., Proteins: Struct., Funct., Gen. 42, 38–48, copyright © 2001. Reprinted by permission of Wiley-Liss, Inc., a subsidiary of John Wiley & Sons, Inc.). The ordinates are (% amino acid in disordered dataset − % amino acid in Globular 3-D)/(% amino acid in Globular 3-D) = (D−O)/O. Negative values indicate that the disordered database has less than the ordered, positive indicates more than the ordered. Error bars are one standard deviation.

The last two amino acids, H and M, are inconsistent across the three datasets, but both are found between the order-promoting and disorder-promoting sets. Thus, these amino acids exhibit intermediate tendencies to be buried or exposed and likewise exhibit intermediate tendencies to promote order or disorder, respectively.

The enrichments and depletions displayed in Figure 1 are concordant with what would be expected if disorder were encoded by the sequence (Williams *et al.*, 2001). Disordered regions are depleted in the hydrophobic amino acids, which tend to be buried, and enriched in the hydrophilic amino acids, which tend to be exposed. Such sequences would be expected to lack the ability to form the hydrophobic cores that stabilize ordered protein structure. Thus, these data strongly support the conjecture that intrinsic disorder is encoded by local amino acid sequence information, and not by a more complex code involving, for example, lack of suitable tertiary interactions.

Others have studied the relationship of amino acid composition and protein structure from different points of view. Karlin identified sequences with unusual compositions (Karlin and Brendel, 1992), while Wootton, who used Shannon's entropy to estimate sequence complexity, showed that nonglobularity was associated with low complexity and found that sequence databases were much richer than PDB in proteins with regions of low complexity (Wootton, 1993, 1994b,a; Wootton and Federhen, 1996). Our extensions of Wootton's work revealed that not one of the more than 2.6×10^6 overlapping 45-residue segments in Globular 3-D contains fewer than 10 different amino acids or a Shannon's entropy value less than 2.9 (Romero *et al.*, 1999). Attempts to select a folded but simplified SH3 domain by phage display (Riddle *et al.*, 1997) yielded a protein with a greatly reduced average value for Shannon's entropy and a reduced average amino acid alphabet size, yet the lowest of the resulting reduced complexity values were very similar to the lowest observed in Globular 3-D. The near coincidence of the lowest complexity values for both laboratory and natural selection suggested the possibility of a lower bound for the sequence complexity of ordered, globular protein structure (Romero *et al.*, 1999).

3. Comparing Sequence Attributes

Another way of studying amino acid sequences is by means of sequence attributes such as hydropathy, net charge, side chain volume, and bulkiness. If $P(S|x)$ is the conditional probability of observing structure type S in a region of sequence having an attribute value of x, graphs of $P(S|x)$ versus x were previously shown to provide useful insight regarding sequence–structure relationships (Arnold *et al.*, 1992).

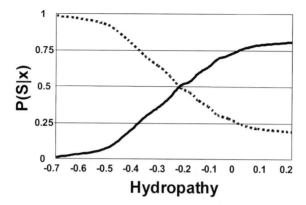

Fig. 2. Conditional probability plot for Sweet and Eisenberg's (1983) hydropathy scale. The black line is the probability (*y*-axis) that a residue is ordered given the hydropathy score indicated on the *x*-axis. The dashed line is the probability of disorder. Negative values for hydropathy indicate hydrophilicity, positive values indicate hydrophobicity. The area between the two curves is divided by the total area of the graph to obtain the area ratio.

For $S =$ order or disorder and for $x =$ average Sweet–Eisenberg hydropathy (Sweet and Eisenberg, 1983) over a window of 21 amino acids, plots of $P(S|x)$ versus x derived from a balanced set of ordered and disordered 21-residue segments gave the data of Figure 2. As expected from the compositional biases displayed in Figure 1, increasing hydropathy values correlate with the formation of ordered structure: The more hydrophilic the sequence, the greater the tendency to be disordered, whereas the more hydrophobic the sequence, the greater the tendency to be ordered.

Dividing the area between the two curves in Figure 2 by the total area gives the area ratio (AR), which provides a means for ranking different attributes (Xie *et al.*, 1998). AR values were used to rank 265 attributes using balanced numbers of ordered and disordered segments having 21 residues each. When this approach was used to compare the X-ray, NMR, and CD–characterized sets of disorder with the same (randomly selected) set of ordered segments, the rankings were very similar but not quite identical across the three disorder datasets, suggesting only very slight differences among the disorder characterized by the three different methods (Williams *et al.*, 2001). These results are consistent with the small compositional differences between the differently characterized disordered regions shown in Figure 1.

Uversky and co-workers recently used a pair of sequence attributes, specifically the Kyte–Doolittle hydropathy scale and net charge, to distinguish between folded and "natively unfolded" proteins (Uversky *et al.,* 1999). With an AR of 0.42, the Kyte–Doolittle scale ranked 79th among the 265 scales, well below five other hydropathy scales. The Sweet–Eisenberg scale, which ranked third overall with an AR value of 0.538 (Williams *et al.,* 2001), was the best of this type for discriminating between the order and disorder in our datasets. Since the Sweet–Eisenberg scale was developed by determining each amino acid's average degree of exposure in a set of structures from PDB, it is interesting that this scale outperforms other hydropathy scales with regard to discriminating between ordered and disordered segments. Furthermore, net charge, with an AR value of 0.236, ranked 174th. Thus, the hydropathy–net charge pair is almost certainly not the best pair of attributes for discriminating ordered and disordered sequences. Nevertheless, the work of Uversky and co-workers was a very important contribution because of its simplicity and the insight it provided. Also, this pair of attributes could be optimal for some particular groups of natively unfolded proteins.

Like the differences in composition, the sequence attribute differences between ordered and disordered sequences are exactly as would be expected if disorder were encoded by the sequence. Attribute values, of course, depend on amino acid compositions, so the composition and attribute results are not independent. Nevertheless, the attribute analysis is useful because it provides a biophysical perspective, allowing insight into the amino acid properties that are important for promoting order or disorder.

4. Comparing Evolutionary Characteristics

A third way of comparing ordered and disordered sequences is by their relative changes over evolutionary time. The requirement to form well-packed protein cores imposes constraints on ordered protein, and these constraints are absent in disordered regions. Disordered regions are therefore generally expected to exhibit higher rates and different patterns of amino acid substitution over evolutionary time as compared to ordered regions. But because function, not structure, is the property subjected to natural selection, special circumstances could mitigate the expected general trend. Such exceptions could potentially provide additional insight regarding the importance of intrinsic disorder. On a related matter, if differences in evolutionary characteristics are found between ordered and disordered regions, such a result would be a strong indicator that disorder exists *in vivo*.

a. Construction of Disordered Protein Families. To test the hypothesis that disordered and ordered proteins differ in both the quantity and quality of their evolutionary change, families of disordered proteins were developed. Families were constructed for proteins from each of the three disordered groups listed in Table I, including proteins that are wholly disordered and proteins with regions of disorder and order. The entire protein sequence was used to identify homologous members of the protein family by BLASTP searches (Altschul *et al.*, 1990,1997) on the nonredundant protein database at NCBI (www.ncbi.nlm.nih.gov). Homologs were aligned using the default settings of CLUSTALW (Thompson *et al.*, 1994) at the Baylor College of Medicine website (Smith *et al.*, 1996). Except for the wholly disordered proteins, aligned sequences were then partitioned into ordered sequences and disordered sequences based on alignment with the structurally characterized sequence. This procedure is outlined in Figure 3.

  b. Comparing Substitution Patterns. To test whether disordered and ordered proteins differ in the pattern of their evolutionary change, substitution matrices were constructed based on 55 aligned disordered protein families. The substitution matrix based on disordered protein families was then compared to the commonly used substitution matrices, which are based on mostly ordered protein families. To develop the disordered

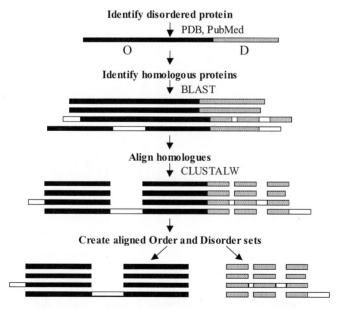

FIG. 3. Procedures for identifying order and disorder in protein families. Black boxes indicate ordered sequences, gray boxes indicate disordered sequences, and open boxes indicate insertions relative to the starting sequence. (From Brown *et al.*, 2002, *J. Mol. Evol.* **55**, 104–110, with permission of Springer-Verlag.)

matrix, initial alignments were performed (Fig. 3) using the BLOSUM62 substitution matrix and the usual first-gap/gap-extension penalties. From the aligned disordered sequences, a new substitution matrix was built. New sequence alignments were developed using the new matrix for disorder. This new-alignment/new-matrix cycle was continued until the change of the matrix in two successive iterations dropped below a prespecified threshold, thus yielding a substitution matrix specifically for regions of intrinsic disorder. The relative improvement of the final matrix over published substitution matrices was confirmed by tests that used Hidden Markov Models of aligned family and nonfamily sequences (Radivojac *et al.*, 2002).

This procedure led to a substitution matrix for aligning disordered protein that was different from the commonly used substitution matrices, such as BLOSUM62 (Fig. 4). The matrix for disordered protein is generally better than order-based matrices for aligning disordered proteins whose sequence identities are between 20 and 50%. These results indicate that disordered and ordered protein can be distinguished by their patterns of evolutionary change.

	C	S	T	P	A	G	N	D	E	Q	H	R	K	M	I	L	V	F	Y	W	B	Z
C	10/-1	-1	-2	-1	1	0	-2	0	0	0	-2	-2	0	-1	-1	0	-2	-1	-2	3	0	0
S	0	3/1	0	-1	0	0	0	0	1	0	0	0	1	1	0	0	0	0	0	0	0	1
T	1	1	4/1	0	0	0	0	0	0	-1	-2	0	-1	0	0	1	0	0	-1	3	0	0
P	-2	0	-1	6/1	0	-1	-1	1	0	0	0	0	0	0	-1	-2	-1	-1	0	-3	1	0
A	-1	1	0	-1	3/1	0	-1	-1	0	0	0	1	0	0.	0	0	0	0	0	2	-1	0
G	-3	0	-2	-1	0	5/1	0	0	0	0	0	-1	0	0	1	1	0	1	1	0	2	0	0
N	-1	1	0	-1	-1	0	4/2	0	0	0	-1	-1	0	0	0	0	0	-1	-1	-1	0	0
D	-3	0	-1	-2	-1	-1	1	4/2	0	0	0	0	0	1	1	0	1	1	1	0	2	0
E	-4	-1	-1	-1	-1	-2	0	2	4/1	2	1	1	1	1	0	0	0	1	1	1	0	1
Q	-3	0	0	-1	-1	-2	1	0	0	5/0	-1	0	1	1	-1	0	0	-1	-1	-1	0	2
H	-1	-1	0	-2	-2	-1	2	-1	-1	1	8/0	0	0	0	-1	-1	-1	-1	0	0	0	1
R	-1	-1	-1	-2	-2	-2	0	-2	-1	1	0	5/0	0	0	-1	0	-1	0	0	-3	0	1
K	-3	-1	0	-1	-1	-2	0	-1	0	0	-1	2	4/1	1	-1	0	0	0	0	0	0	1
M	0	-2	-1	-2	-1	-4	-2	-4	-3	-1	-2	-1	-2	7/-2	0	0	0	-1	0	0	1	1
I	0	-2	-1	-2	-1	-5	-3	-4	-3	-2	-2	-2	-2	1	4/0	0	0	-1	-1	-1	1	0
L	-1	-2	-2	-1	-1	-4	-3	-4	-3	-2	-2	-2	-2	2	2	4/0	0	-1	-1	0	0	0
V	1	-2	0	-1	0	-4	-3	-4	-2	-2	-2	-2	-2	1	3	1	4/0	-1	0	1	1	0
F	-1	-2	-2	-3	-2	-4	-2	-4	-4	-2	0	-3	-3	1	1	1	0	7/-1	-1	2	1	1
Y	0	-2	-1	-3	-2	-3	-1	-4	-3	0	2	-2	-2	-1	0	0	-1	4	8/-1	3	1	1
W	-5	-3	-5	-1	-5	-4	-3	-4	-4	-1	-2	0	-3	-1	-2	-2	-4	-1	3	13/-2	0	1
B	-3	0	-1	-2	-1	-1	1	4	2	0	-1	-2	-1	-4	-4	-4	-4	-4	-4	-4	4/2	0
Z	-4	-1	-1	-1	-1	-2	0	2	4	0	-1	-1	0	-3	-3	-3	-2	-4	-3	-4	2	4/1

FIG. 4. Substitution matrix based on disordered protein families. Below the diagonal are the scores for each amino acid substitution. Above the diagonal are the differences between BLOSUM 62 and the disorder matrix. On the diagonal are the scores/differences. (From Radivojac *et al.*, 2002, *PSB 2002* **7**, 589–600, with permission of World Scientific Publishing Co. Pte Ltd.)

c. Comparing Substitution Rates. To test whether disordered protein evolves more rapidly than ordered protein, comparisons were made between the ordered and disordered regions of 26 protein families having both order and long regions of disorder (Brown *et al.*, 2002). Twenty-four of the families had been structurally characterized by NMR or X-ray crystallography. The ordered and disordered regions in the two CD-characterized proteins had been dissected by limited proteolysis. The pairwise genetic distance for each ordered region was compared to the pairwise genetic distance for the corresponding disordered region from the same protein pairs. The average difference between these pairwise genetic distances (Δ) was calculated for each protein family (Table III). An appropriate statistical test was designed to determine whether Δ was significantly different from zero (i.e., if ordered and disordered regions differed in their rates of evolution). For five families, there were no significant differences in pairwise genetic distances between ordered and disordered sequences. The disordered region evolved significantly faster than the ordered region for 19 of the 26 families. The functions of these disordered regions are diverse, including binding sites for protein, DNA, or RNA as well as flexible linkers. The functions of some of these regions are unknown. The disordered regions evolved significantly more slowly than the ordered regions for the two remaining families. The functions of these more slowly evolving disordered regions include sites for DNA binding. These results indicate that, in general, disordered regions of proteins evolve more rapidly than their ordered regions. Understanding the exceptions to these rules may help to further characterize the roles of disorder in protein function.

II. Prediction of Order and Disorder from the Amino Acid Sequence

The results described in Section I suggest that amino acid sequence codes for intrinsic protein disorder. In this circumstance, constructing a predictor of order and disorder would be useful as a means to extend and generalize from the current experimental results.

The steps involved in building disordered predictors are the following: (1) Develop datasets of ordered and disordered protein. (2) Identify a set of features or attributes for discriminating between order and disorder. (3) Use an appropriate set of features or attributes as the basis for predictors of intrinsic order and disorder, while taking care to use disjoint subsets of sequences for training and testing. Each of these steps has been investigated in some detail as described in Section I,A or as reported previously (Romero *et al.*, 1997a,b; Li *et al.*, 1999; Romero *et al.*, 2000; Vucetic *et al.*, 2001).

TABLE III
*Average Difference in Genetic Distance (Δ) between Ordered and Disordered Regions
of 26 Protein Families[a]*

Protein family	Reference	Detection method[b]	Number of sequences	Δ[c]	p Value[d]
Replication protein A	(Jacobs et al., 1999)	NMR	7	1.92	0.001
NF-KB p65	(Schmitz et al., 1994)	NMR	4	1.18	0.001
Glycyl-tRNA synthetase	(Logan et al., 1995)	X-ray	24	1.69	0.002
Regulator of G-protein signaling 4	(Tesmer et al., 1997)	X-ray	17	0.96	0.001
Topoisomerase II	(Berger et al., 1996)	X-ray	28	0.87	0.001
Calcineurin	(Kissinger et al., 1995)	X-ray	23	0.84	0.001
c-Fos	(Campbell et al., 2000)	NMR	23	0.82	0.001
Thyroid transcription factor	(Tell et al., 1998)	CD, LP	12	0.76	0.001
Sulfotransferase	(Bidwell et al., 1999)	X-ray	12	0.74	0.013
Phenylalanine-tRNA synthetase	(Mosyak et al., 1995)	X-ray	14	0.69	0.001
Coat protein, tomato bushy stunt virus	(Hopper et al., 1984)	X-ray	7	0.63	0.001
Gonadotropin	(Lapthorn et al., 1994)	X-ray	9	0.61	0.001
Coat protein, sindbis virus	(Choi et al., 1991)	X-ray	6	0.60	0.025
Histone H5	(Aviles et al., 1978)	NMR	9	0.41	0.001
Small heat-shock protein	(Kim et al., 1998)	X-ray	6	0.36	0.457
Telomere binding protein	(Horvath et al., 1998)	X-ray	8	0.29	0.001
Cytochrome BC1	(Iwata et al., 1998)	X-ray	7	0.27	0.034
DNA-lyase	(Gorman et al., 1997)	X-ray[e]	8	0.18	0.001
Bcl-x$_L$	(Muchmore et al., 1996)	X-ray, NMR	7	0.13	0.001
Coat protein, southern bean mosaic virus	(Silva and Rossmann, 1985)	X-ray	6	0.09	0.100
α-Tubulin	(Jimenez et al., 1999)	NMR	80	0.06	0.034
Epidermal growth factor	(Louie et al., 1997)	X-ray	10	0.03	0.736
Prion	(Riek et al., 1997)	NMR	72	0.03	0.636
Glycine N-methyl-transferase	(Huang et al., 2000)	X-ray	11	0.09	0.095
ssDNA binding protein	(Tucker et al., 1994)	X-ray	20	0.37	0.010
Flagellin	(Vonderviszt et al., 1989)	LP	34	0.66	0.023

[a] From Brown et al. (2002). *J. Mol. Evol.* **55,** 104–110, with permission of Springer-Verlag.
[b] Disordered state detected by NMR, nuclear magnetic resonance; X-ray, X-ray crystallography; CD, circular dichroism; LP, limited proteolysis.
[c] Negative values of Δ indicate disordered regions are evolving faster than ordered.
[d] *p*-Values for a two-sided test of the null hypothesis.
[e] Useful crystallization only in the absence of most of the disordered region.

A. Predictors of Natural Disordered Regions (PONDRs)

Applying standard machine learning algorithms and approaches to databases and sequence information as described in Section I, a series of predictors of intrinsic disorder and order called PONDRs have been developed. These predictors use amino acid sequence as inputs and give numerical outputs, with 0.0 to <0.5 indicating order and ≥0.5 to 1.0 indicating disorder. Various data representations for the inputs have been tried, such as the compositions of selected amino acids or the averaged values of selected sequence attributes including hydropathy, net charge, and aromaticity. In addition, the data representation typically involved calculating the inputs over sliding windows, and some experimentation to optimize window size has been carried out. Finally, various operations on the input data have been tried, including linear data modeling (such as logistic regression) and nonlinear modeling (such as artificial neural networks), to yield predictors of intrinsic order and disorder (Li et al., 1999; Vucetic et al., 2001). The various experiments suggested that data representation, window size, and linear or nonlinear data modeling make relatively small differences in prediction accuracies.

To distinguish among the several PONDRs that resulted from the various experiments described above, a two-letter extension and a version number were added. The first letter, X, N, or V, indicates the method of structural characterization of the training data, namely, X-ray diffraction, NMR, or various means, respectively. The second letter, S, M, L, N, C or T, indicates the length or position of the disordered regions in the training data, with S for short (length = 9–20 residues), M for medium (length = 21–39 residues), L for long (length ≥ 40 residues), N for amino termini, C for carboxyl termini, and T for both termini (length = 5–14 residues). Although predictors were initially developed for the three indicated length classes (Romero et al., 1997b), since then the focus has been on long regions of disorder (Romero et al., 1998). PONDRs have also been developed that use disorder information from a single family of proteins; the extension for these PONDRs is the abbreviation of the name of the representative family member. For example, the putative disordered regions from 13 calcineurin (CaN) proteins were used to construct PONDR CaN (Romero et al., 1997a). Finally, as stated above, for each type of predictor, a version number is specified.

One recent predictor is called PONDR VL-XT (Romero et al., 2001) because it is a merger of PONDR VL1 (training set = variously characterized, long regions of disorder) with PONDR XN and PONDR XC, i.e., PONDR XT (training set = X-ray characterized chain termini) (Li et al., 2000), while another is called PONDR VL2 (training set = variously characterized, long regions of disorder, version 2) (Vucetic et al., 2002).

TABLE IV
Accuracies of Predictors of Natural Disordered Regions (PONDR)

Name	Training set	Number of disordered residues	Accuracy(%)		
			Train	Order	Disorder
XL1	7 X-ray	502	73 ± 4	71	47
CaN	13 CaN	1175	83 ± 5	84	29
VL1	7 NMR 8 X-ray	1366	83 ± 2	83	45
VL-XT	Merger of VL1, XN and XC			80	60
VL2	35 NMR 52 X-ray 56 CD	16,854	79 ± 4	83	75.5

B. PONDR Accuracies

The success of the initial PONDRs based on small databases of disordered protein motivated attempts to improve predictor accuracy. The main limitation for such attempts has been and continues to be the lack of low-noise structural data for both ordered and disordered protein, where noise means ordered regions misclassified as disordered and *vice versa.*

The accuracies of the various PONDRs were estimated (Table IV) by applying them to the ordered sequences in O_PDBS25 as summarized in Table II and to the merged set of disordered proteins described in Table I. Overall, the prediction accuracy of each PONDR was much better on the 222,116 ordered residues of O_PDBS25 than on the 18,833 residues of the merged disorder set. Thus, prediction of order generalized much better than prediction of disorder.

Even when as few as 502 ordered and 502 disordered residues were used for predictor development, the accuracy of $73 \pm 4\%$ estimated from 5-cross validation during training matched within 1 standard deviation the accuracy of 71% observed when the predictor was applied to O_PDBS25, which had 222,116 ordered residues. Since O_PDBS25 contains one representative from each protein family in PDB, this database spans the information on ordered protein structure. Such a good generalization from such a small training set was totally unexpected. Every PONDR so far tested, including some others not among the five examples given in Table IV, show comparably good generalization for prediction of protein order.

The poorer generalization of the prediction of disorder probably arises from two or perhaps three sources. These relate to differences in the training sets, possible differences in the volumes of sequence space occupied by ordered and disordered protein, and differences in the levels of noise in the ordered and disordered testing data.

The ordered training data involved nonoverlapping ordered segments randomly selected from different proteins, whereas the disordered training data involved overlapping windows that were generated by sliding windows in single-residue steps. This strategy was chosen to maximize use of the limited amount of disordered data. Thus, the disordered training data spanned a smaller volume of sequence space than did the ordered data.

The volume of sequence space occupied by natively disordered proteins might simply be much larger than the corresponding region for ordered proteins. Lattice-model approximations of protein folding suggest that only a small fraction of random sequences fold into specific structure (Abkevich et al., 1996). A random library containing 80-residue sequences designed to have helical periodicities with an average hydrophobicity level similar to that of natural proteins yielded only a small fraction that exhibited cooperative folding behavior (Davidson et al., 1995), providing further evidence that folding sequences represent a small proportion of all sequences. If the ordered and disordered sequences utilized in nature reflect their statistical abundances, disordered protein sequence space would be much larger than ordered protein sequence space. In this case, a greater number of disorder examples would be needed to achieve the level of generalization observed for the ordered data. Note the especially poor performance of PONDR CaN on the disordered set; the poor performance probably relates substantially to the small region of disordered sequence space sampled by the CaN regions of disorder.

A third possible origin of the reduced performance of disorder prediction is that the disordered data may simply be noisier than the ordered data; i.e., the disordered data might simply contain more ordered residues that are misclassified as disordered than vice versa. NMR probably yields the best characterization of disorder, yet NMR-characterized regions of disorder very likely contain significant regions of misclassified order (Garner et al., 1999). The misclassification problem is likely to be worse for both the CD- and X-ray-characterized data as compared to the NMR data. Because CD spectra provide estimates of the structure averaged over the entire sequence, an ordered domain within a sea of disorder could easily be missed in the CD spectrum. Long regions of missing coordinates in X-ray structures could be structured but wobbly domains rather than true disorder.

Despite obtaining all the ordered data from crystal structures, which give accurate structural information, and despite the removal of the unobserved (disordered) residues, it still cannot be assumed that the ordered data in Globular 3-D, O_PDBS25, and CO_PDBS25 are noise-less. For example, examination of segments from O_PDBS25 having the longest consecutive (putatively false positive) predictions of disorder shows many of these segments to be associated with DNA, with other large ligands including other proteins, or with the contacts that form the crystal lattice. Since disorder-to-order transitions upon complex formation or upon crystallization can occur, it is difficult to know whether such segments are ordered or disordered in the absence of the intermolecular association. Thus, these segments very possibly correspond to segments that are actually disordered when not ligand-associated or when not in the crystal, in which case these segments would represent noise in the ordered data.

The next phase of testing PONDR accuracy will be the careful comparison of experiment and prediction on individual proteins. Figure 5 shows results from one such analysis in which proteolysis was used to test PONDR VL-XT accuracy. This test was based on the knowledge that disordered protein is orders of magnitude more sensitive to protease digestion than is ordered protein (Fontana *et al.*, 1997). The sequence of *Xeroderma pigmentosum* group A (XPA), which is a DNA damage-recognition protein, was the test case. The predictions indicated that the XPA sequence has long regions of intrinsic disorder at both ends. Identification of hypersensitive trypsin digestion sites by mass spectrometry revealed a remarkable agreement between predicted disorder and protease sensitivity, showing that the PONDR indications of disorder were very accurate for this protein. These results suggested that the combination of PONDR predictions + protease digestion + mass spectrometry offers a useful approach for the analysis of protein disorder (Iakoucheva *et al.*, 2001).

III.   PONDR Estimations of the Commonness of Intrinsically Disordered Proteins

More than 30 proteomes have been PONDRed to estimate the commonness of putative disorder. We have used 40 or more consecutive predictions of disorder (e.g., a putative long disordered region, or LDR) as a convenient indicator, then scored each proteome by estimating the fraction of proteins predicted to contain at least one LDR. As shown in Table V, proteins with putative LDRs are quite common. The wide range of putative LDRs among the proteomes from eubacteria and archaea was quite surprising. A second surprise was the large jump in putative LDRs in the eukaryota compared to the eubacteria and archaea.

Further discussion of prediction errors provides more insight into these estimates of the commonness of disorder. Based on O_PDBS25,

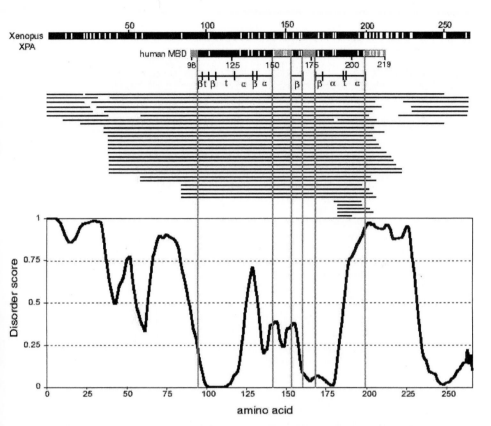

FIG. 5. Comparison of proteolysis data for *Xeroderma pigmentosum* group A (XPA) with PONDR predictions and NMR structure (From Iakoucheva *et al.*, 2001, *Protein Science* **10**, 560–571, with permission of Cold Spring Harbor Laboratory Press). (Top) Full-length *Xenopus laevis* XPA is depicted as a bar, with all possible trypsin sites indicated by white vertical lines (*X. laevis* numbering). The line below represents the human Minimal Binding Domain of XPA in the same format, the structure of which has been determined by NMR. Four regions with low certainty of assignment or high flexibility are indicated in gray. Each of the unique experimentally observed trypsin proteolysis fragments are drawn as horizontal lines below; the endpoints of these lines indicate the trypsin sensitive sites. (Below) Disorder prediction for *X. laevis* XPA. Each residue (*x*-axis) is assigned a disorder score (*y*-axis) by PONDR VL-XT. Disorder scores ≥0.5 signify disorder. Note the coincidence between predictions of disorder and the observed cut sites, and note also the coincidence between predictions of order and lack of observed cut sites.

VL-XT gave a false positive prediction of disorder of ~20% on a per-residue basis (Table IV); this error decreases to ~0.4% for consecutive predictions of 40 or longer (Romero *et al.*, 2001). These error rates lead to ~6% of the nonredundant proteins from PDB having consecutive false positive predictions of disorder ≥40 residues in length. As discussed in

TABLE V

Prevalence of Predicted Disorder in Genomes of Various Species[a]

Kingdom	Species	Number of sequences	Disorder lengths $\geq 40^b$ (%)
Archaea	*Methanococcus jannaschii*	1714	9
	Pyrococcus horikoshii	2062	16
	Pyrococcus abyssi	1764	19
	Archaeoglobus fulgidus	2402	20
	Methanobacterium thermoautotrophicum	1869	34
	*Halobacterium sp.*NRC-1	2057	35
	Aeropyrum pernix K1	2694	37
Bacteria	*Ureaplasma urealyticum*	611	7
	Rickettsia prowazekii	834	6
	Borrelia burgdorferi	845	7
	Campylobacter jejuni	2309	6
	Mycoplasma genitalium	480	8
	Helicobacter pylori	1532	9
	Aquifex aeolicus	1522	15
	Haemophilus influenzae	1708	13
	Bacillus subtilis	4093	15
	Escherichia coli	4281	17
	Vibrio cholerae	3815	16
	Mycoplasma pneumoniae	675	14
	Xylella fastidiosa	2761	17
	Thermotoga maritime	1842	18
	Neisseria meningitidis MC58	2015	17
	Chlamydia pneumoniae	1052	18
	Synechocystis sp	3167	20
	Chlamydia trachomatis	894	19
	Treponema pallidum	1028	11
	Pseudomonas aeruginosa	5562	24
	Mycobacterium tuberculosis	3916	31
	Deinococcus radiodurans chr 1	2580	33
Eukaryota	*Plasmodium falciparum* chr II, III	422	35
	Caenorhabditis elegans	17,049	36
	Arabodiopsis thaliana	7849	41
	Saccharomyces cerevisiae	6264	40
	Drosophila melanogaster	13,885	51

[a] From Dunker *et al.* (2000). *Genome Informatics* **11,** 161–171, with permission from Japanese Society for Bioinformatics.

[b] The percentages of proteins in the indicated genomes predicted to have at least one region of disorder of $\geq$40 amino acids. Predictions were made by PONDR VL-XT (Li *et al.*, 1999; Romero *et al.*, 2001).

Section II,B, this may be an overestimate of the false positive error rate because many of the apparent consecutive errors correspond to regions of disorder that are ordered in the crystal due to ligand binding or crystal contacts. Also, because disordered regions of length $\geq$40 residues are often missed due to false negative predictions of order, the data in

Table V probably represent lower bounds on the amount of disorder per genome.

IV. FUNCTIONS OF INTRINSICALLY DISORDERED REGIONS

We are attempting to understand the biological significance of the large variations in frequency of putative LDRs, whether between different types of bacteria or archaea, or between pro- and eukaryota. We have carefully studied the literature of more than 90 example proteins selected from our disordered protein databases and found reports on the functions of most of the disordered regions (Dunker *et al.*, 2002). The observed functions and the number of examples in each functional class are given in Table VI. As indicated, four major functional classes were found: molecular recognition, molecular assembly or disassembly, protein modification, and entropic chains.

For two of the categories, molecular recognition and protein modification, the proteins in Table VI are mostly involved in signaling, control, or regulation. Thus, our current hypothesis is that an increased requirement for signaling, regulation, and control is the underlying cause for the significantly larger fractions of proteins with long regions of intrinsic disorder in some organisms as compared to others (Dunker and Obradovic, 2001). Consistent with our current hypothesis but developed without regard to the role of protein disorder, recent proteomic comparisons indicate that eukaryotes are much richer in regulatory proteins than are the prokaryotes and archaea (Liu and Rost, 2001). Tests are underway to determine whether the increased amount of predicted disorder is indeed associated with an increased number of regulatory proteins in the various proteomes.

TABLE VI
Functional Categories for Disordered Regions of Proteins

Category	Number	Transition	Description
Molecular recognition	113	D → O	Protein, ssDNA, dsDNA, tRNA, rRNA, mRNA, nRNA, bilayers, ligands, cofactors, metals, autoregulatory
Molecular assembly/ Disassembly	>13	D → O O → D	Heterocomplexes, linear polymers, phages, viruses
Protein modification	36	Variable	Acetylation, fatty acylation, glycosylation, methylation, phosphorylation, ADP-ribosylation, ubiquitination, proteolytic digestion
Entropic chains	17	None	Linkers, spacers, bristles, clocks, springs, detergents, self-transport

V. Conclusions

The experiments and data presented herein support the proposals that protein disorder is encoded by the amino acid sequence and that protein disorder is essential for many important biological functions. Thus, intrinsic disorder ought to be considered a distinct category of *native* protein structure. Categorization is, of course, an essential step in the development of knowledge (Lakoff, 1987), so the concept that intrinsic disorder represents a *category of native protein structure,* rather than being simply an intermediate on the way to the native structure, has important implications (Dunker *et al.,* 1997, 2001, 2002). The association of protein disorder with function is not a new idea. Experiments reported more than 50 years ago gave strong indications that an ensemble of structures enabled serum albumin to bind a structurally diverse set of ligands (Karush, 1950), and numerous additional papers were published on disorder/function relationships about 20 years ago for several other proteins (Stubbs *et al.,* 1977; Bloomer *et al.,* 1978; Bode *et al.,* 1978; Jardetzky *et al.,* 1978; Huber, 1979; Schulz, 1979; Blow, 1982; Holmes, 1983; Bennett and Huber, 1984). Our extensions of current disorder knowledge by predictions on whole proteomes lead to the conclusion that functions carried out by disordered regions are likely to be very common, especially with regard to signaling and regulatory functions in eukaryotic cells.

References

Abkevich, V. I., Gutin, A. M., and Shakhnovich, E. I. (1996). *Proc. Natl. Acad. Sci. USA* **93,** 839–844.

Altschul, S. F., Gish, W., Miller, W., Myers, E. W., and Lipman, D. J. (1990). *J. Mol. Biol.* **215,** 403–410.

Altschul, S. F., Madden, T. L., Schaffer, A. A., Zhang, J., Zhang, Z., Miller, W., and Lipman, D. J. (1997). *Nucleic Acids Res.* **25,** 3389–3402.

Arnold, G. E., Dunker, A. K., Johns, S. J., and Douthart, R. J. (1992). *Proteins: Struct., Funct. Gen.* **12,** 382–399.

Aviles, F. J., Chapman, G. E., Kneale, G. G., Crane-Robinson, C., and Bradbury, E. M. (1978). *Eur. J. Biochem.* **88,** 363–371.

Bennett, W. S., and Huber, R. (1984). *Crit. Rev. Biochem.* **15,** 291–384.

Berger, J. M., Gamblin, S. J., Harrison, S. C., and Wang, J. C. (1996). *Nature* **379,** 225–232.

Bidwell, L. M., McManus, M. E., Gaedigk, A., Kakuta, Y., Negishi, M., Pedersen, L., and Martin, J. L. (1999). *J. Mol. Biol.* **293,** 521–530.

Bloomer, A. C., Champness, J. N., Bricogne, G., Staden, R., and Klug, A. (1978). *Nature* **276,** 362–368.

Blow, D. M. (1982). *Nature* **297,** 454.

Bode, W., Schwager, P., and Huber, R. (1978). *J. Mol. Biol.* **118,** 99–112.

Brown, C. J., Takayama, S., Campen, A. M., Vise, P., Marshall, T., Oldfield, C. J., Williams, C. J., and Dunker, A. K. (2002). *J. Mol. Evol.* **55,** 104–110.

Campbell, K. M., Terrell, A. R., Laybourn, P. J., and Lumb, K. J. (2000). *Biochemistry* **39,** 2708–2713.

Choi, H. K., Tong, L., Minor, W., Dumas, P., Boege, U., Rossmann, M. G., and Wengler, G. (1991). *Nature* **354,** 37–43.

Davidson, A. R., Lumb, K. J., and Sauer, R. T. (1995). *Nat. Struct. Biol.* **2,** 856–864.

Demchenko, A. P. (2001). *J. Mol. Recognit.* **14,** 42–61.

Dunker, A. K., Brown, C. J., Lawson, J. D., Iakoucheva, L. M., and Obradovic, Z. (2002). *Biochemistry* **41,** 6573–6582.

Dunker, A. K., Lawson, J. D., Brown, C. J., Williams, R. M., Romero, P., Oh, J. S., Oldfield, C. J., Campen, A. M., Ratliff, C. M., Hipps, K. W., Ausio, J., Nissen, M. S., Reeves, R., Kang, C., Kissinger, C. R., Bailey, R. W., Griswold, M. D., Chiu, W., Garner, E. C., and Obradovic, Z. (2001). *J. Mol. Graph. Model.* **19,** 26–59.

Dunker, A. K., and Obradovic, Z. (2001). *Nat. Biotech.* **19,** 805–806.

Dunker, A. K., Obradovic, Z., Romero, P., Garner, E. C., and Brown, C. J. (2000). *Genome Informatics* **11,** 161–171.

Dunker, A. K., Obradovic, Z., Romero, P., Kissinger, C., and Villafranca, E. J. (1997). *PDB Newsletter* **81,** 3–5.

Fontana, A., Zambonin, M., Polverino de Laureto, P., De Filippis, V., Clementi, A., and Scaramella, E. (1997). *J. Mol. Biol.* **266,** 223–230.

Garner, E., Romero, P., Dunker, A. K., Brown, C. J., and Obradovic, Z. (1999). *Genome Informatics* **10,** 41–50.

Gorman, M. A., Morera, S., Rothwell, D. G., de La Fortelle, E., Mol, C. D., Tainer, J. A., Hickson, I. D., and Freemont, P. S. (1997). *EMBO J.* **16,** 6548–6558.

Hobohm, U., and Sander, C. (1994). *Protein Sci.* **3,** 522–524.

Holmes, K. C. (1983). *CIBA Found. Symp.* **93,** 116–138.

Hopper, P., Harrison, S. C., and Sauer, R. T. (1984). *J. Mol. Biol.* **177,** 701–713.

Horvath, M. P., Schweiker, V. L., Bevilacqua, J. M., Ruggles, J. A., and Schultz, S. C. (1998). *Cell* **95,** 963–974.

Huang, Y., Komoto, J., Konishi, K., Takata, Y., Ogawa, H., Gomi, T., Fujioka, M., and Takusagawa, F. (2000). *J. Mol. Biol.* **298,** 149–162.

Huber, R. (1979). *Nature* **280,** 538–539.

Iakoucheva, L. M., Kimzey, A. L., Masselon, C. D., Bruce, J. E., Garner, E. C., Brown, C. J., Dunker, A. K., Smith, R. D., and Ackerman, E. J. (2001). *Protein Sci.* **10,** 560–571.

Iwata, S., Lee, J. W., Okada, K., Lee, J. K., Iwata, M., Rasmussen, B., Link, T. A., Ramaswamy, S., and Jap, B. K. (1998). *Science* **281,** 64–71.

Jacobs, D. M., Lipton, A. S., Isern, N. G., Daughdrill, G. W., Lowry, D. F., Gomes, X., and Wold, M. S. (1999). *J. Biomol. NMR* **14,** 321–331.

Jardetzky, O., Akasaka, K., Vogel, D., Morris, S., and Holmes, K. C. (1978). *Nature* **273,** 564–566.

Jimenez, M. A., Evangelio, J. A., Aranda, C., Lopez-Brauet, A., Andreu, D., Rico, M., Lagos, R., Andreu, J. M., and Monasterio, O. (1999). *Protein Sci.* **8,** 788–799.

Karlin, S., and Brendel, V. (1992). *Science* **257,** 39–49.

Karush, F. (1950). *J. Am. Chem. Soc.* **72,** 2705–2713.

Kim, K. K., Kim, R., and Kim, S. H. (1998). *Nature* **394,** 595–599.

Kissinger, C. R., Parge, H. E., Knighton, D. R., Lewis, C. T., Pelletier, L. A., Tempczyk, A., Kalish, V. J., Tucker, K. D., Showalter, R. E., Moomaw, E. W., Gastinel, L. N., Habuka, N., Chen, X., Maldonado, F., Barker, J. E., Bacquet, R., and Villafranca, J. E. (1995). *Nature* **378,** 641–644.

Lakoff, G. (1987). *Women, Fire and Dangerous Things: What Categories Reveal about the Human Mind.* University of Chicago Press, Chicago.

Lapthorn, A. J., Harris, D. C., Littlejohn, A., Lustbader, J. W., Canfield, R. E., Machin, K. J., Morgan, F. J., and Isaacs, N. W. (1994). *Nature* **369,** 455–461.

Li, X., Obradovic, Z., Brown, C.J., Garner, E. C., and Dunker, A. K. (2000). *Genome Informatics* **11,** 172–184.

Li, X., Romero, P., Rani, M., Dunker, A. K., and Obradovic, Z. (1999). *Genome Informatics* **10,** 30–40.

Liu, J., and Rost, B. (2001). *Protein Sci.* **10,** 1970–1979.

Logan, D. T., Mazauric, M. H., Kern, D., and Moras, D. (1995). *EMBO J.* **14,** 4156–4167.

Louie, G. V., Yang, W., Bowman, M. E., and Choe, S. (1997). *Mol. Cell* **1,** 67–78.

Mosyak, L., Reshetnikova, L., Goldgur, Y., Delarue, M., and Safro, M. G. (1995). *Nat. Struct. Biol.* **2,** 537–547.

Muchmore, S. W., Sattler, M., Liang, H., Meadows, R. P., Harlan, J. E., Yoon, H. S., Nettesheim, D., Chang, B. S., Thompson, C. B., Wong, S. L., Ng, S. L., and Fesik, S. W. (1996). *Nature* **381,** 335–341.

Nakashima, H., Nishikawa, K., and Ooi, T. (1986). *J. Biochem.* (*Tokyo*) **99,** 153–162.

Pattabiraman, N., Namboodiri, K., Lowrey, A., and Gaber, B. P. (1990). *Prot. Seq. Data Anal.* **3,** 387–405.

Presta, L. G., and Rose, G. D. (1988). *Science* **240,** 1632–1641.

Radivojac, P., Obradovic, Z., Brown, C. J., and Dunker, A. K. (2002). *PSB 2002* **7,** 589–600.

Richardson, J. S., and Richardson, D. C. (1988). *Science* **240,** 1648–1652.

Riddle, D. S., Santiago, J. V., Bray-Hall, S. T., Doshi, N., Grantcharova, V. P., Yi, Q., and Baker, D. (1997). *Nat. Struct. Biol.* **4,** 805–809.

Riek, R., Hornemann, S., Wider, G., Glockshuber, R., and Wuthrich, K. (1997). *FEBS Lett.* **413,** 282–288.

Romero, P., Obradovic, Z., and Dunker, A. K. (1997a). *Genome Informatics* **8,** 110–124.

Romero, P., Obradovic, Z., and Dunker, A. K. (1999). *FEBS Lett.* **462,** 363–367.

Romero, P., Obradovic, Z., and Dunker, A. K. (2000). *Artificial Intelligence Rev.* **14,** 447–484.

Romero, P., Obradovic, Z., Kissinger, C. R., Villafranca, J. E., and Dunker, A. K. (1997b). *Proc. IEEE Intl. Conf. Neural Networks* **1,** 90–95.

Romero, P., Obradovic, Z., Kissinger, C. R., Villafranca, J. E., Guilliot, S., Garner, E., and Dunker, A. K. (1998). *PSB 1998* **3,** 437–448.

Romero, P., Obradovic, Z., Li, X., Garner, E. C., Brown, C. J., and Dunker, A. K. (2001). *Proteins: Struct., Funct., Gen.* **42,** 38–48.

Schmitz, M. L., dos Santos Silva, M. A., Altmann, H., Czisch, M., Holak, T. A., and Baeuerle, P. A. (1994). *J. Biol. Chem.* **269,** 25613–25620.

Schulz, G. E. (1979). In *Molecular Mechanism of Biological Recognition* (M. Balaban, ed.), pp. 79–94. Elsevier/North-Holland Biomedical Press, New York.

Silva, A. M., and Rossmann, M. G. (1985). *ACTA Cryst.* **B41,** 147–157.

Smith, R. F., Wiese, B. A., Wojzynski, M. K., Davison, D. B., and Worley, K. C. (1996). *Genome Res.* **6,** 454–462.

Stubbs, G., Warren, S., and Holmes, K. (1977). *Nature* **267,** 216–221.

Sweet, R. M., and Eisenberg, D. (1983). *J. Mol. Biol.* **171,** 479–488.

Tell, G., Perrone, L., Fabbro, D., Pellizzari, L., Pucillo, C., De Felice, M., Acquaviva, R., Formisano, S., and Damante, G. (1998). *Biochem. J.* **329,** 395–403.

Tesmer, J. J., Berman, D. M., Gilman, A. G., and Sprang, S. R. (1997). *Cell* **89,** 251–261.

Thompson, J. D., Higgins, D. G., and Gibson, T. J. (1994). *Nucleic Acids Res.* **22,** 4673–4680.

Tucker, P. A., Tsernoglou, D., Tucker, A. D., Coenjaerts, F. E., Leenders, H., and van der Vliet, P. C. (1994). *EMBO J.* **13,** 2994–3002.

Uversky, V. N. (2002). *Eur. J. Biochem.* **269,** 2–12.

Uversky, V. N., Gillespie, J. R., Millett, I. S., Khodyakova, A. V., Vasiliev, A. M., Chernovskaya, T. V., Vasilenko, R. N., Kozlovskaya, G. D., Dolgikh, D. A., Fink, A. L., Doniach, S., and Abramov, V. M. (1999). *Biochemistry* **38,** 15009–15016.

Vihinen, M., Torkkila, E., and Riikonen, P. (1994). *Proteins* **19,** 141–149.

Vonderviszt, F., Kanto, S., Aizawa, S., and Namba, K. (1989). *J. Mol. Biol.* **209,** 127–133.

Vucetic, S., Brown, C. J., Dunker, A. K., and Obradovic, Z. (2002). *Bioinformatics.* Submitted.

Vucetic, S., Radivojac, P., Obradovic, Z., Brown, C. J., and Dunker, A. K. (2001). *Intl. Joint INNS-IEEE Conf. Neural Networks* **4,** 2718–2723.

Williams, R. J. P. (1989). *Eur. J. Biochem.* **183,** 479–497.

Williams, R. M., Obradovic, Z., Mathura, V., Braun, W., Garner, E. C., Young, J., Takayama, S., Brown, C. J., and Dunker, A. K. (2001). *PSB 2001* **6,** 89–100.

Wootton, J. C. (1993). *Comput. Chem.* **17,** 149–163.

Wootton, J. C. (1994a). *Comput. Chem.* **18,** 269–285.

Wootton, J. C. (1994b). *Curr. Opin. Struct. Biol.* **4,** 413–421.

Wootton, J. C., and Federhen, S. (1996). *Methods Enzymol.* **266,** 554–571.

Wright, P. E., and Dyson, H. J. (1999). *J. Mol. Biol.* **293,** 321–331.

Xie, Q., Arnold, G. E., Romero, P., Obradovic, Z., Garner, E., and Dunker, A. K. (1998). *Genome Informatics* **9,** 193–200.

UNFOLDED PROTEINS STUDIED BY RAMAN OPTICAL ACTIVITY

By L. D. BARRON, E. W. BLANCH, and L. HECHT

Department of Chemistry, University of Glasgow, Glasgow G12 8QQ, United Kingdom

I. INTRODUCTION

Interest in unfolded states of proteins has been growing because of their importance in studies of protein folding, stability, and function (Creighton, 1992; Pain, 1994). Until recently, unfolded proteins were usually considered in the context of "nonnative" or "denatured" protein states embracing a plethora of structures ranging from the ideal random coil at one extreme (Smith *et al.*, 1996) to molten globules at the other (Dobson, 1994; Ptitsyn, 1995; Arai and Kuwajima, 2000). Although it has been appreciated for some time that the molten globule state, which remains partially collapsed and retains a significant amount of secondary structure, may play an important role in a number of physiological processes (Bychkova *et al.*, 1988), more and more examples of proteins are coming to light which are completely unfolded in their native biologically active state (Uversky *et al.*, 2000). The fact that such "natively unfolded" proteins can have important biological functions has necessitated a reassessment of the structure–function paradigm (Wright and Dyson, 1999). Examples of natively unfolded proteins include the

ADVANCES IN
PROTEIN CHEMISTRY, Vol. 62

casein milk proteins (Holt and Sawyer, 1993), the synuclein and tau brain proteins (Goedert *et al.*, 2001), and the gluten cereal proteins (Shewry and Tathum, 1990). In addition to their role in normal function, un-folded sequences in both native and nonnative states are also of interest because of their susceptibility to the type of aggregation found in many protein misfolding diseases such as amyloidosis and neurodegenerative disease (Dobson *et al.*, 2001).

Unfolded proteins are characterized by disordered amino acid se-quences containing little or no extended secondary structure such as α-helix or β-sheet. Because the experimental characterization of disor-dered polypeptide structure is difficult, all such structure is often called random coil. The random coil state was originally conceived as the col-lection of an enormous number of possible random conformations of an extremely long molecule in which chain flexibility arises from internal rotation (with some degree of hindrance) around the covalent backbone bonds (Poland and Scheraga, 1970). The random coil state is therefore the delimiting case of a dynamic disorder in which there is a distribu-tion of Ramachandran ϕ,ψ angles for each amino acid residue, giving rise to an ensemble of rapidly interconverting conformers. However, a second, equally important delimiting case is that of a static disorder cor-responding to that found in loops and turns within native proteins with well-defined tertiary folds that contain sequences of residues with fixed but nonrepetitive ϕ,ψ angles. Unfolded proteins are expected to exhibit types of disorder somewhere between these two delimiting cases.

Natively unfolded proteins are loose structures that simply become looser through a continuous transition on heating which takes them closer to the true random coil. They are particular examples of "native proteins with nonregular structures," which is a broader term also en-compassing proteins with fixed nonregular folds comprising statically disordered sequences stabilized, for example, by cooperative side chain interactions, multiple disulfide links, or multiple metal ions. Examples of proteins with fixed nonregular folds include Bowman–Birk protease inhibitors (de la Sierra *et al.*, 1999) and metallothioneins (Braun *et al.*, 1992). These fixed folds may often be evidenced by a first-order thermal transition observed using differential scanning calorimetry (DSC) and are sometimes accessible through X-ray crystallography or multidimen-sional solution NMR.

To further our understanding of the behavior of unfolded proteins, it is necessary to employ experimental techniques able to discriminate between the dynamic true random coil state and more static types of disorder, including situations in which some ordered secondary struc-ture may also be present. One such technique is a novel chiroptical

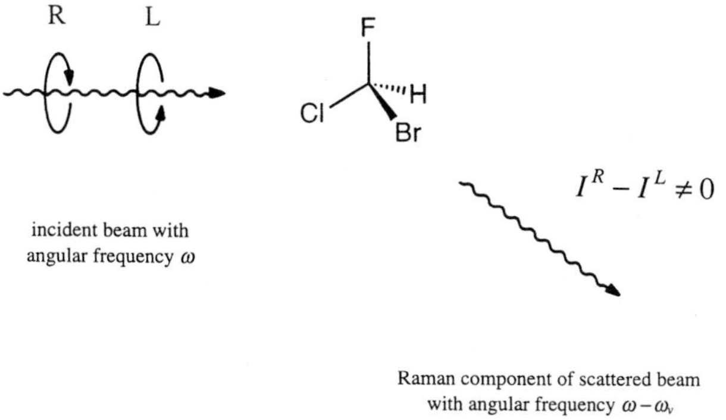

Raman component of scattered beam
with angular frequency $\omega - \omega_v$

Fig. 1. The basic ROA experiment measures a small difference in the intensity of Raman scattering from chiral molecules in right- and left-circularly polarized incident light. Reprinted from Barron *et al.*, 2000, *Prog. Biophys. Mol. Biol.* **73**, 1–49, with permission from Elsevier Science.

spectroscopy called Raman optical activity (ROA). The power of ROA in this area derives from the fact that, like the complementary technique of vibrational circular dichroism (VCD), it is a form of vibrational optical activity (Barron, 1982; Diem, 1993; Polavarapu, 1998; Nafie and Freedman, 2000) and so is sensitive to chirality associated with all the $3N - 6$ fundamental molecular vibrational transitions, where N is the number of atoms.

Vibrational optical activity in typical chiral molecules in the liquid phase was first observed by Barron *et al.* (1973) using one version of an ROA technique in which, as depicted in Figure 1, a small difference in the intensity of Raman scattering is measured using right and left circularly polarized incident laser light (Nafie and Freedman, 2000; Barron and Hecht, 2000). Raman spectroscopy itself provides molecular vibrational spectra by means of inelastic scattering of visible light. During the Stokes Raman scattering event, the interaction of the molecule with the incident visible photon of energy $\hbar\omega$, where ω is its angular frequency, can leave the molecule in an excited vibrational state of energy $\hbar\omega_v$, with a corresponding energy loss, and hence a shift to lower angular frequency $\omega-\omega_v$, of the scattered photon. Therefore, by analyzing the scattered light with a visible spectrometer, a complete vibrational spectrum may be obtained. Until recently, lack of sensitivity restricted ROA studies to favorable samples such as small chiral organic molecules. However, major advances in instrumentation in recent years (Hecht and Barron, 1994; Hug, 1994; Vargek *et al.*, 1997; Hug and Hangartner, 1999; Hecht *et al.*, 1999) have now rendered biomolecules in aqueous solution accessible to ROA studies (Barron *et al.*, 2000).

Conventional Raman spectroscopy has a number of favorable characteristics that have led to many applications in biochemistry (Carey, 1982; Miura and Thomas, 1995; Thomas, 1999). In particular, the complete vibrational spectrum from $\sim$100 to 4000 cm^{-1} is accessible on one simple instrument and both H_2O and D_2O are excellent solvents for Raman studies. Since ROA is sensitive to chirality, it is able to build on these advantages by adding to Raman spectroscopy an extra sensitivity to the details of the three-dimensional structure, thus opening a new window on the structure and behavior of biomolecules. Although incapable of providing the atomic resolution of X-ray crystallography and multidimensional NMR, its simple routine application to aqueous solutions, with no restrictions on the size of the biomolecules, makes ROA ideal for studying many timely problems.

The application of ROA to studies of unfolded and partially folded proteins has been especially fruitful. As well as providing new information on the structure of disordered polypeptide and protein sequences, ROA has provided new insight into the complexity of order in denatured proteins and the structure and behavior of proteins involved in misfolding diseases. All the ROA data shown in this chapter have been measured in our Glasgow laboratory because, at the time of writing, ROA data on typical large biomolecules had not been published by any other group. We hope that this review will encourage more widespread use of ROA in protein science.

II.   RAMAN OPTICAL ACTIVITY THEORY AND EXPERIMENT

A.   *The Raman Optical Activity Observables*

The fundamental scattering mechanism responsible for ROA was discovered by Atkins and Barron (1969), who showed that interference between the waves scattered via the polarizability and optical activity tensors of the molecule yields a dependence of the scattered intensity on the degree of circular polarization of the incident light and to a circular component in the scattered light. Barron and Buckingham (1971) subsequently developed a more definitive version of the theory and introduced a definition of the dimensionless circular intensity difference (CID),

$$\Delta = (I^R - I^L)/(I^R + I^L) \qquad (1)$$

as an appropriate experimental quantity, where I^R and I^L are the scattered intensities in right- and left-circularly polarized incident light. In terms of the electric dipole–electric dipole molecular polarizability tensor $\alpha_{\alpha\beta}$ and the electric dipole–magnetic dipole and electric dipole–electric quadrupole optical activity tensors $G'_{\alpha\beta}$ and $A_{\alpha\beta\gamma}$ (Buckingham, 1967; Barron, 1982; Polavarapu, 1998), the CIDs for *forward* (0°) and *backward* (180°) scattering from an isotropic sample for incident transparent wavelengths much greater than the molecular dimensions are as follows (Barron, 1982; Hecht and Barron, 1990; Nafie and Che, 1994; Polavarapu, 1998):

$$\Delta(0°) = \frac{4[45\alpha G' + \beta(G')^2 - \beta(A)^2]}{c[45\alpha^2 + 7\beta(\alpha)^2]} \tag{2a}$$

$$\Delta(180°) = \frac{24[\beta(G')^2 + \frac{1}{3}\beta(A)^2]}{c[45\alpha^2 + 7\beta(\alpha)^2]} \tag{2b}$$

where the isotropic invariants are defined as

$$\alpha = \tfrac{1}{3}\alpha_{\alpha\alpha}, \qquad G' = \tfrac{1}{3}G'_{\alpha\alpha}$$

and the anisotropic invariants as

$$\beta(\alpha)^2 = \tfrac{1}{2}(3\alpha_{\alpha\beta}\alpha_{\alpha\beta} - \alpha_{\alpha\alpha}\alpha_{\beta\beta})$$
$$\beta(G')^2 = \tfrac{1}{2}(3\alpha_{\alpha\beta}G'_{\alpha\beta} - \alpha_{\alpha\alpha}G'_{\beta\beta})$$
$$\beta(A)^2 = \tfrac{1}{2}\omega\alpha_{\alpha\beta}\varepsilon_{\alpha\gamma\delta}A_{\gamma\delta\beta}$$

We are using a Cartesian tensor notation in which a repeated Greek suffix denotes summation over the three components, and $\varepsilon_{\alpha\beta\gamma}$ is the third-rank unit antisymmetric tensor. Additional CID expressions can be developed for right-angle scattering but are not given her because, as explained below, backscattering is the optimum strategy for ROA studies of biomolecules in aqueous solution. These results apply specifically to *Rayleigh* (elastic) scattering: For *Raman* (inelastic) scattering, the same basic CID expressions apply but with the molecular property tensors replaced by corresponding vibrational Raman transition tensors. Raman optical activity is a very weak phenomenon, with CID values usually less than 10^{-3}, this being the approximate ratio of the magnitude of a magnetic dipole transition moment to that of an electric dipole transition moment.

Using a simple bond polarizability theory of ROA for the case of a molecule composed entirely of idealized axially symmetric bonds, the relationships $\beta(G')^2 = \beta(A)^2$ and $\alpha G' = 0$ are found (Barron and

Buckingham, 1974; Barron, 1982). Within this model, therefore, isotropic scattering makes zero contribution to the ROA intensity, which
is generated exclusively by anisotropic scattering, and the CID Eqs. (2)
reduce to

$$\Delta(0°) = 0 \qquad (3a)$$

$$\Delta(180°) = \frac{32\beta(G')^2}{c[45\alpha^2 + 7\beta(\alpha)^2]} \qquad (3b)$$

Hence, unlike the conventional Raman intensity which is the same in
the forward and backward directions, the ROA intensity is maximized
in backscattering and is zero in forward scattering. The CID components for right-angle scattering take intermediate values, and it has been
shown that a given ROA signal-to-noise ratio can be achieved at least an
order of magnitude faster in backscattering than in right-angle scattering (Hecht *et al.*, 1989; Hecht and Barron, 1994). These considerations
lead to an important conclusion: *Backscattering boosts the ROA signal relative to the background Raman intensity and is therefore the best experimental
strategy for most ROA studies of biomolecules in aqueous solution.* Because the
numerators of the general CID Eqs. (2) reveal that ROA in backscattering is generated exclusively by anisotropic scattering (within the approximations used in the derivation), and anisotropic scattering usually
generates the largest ROA signals (as in vibrations based on bond angle deformations), this conclusion in fact holds independently of any
model.

Insight into these results is provided by a simple two-group model of
ROA in which two achiral axially symmetric groups or bonds are held
together in a twisted chiral structure (Barron and Buckingham, 1974;
Barron, 1982). However, although they have conceptual value, such simple models have been found to have little practical value in predicting
observed ROA features because of the complexity of the normal modes
in typical chiral molecular structures. *Ab initio* methods are manda-
tory for realistic ROA computations. These have been developed by
Polavarapu (1990, 1998) and Helgaker *et al.* (1994) with some preliminary applications made to model peptide and protein structures
(Polavarapu and Deng, 1994; Bour 2001). At the time of writing, *ab initio*
computations had not been developed sufficiently to assist the interpretation of the observed polypeptide and protein ROA spectra presented
in this chapter.

B. Enhanced Sensitivity of Raman Optical Activity to Structure and Dynamics of Chiral Biomolecules

The normal modes of vibration of biomolecules can be highly complex, with contributions from local vibrational coordinates in both the backbone and the side chains. The advantage of ROA is that it is able to cut through the complexity of the corresponding vibrational spectra since the largest ROA signals are often associated with vibrational coordinates from the most rigid and chiral parts of the structure. These are usually within the backbone and often give rise to ROA band patterns characteristic of the backbone conformation. Polypeptides in the standard conformations defined by characteristic Ramachandran ϕ,ψ angles found in secondary, loop and turn structure within native proteins are particularly favorable in this respect because signals from the peptide backbone (Fig. 2) usually dominate the ROA spectra, unlike the parent conventional Raman spectra in which bands from the amino acid side chains often obscure the peptide backbone bands.

Because the time scale of the Raman scattering event ($\sim 3.3 \times 10^{-14}$ s for a vibration with wavenumber shift 1000 cm^{-1} excited in the visible) is much shorter than that of the fastest conformational fluctuations in biomolecules, the ROA spectrum is a superposition of "snapshot" spectra from all the distinct chiral conformers present in the sample. Together with the dependence of ROA on chirality, this leads to an enhanced sensitivity to the dynamic aspects of biomolecular structure. The two-group model provides a qualitative explanation since it predicts ROA intensities that depend on absolute chirality in the form of a sin χ dependence

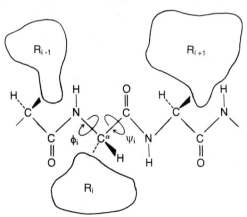

FIG. 2.   Sketch of the polypeptide backbone illustrating the Ramachandran ϕ,ψ angles and the amino acid side chains R. Reprinted from Barron *et al.*, 2000, *Prog. Biophys. Mol. Biol.* **73**, 1–49, with permission from Elsevier Science.

on the torsion angle χ between the two groups for the simple geometry where the angles between the group axes and the connecting bond are 90° (Barron *et al.*, 2000). This leads to cancelation of contributions from enantiomeric structures, such as two-group units with torsion angles of approximately equal magnitude but opposite sign, which can arise as the mobile structure explores the set of accessible conformations. In contrast, observables that are "blind" to chirality, such as conventional Raman band intensities, tend to be additive and hence much less sensitive to this type of mobility.

C. Measurement of Raman Optical Activity

For the reasons given above, a backscattering geometry is essential for the routine measurement of the ROA spectra of typical biomolecules in aqueous solution, especially since these samples often show a high background in the conventional Raman spectrum from the solvent water together with residual fluorescence. In the current Glasgow backscattering ROA instrument, a visible argon-ion laser beam is focused into the sample, which is held in a small rectangular cell. The cone of backscattered light is collected by means of a 45° mirror with a small central hole to allow passage of the incident laser beam. The Raman light is thereby reflected into the collection optics of a single grating spectrograph. The charge-coupled device (CCD) camera used as a multichannel detector permits measurement of the full spectral range in a single acquisition. To measure the tiny ROA signals, the spectral acquisition is synchronized with an electro-optic modulator (EOM) used to switch the polarization of the incident laser beam between right and left circular at a suitable rate. Further details may be found in Hecht and Barron (1994) and Hecht *et al.* (1999).

Typical sample concentrations are in the range 20–100 mg/ml, but concentrations of virus samples can be as low as 5 mg/ml. The green 514.5-nm line of the argon-ion laser is used because it provides a compromise between reduced fluorescence from sample impurities with increasing wavelength and the increased scattering power with decreasing wavelength due to the Rayleigh λ^{-4} law. A typical laser power at the sample is ~700 mW. Under these conditions, high-quality ROA spectra of small peptides may be collected in an hour or two, and polypeptides and proteins in ~5–20 h.

In order to study dynamic aspects of biomolecular structure, it is necessary to perform measurements over an appropriate temperature range. This is accomplished by directing dry air downward over the sample cell from the nozzle of a device used to cool protein crystals in X-ray diffraction experiments, thus allowing the temperature to be varied in the range 0–70°C.

Several novel design features of a new ROA instrument developed by Hug and Hangartner (1999) will be especially valuable for biomolecular studies. It is based on measurement of the circularly polarized component in the Raman-scattered light, instead of the Raman intensity difference in right- and left-circularly polarized incident light employed in the Glasgow instrument. This allows "flicker noise" arising from dust particles, density fluctuations, etc, to be eliminated since the intensity differences required to extract the circularly polarized component of the scattered beam are taken between two components of the scattered light measured during the same acquisition period. The flicker noise therefore cancels out, resulting in superior signal-to-noise characteristics. A commercial instrument based on this design will be available shortly.

III. SURVEY OF POLYPEPTIDE AND PROTEIN RAMAN OPTICAL ACTIVITY

A. General

Raman optical activity is an excellent technique for studying polypeptide and protein structure in aqueous solution since, as mentioned above, their ROA spectra are often dominated by bands originating in the peptide backbone that directly reflect the solution conformation. Furthermore, the special sensitivity of ROA to dynamic aspects of structure makes it a new source of information on order–disorder transitions.

Vibrations of the backbone in polypeptides and proteins are usually associated with three main regions of the Raman spectrum (Tu, 1986; Miura and Thomas, 1995). These are the backbone skeletal stretch region $\sim$870–1150 cm^{-1} originating in mainly C_{α}–C, C_{α}–C_{β}, and C_{α}–N stretch coordinates; the amide III region $\sim$1230–1310 cm^{-1}, which is often thought to involve mainly the in-phase combination of largely the N–H in-plane deformation with the C_{α}–N stretch; and the amide I region $\sim$1630–1700 cm^{-1}, which arises mostly from the C=O stretch. However, Diem (1993) has shown that, in small peptides, the amide III region involves much more mixing between the N–H and C_{α}–H deformations than previously thought, and should be extended to at least 1340 cm^{-1}. This extended amide III region is particularly important for ROA studies because, as first shown in an early ROA study of alanyl peptide oligomers (Ford et al., 1994), the coupling between N–H and C_{α}–H deformations is very sensitive to geometry and generates a rich and informative ROA band structure. Schweitzer-Stenner (2001) has provided a detailed review of the extended amide III and other vibrational modes of small peptides based on conventional Raman studies. Side-chain vibrations also generate many characteristic Raman bands (Tu, 1986; Miura and Thomas, 1995; Overman and Thomas, 1999): Although less prominent in ROA spectra, some side-chain vibrations do give rise to useful ROA signals.

Poly-L-lysine and poly-L-glutamic acid adopt well-defined conformations under certain conditions of temperature and pH and have long been used as models for the spectroscopic identification of secondary structure sequences in proteins. Poly-L-lysine at alkaline pH and poly-L-glutamic acid at acid pH have neutral side chains and so are able to support α-helical conformations stabilized both by internal hydrogen bonds and by hydrogen bonds to the solvent, whereas poly-L-lysine at neutral and acid pH and poly-L-glutamic acid at neutral and alkaline pH have charged side chains that repel one another, thereby encouraging a disordered structure (Bergethon, 1998). The backscattered Raman and ROA spectra of these samples are shown in Figure 3 (Wilson *et al.*, 1996a; Hecht *et al.*, 1999). There are clearly many similarities between the ROA spectra of the α-helical conformations of the two polypeptides and also between those of the disordered conformations; but there are sufficiently large differences between the ROA spectra of the corresponding α-helical and disordered conformations to enable ROA to distinguish between these two states. Until very recently it has not been possible to obtain ROA spectra of β-sheet polypeptides due to experimental difficulties, but this has now been achieved for β-sheet poly-L-lysine (I. H. McColl, E. W. Blanch, L. Hecht and L. D. Barron, unpublished work).

Preliminary accounts of the assignment of polypeptide and protein ROA bands to the various types of structural elements were given in earlier articles (Wilson *et al.*, 1996a; Barron *et al.*, 1996; Teraoka *et al.*, 1998). However, with the measurement of many more ROA spectra, some of these early assignments have been revised (Barron *et al.*, 2000) and are still being refined as more data accumulate.

B. *Raman Optical Activity Band Assignments for Folded Proteins*

Before discussing the ROA band signatures and general spectral characteristics of the disordered types of structure found in unfolded proteins, it is helpful to review the ROA band signatures of α-helix and β-sheet together with those of loops, turns, and side chains, as shown by folded proteins containing significant amounts of extended secondary structure in order to demonstrate that ROA is able to discriminate adequately between ordered and disordered polypeptide sequences. Typical

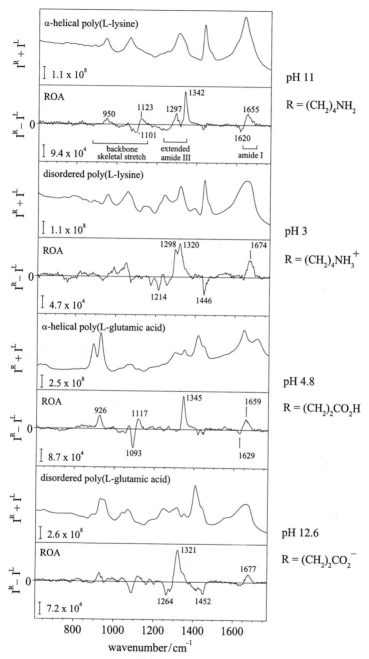

FIG. 3. Backscattered Raman ($I^R + I^L$) and ROA ($I^R - I^L$) spectra of poly-L-lysine in α-helical (top pair) and disordered (second pair) conformations, and of poly-L-glutamic acid in α-helical (third pair) and disordered (bottom pair) conformations in aqueous solution. Reprinted from Barron *et al.*, 2000, *Prog. Biophys. Mol. Biol.* **73,** 1–49, with permission from Elsevier Science.

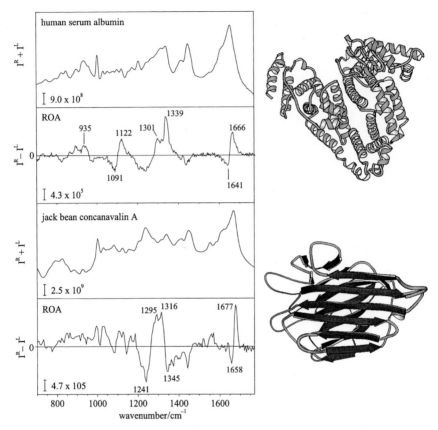

FIG. 4. Backscattered Raman and ROA spectra of the α-helical protein human serum albumin in H$_2$O (top pair) and the β-sheet protein jack bean concanavalin A in acetate buffer solution at pH ∼5.4, together with MOLSCRIPT diagrams (Kraulis, 1991) of their X-ray crystal structures (PDB codes 1ao6 and 2cna).

examples of proteins containing large amounts of exclusively α-helix or β-sheet as their secondary structure components, with the rest made up of nonregular (disordered) structure in the form of loops and turns, are human serum albumin and jack bean concanavalin A, respectively. Their backscattered Raman and ROA spectra are shown in Figure 4.

1. α-Helix

According to the PDB X-ray crystal structure 1 ao6, human serum albumin contains 69.2% α-helix and 1.7% 3$_{10}$-helix, the rest being made up of turns and long loops. The amide I ROA couplet centered at ∼1650 cm^{-1} (Fig. 4), which is negative at low wavenumber and positive at high, is a good signature of α-helix in proteins and accords with the wavenumber range $\sim$1645–1655 cm^{-1} for α-helix bands in conventional Raman spectra (Miura and Thomas, 1995). The two polypeptides in α-helical conformations show similar amide I ROA couplets (Fig. 3) but shifted to lower wavenumber, which may be due to the particular side chains or due to the absence of disordered structure. Positive ROA intensity in the range $\sim$870–950 cm^{-1} also appears to be a signature of α-helix. Again, the two α-helical polypeptides show positive ROA bands in this region. The conservative ROA couplet centered at $\sim$1103 cm^{-1}, negative at low wavenumber and positive at high, may also be due to α-helix since a similar feature appears in the ROA spectra of the two α-helical polypeptides. However, other structural elements may be involved since disordered poly-L-glutamic acid shows a similar ROA feature (Fig. 3).

The ROA band structure of human serum albumin in the extended amide III region is especially rich and interesting. Before X-ray crystal structures of serum albumins were available in the PDB, it was suggested that the intense positive ROA band at $\sim$1340 cm^{-1} (Fig. 4) originated in 3_{10}-helix (Barron et al., 1996; Teraoka et al., 1998). However, evidence subsequently emerged for reassigning this band to a hydrated form of α-helix (Barron et al., 2000). For example, this band usually disappears rapidly when the protein is dissolved in D$_2$O. This indicates both that the sequence is highly accessible to solvent and that N—H deformations of the peptide backbone make a significant contribution to the generation of this ROA band (because the corresponding N—D deformations contribute to normal modes in a spectral region several hundred wavenumbers lower). Furthermore, the two polypeptides in α-helical conformations, which are exposed to water all around in aqueous solution, also show strong positive ROA bands at $\sim$1342–1345 cm^{-1} (Fig. 3). A weak conventional Raman band near this wavenumber in proteins has recently been assigned to α-helix by studying isotopic variants of filamentous bacteriophages (Tsuboi et al., 2000), the major coat proteins of that comprise mainly extended α-helix. ROA data on these bacteriophages have similarly proved valuable for α-helix ROA band assignments (Blanch et al., 1999a).

Positive ROA bands in the range $\sim$1297–1312 cm^{-1} are also characteristic of α-helix. These are observed at $\sim$1300 cm^{-1} in human serum albumin (Fig. 4) and at $\sim$1297 cm^{-1} in α-helical poly-L-lysine (Fig. 3). These additional bands appear to be associated with α-helix in a more hydrophobic environment (Barron et al., 2000). The striking absence of a positive ROA band in the range $\sim$1297–1312 cm^{-1} in α-helical poly-L-glutamic acid would then suggest that only the hydrated form of α-helix is present, possibly due to the shorter side chains relative to poly-L-lysine, which allows easier penetration of water to the peptide backbone. In conventional Raman spectroscopy, α-helix is thought to contribute to amide III bands in the range $\sim$1260–1310 cm^{-1} (Miura and Thomas, 1995).

Further evidence for these α-helix ROA band assignments in the extended amide III region comes from the ROA spectrum of poly-L-alanine dissolved in a mixture of chloroform (70%) and dichloracetic acid (30%), known to promote α-helix formation (Fasman, 1987), which shows strong positive ROA bands at $\sim$1305 and 1341 cm^{-1} (unpublished results), and of the α-helix forming alanine-rich peptide AK21 (sequence Ac–AAKAAAAKAAAAKAAAAKAGY–NH$_2$) in aqueous solution which shows strong positive ROA bands at $\sim$1309 and 1344 cm^{-1} (Blanch *et al.*, 2000).

Recent infrared and Raman studies of α-helical poly-L-alanine have provided definitive assignments of a number of the normal modes of vibration in the extended amide III region over the range in which we have identified α-helix ROA bands (Lee and Krimm, 1998a,b). These normal modes variously transform as the A, E_1, and E_2 symmetry species of the point group of a model infinite regular helix. The polypeptide and protein α-helix ROA peaks observed in this region may be related to a number of these normal modes, the ROA intensities being a function of the perturbations (geometric and/or due to various types of hydration) to which the particular helical sequence is subjected.

New insight into the nature of the hydrophobic and hydrated variants of α-helix identified by ROA has been provided by recent electron spin resonance studies of double spin-labeled alanine-rich peptides which identified a new, more open conformation of the α-helix (Hanson *et al.*, 1998; Bolin and Millhauser, 1999). Computer modeling suggests that the more open geometry leaves the hydrogen bonding network intact but with a changed C–O $\cdots$ N angle which results in a splaying of the backbone amide carbonyls away from the helix axis and into solution. Hence the more open structure may be the preferred conformation in aqueous solution since it allows main-chain hydrogen bonding with water molecules. If so, an equilibrium would exist between the canonical form of α-helix, which may be responsible for the positive $\sim$1300 cm^{-1} ROA band, and the more open form, which may be responsible for the positive $\sim$1340 cm^{-1} band, the former being favored in a hydrophobic environment and the latter in an aqueous (or hydrogen-bonding) environment.

2. β-Sheet

According to the PDB X-ray crystal structure 2cna, jack bean concanavalin A contains 43.5% β-strand, 1.7% α-helix, and 1.3% 3_{10}-helix, the rest being made up of hairpin bends and long loops. The strong sharp negative ROA band at $\sim$1241 cm^{-1} (Fig. 4) appears to be a good signature of β-strand. Among many other examples, this assignment is reinforced by ROA date on the protein capsid of the icosahedral virus MS2, the coat proteins of which contain a large amount of extended antiparallel β-sheet and which shows a very strong sharp negative ROA band at $\sim$1247 cm^{-1} (Blanch et al., 2002a). In conventional Raman spectroscopy, amide III bands from β-sheet are assigned to the region $\sim$1230–1245 cm^{-1} (Miura and Thomas, 1995). Some proteins containing β-sheet also show a negative ROA band at $\sim$1220 cm^{-1}, which appears to be the signature of a distinct variant of β-strand or sheet, possibly hydrated (Barron et al., 2000).

The ROA couplet in the amide I region, negative at low wavenumber, positive at high, and centered at $\sim$1665 cm^{-1} (Fig. 4), is a good signature of β-sheet. This can be distinguished from the similar couplet shown by α-helix since it is $\sim$5–19 cm^{-1} higher. In conventional Raman spectroscopy, amide I β-sheet bands are assigned to the range $\sim$1665–1680 cm^{-1} (Miura and Thomas, 1995).

The backbone skeletal stretch region is rather ambiguous for β-sheet assignments. As may be seen in the ROA spectrum of jack bean concanavalin A (Fig. 4), there are many bands in this region, mostly positive. These bands can be rather variable in intensity, width, and wavenumber in β-sheet proteins, which may reflect the various types of distortions found in β-sheet.

Although the ROA spectra of typical β-sheet proteins share some of the features observed in β-sheet poly-L-lysine, there are also some differences, especially in the amide I region. This is because the β-sheet in proteins tends to be twisted and irregular, whereas that in polypeptides tends to be extended, multistranded and relatively flat.

3. Loops and Turns

Loops and turns in proteins are polypeptide sequences of nonrepetitive conformation connecting structural elements with well-defined secondary structure (Donate et al., 1996). They are sometimes called linkers of secondary structure (Geetha and Munson, 1997). The ϕ,ψ angles of individual residues within loop and turn sequences in proteins tend to cluster in well-defined regions of the Ramachandran surface, with those regions where residues in α-helix, β-sheet, and polyproline II (PPII) helix are usually found being especially favored (Adzhubei et al., 1987; Swindells et al., 1995; Donate et al., 1996; Geetha and Munson, 1997). In addition to ROA bands characteristic of α-helix and β-sheet, the extended amide III region often contains sharp ROA bands that appear to originate in loops and turns, possibly because the short-range nature of the vibrational couplings responsible for ROA reflect the local geometry of the individual residues.

Proteins containing a large amount of antiparallel β-sheet usually show negative ROA bands in the range $\sim$1340–1380 cm^{-1}, especially if the β-sheet is extended as in a sandwich or barrel, which appear to originate in tight turns of the type found in β-hairpins. One example is the band at $\sim$1345 cm^{-1} in jack bean concanavalin A (Fig. 4). These bands do not appear in the ROA spectra of α/β proteins, since these contain only parallel β-sheet with the ends of the parallel β-strands connected by α-helical sequences rather than tight turns (Barron *et al.*, 2000).

Proteins containing a significant amount of β-sheet also often show a sharp positive ROA band at $\sim$1314–1325 cm^{-1} similar to a prominent positive ROA band shown by unordered poly-L-lysine and poly-L-glutamic acid at $\sim$1320 cm^{-1} (Fig. 3). Since these unordered polypeptides may contain significant amounts of the PPII helical conformation (*vide infra*), this ROA band in proteins may be the signature of PPII-helical elements known from X-ray crystal structures to be present in some of the longer loops (Adzhubei and Sternberg, 1993; Stapley and Creamer, 1999). An example is observed at $\sim$1316 cm^{-1} in the ROA spectrum of jack bean concanavalin A (Fig. 4). Other ROA bands appear in the range $\sim$1260–1295 cm^{-1} which may also originate in loop and turn structure.

4. Side Chains

Bands from side chains are usually not very prominent in the ROA spectra of polypeptides and proteins. Side-chain differences may be responsible for some of the small variations observed in the characteristic helix, sheet, and turn ROA bands. There are, however, several distinct regions where side chain vibrations appear to be largely responsible for the observed ROA features. In particular, ROA bands in the range $\sim$1400–1480 cm^{-1} originate in CH_2 and CH_3 side-chain deformations and also in tryptophan vibrations; ROA bands in the range $\sim$1545–1560 cm^{-1} originate in tryptophan vibrations; and ROA bands in the range $\sim$1600–1630 cm^{-1} originate in vibrations of aromatic side chains, especially tyrosine (Barron *et al.*, 2000).

The absolute stereochemistry of the tryptophan conformation, in terms of the sign and magnitude of the torsion angle $\chi^{2,1}$ around the bond connecting the indole ring to the C_β atom, may be obtained from the $\sim$1545–1560 cm^{-1} tryptophan ROA band, assigned to a W3-type vibration of the indole ring (Miura and Thomas, 1995). This was discovered from observations of W3 tryptophan ROA bands with equal magnitudes but opposite signs in two different filamentous bacterial viruses with coat protein subunits containing a single tryptophan, which suggested that the tryptophans adopt quasi-enantiomeric conformations in the two viruses (Blanch *et al.*, 2001a). The W3 ROA band may also be used as a probe of conformational heterogeneity among a set of tryptophans in disordered protein sequences since cancelation from ROA contributions with opposite signs will result in a loss of ROA intensity. Examples are observed in molten globule states of equine lysozyme and human lysozyme, as described in Section IV,A. Tryptophan ROA is similar in this respect to the near-UVCD bands from aromatic side chains, which disappear when tertiary structure is lost on partial denaturation (Woody and Dunker, 1996). These two techniques provide complementary perspectives because ROA probes the intrinsic skeletal chirality of the tryptophan side chain whereas UVCD probes the chirality in the general environment of the chromophore.

5. Conformational Plasticity

Because of its ability to detect loop and turn structure separately from secondary structure, ROA is able to monitor delicate changes in the fold, something inaccessible to conventional solution phase techniques such as UVCD and NMR (Dalal *et al.*, 1997). The pH-dependent conformational changes in bovine β-lactoglobulin provide a good example. This protein undergoes a reversible transition (Tanford *et al.*, 1959) which occurs near pH 7.5 at 25°C and which may be of functional importance since this is close to physiological pH (Qin *et al.*, 1998). ROA spectra measured at pH 2.0, 6.8, and 9.0 reveal that the basic β-barrel core is preserved over the entire pH range (Blanch *et al.*, 1999b), in agreement with other studies. However, from the shift of a sharp positive ROA band at $\sim$1268 cm^{-1} to $\sim$1294 cm^{-1} on going from pH values below that of the Tanford transition to values above, the transition appears to be associated with changes in the local conformations of residues in loop sequences possibly corresponding to a migration into the α-helical region from a nearby region. These changes may be related to those detected in X-ray crystal structures (Qin *et al.*, 1998) which revealed that the Tanford transition is associated with conformational changes in loops that form a doorway to the interior of the protein. Migration of residue conformations between different regions of the Ramachandran surface corresponds to what has been called "plasticity" associated with fluctuations between discrete energy wells associated with distinct conformational substates (Rader and Agard, 1997).

Wilson *et al.* (1997) reported what appeared to be a new type of sharp cooperative transition in native hen lysozyme from temperature-dependent ROA measurements. It occurred at ~12°C and seemed to involve loss of residual mobility in tertiary loop and secondary structure together with tryptophan side chains. Since differential scanning calorimetry (DSC) failed to reveal any latent heat, it did not appear to be a simple first-order transition. However, our later attempts to repeat these observations failed to reproduce the transition, which must therefore be regarded as unsubstantiated at present, and may be due to an instrumental artifact. On the other hand, a related, readily reproducible result is a significant increase in the intensity of a positive ~1344 cm^{-1} ROA band in hen lysozyme on binding to the trimer of *N*-acetylglucosamine (NAG$_3$) (Ford *et al.*, 1995; Wilson *et al.*, 1997). This may be due to disordered (mobile) structure around the binding site coalescing into hydrated α-helix.

IV. UNFOLDED PROTEINS

A. *Partially Unfolded Denatured Proteins*

So far it has not been possible to obtain useful ROA data on the fully unfolded denatured states of proteins that have well-defined tertiary folds in their native states. This is because the use of chemical denaturants such as urea or guanadinium hydrochloride at high concentration precludes ROA measurements owing to intense Raman bands from the denaturants, and thermally unfolded proteins often show considerable light scattering due to the formation of aggregates. The few data we have on such states appear to show just a few weak unstructured ROA bands. However, interesting ROA results have been obtained on the partially unfolded denatured states associated with reduced proteins and molten globules.

1. *Reduced Proteins*

The ROA spectra of partially unfolded denatured hen lysozyme and bovine ribonuclease A, prepared by reducing all the disulfide bonds and keeping the sample at low pH, together with the ROA spectra of the corresponding native proteins, are displayed in Figure 5. As pointed out in Section II,B, the short time scale of the Raman scattering event means that the ROA spectrum of a disordered system is a superposition of "snapshot" ROA spectra from all the distinct conformations present at equilibrium. Because of the reduced ROA intensities and large

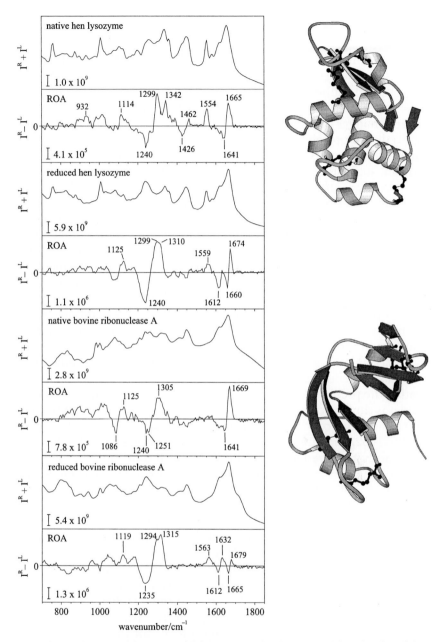

FIG. 5. Backscattered Raman and ROA spectra of native (top pair) and reduced (second pair) hen lysozyme, and of native (third pair) and reduced (bottom pair) bovine ribonuclease A, together with MOLSCRIPT diagrams of the crystal structures (PDB codes 1lse and 1rbx) showing the native disulfide links. The native proteins were in acetate buffer at pH 5.4 and the reduced proteins in citrate buffer at pH 2.4. The spectra were recorded at ~20°C.

fluctuations in the bulk sample, these are difficult samples for ROA measurements: The previously published versions (Ford *et al.*, 1995; Wilson *et al.*, 1996b), although generally similar, are not as reliable in some details as the newly acquired spectra shown in Figure 5. There are significant changes compared with the ROA spectra of the native states. In both reduced proteins, ROA in the backbone skeletal stretch region is generally suppressed as compared with the native state, with differences of detail, suggesting that much fixed structure has been lost. Much of the ROA band structure in the extended amide III region has disappeared in both proteins, being replaced by a large broad couplet. Hints of structure are present throughout these couplets, suggesting that they are generated by a large number of conformations with a range of local residue ϕ,ψ angles clustering in the same general regions of the Ramachandran surface as in the native proteins. This is expected from other studies, including X-ray crystallography (Adzhubei *et al.*, 1987; Swindells *et al.*, 1995) and NMR (Fiebig *et al.*, 1996). However, the absence of positive ROA intensity at $\sim$1340 cm^{-1} reveals that none of the significant amount of hydrated α-helix present in native hen lysozyme is present in the reduced protein. Compared with the native states, the amide I couplets have shifted by $\sim$10 cm^{-1} to higher wavenumber and, in the case of reduced lysozyme, this couplet has become quite sharp, indicating the presence of a significant amount of β-sheet since its amide I couplet is similar to that shown by β-sheet proteins (see the ROA spectrum of jack bean concanavalin A in Fig. 4). The side-chain bands marked at 1426, 1462, and 1554 cm^{-1} in native hen lysozyme have all degraded significantly in the reduced protein, presumably owing to increased conformational heterogeneity. Although of similar overall appearance, the ROA spectra of unfolded hen lysozyme and unfolded bovine ribonuclease A are clearly different in detail, which may reflect the different residue compositions and their different ϕ,ψ propensities. In conventional Raman spectroscopy, disordered structure is assigned to the range $\sim$1245–1270 cm^{-1} in the amide III region and to $\sim$1655–1665 cm^{-1} in the amide I region (Miura and Thomas, 1995).

We have been unable to reproduce the previously reported sharp positive ROA peaks at $\sim$1300 and 1314 cm^{-1} in reduced hen lysozyme and their coalescence with increasing temperature (Wilson *et al.*, 1996b). This casts doubt on the experimental evidence for the suggestion that individual residue conformations in disordered states of polypeptides "flicker" between distinct regions of the Ramachandran surface at rates $\sim$10^{12} s^{-1} (Barron *et al.*, 1997). Nonetheless, the spectra have a rather different general appearance from those of disordered poly-L-lysine and poly-L-glutamic acid shown in Figure 3, which suggests that some sequences may be more mobile. In particular, neither of the reduced proteins shows the strong sharp positive $\sim$1320 cm^{-1} ROA band assigned to PPII structure (*vide infra*) that dominates the ROA spectra of the two polypeptides. This indicates that, unlike the disordered polypeptides and the natively unfolded proteins discussed in Section IV,C, neither has a structure based on large amounts of PPII helix (although the positive $\sim$1315 cm^{-1} ROA band shown by reduced ribonuclease A suggests that some PPII-type structure may be present).

2. Molten Globules

Raman optical activity is proving to be valuable in studies of molten globule protein states. These are denatured entities, stable at equilibrium, which have well-defined secondary structure but lack the specific tertiary interactions characteristic of the native protein (Ptitsyn, 1995). It has been proposed that there could be a close relationship between molten globules and kinetic intermediates on the folding pathway of many proteins (Karplus and Shakhnovich, 1992; Ptitsyn, 1995). Molten globules may also have an important physiological role in the living cell (Bychkova *et al.*, 1988; Ptitsyn, 1995) and may be implicated in a number of genetic diseases due to mutant versions of key proteins failing to fold beyond the molten globule stage (Bychkova and Ptitsyn, 1995). A much-studied molten globule state is that supported by α-lactalbumins at low pH and called the A-state. The ROA spectrum of A-state bovine α-lactalbumin (Fig. 6) is clearly different from that of the native protein. Apart from a few differences of detail, the ROA spectrum of A-state α-lactalbumin is similar to that of reduced hen lysozyme (Fig. 5), which suggests that the structural elements present in the two partially denatured proteins are similar.

It is interesting to compare the ROA spectrum of A-state bovine α-lactalbumin with that of equine lysozyme, also shown in Figure 6 (Blanch *et al.*, 1999c). Equine lysozyme is unusual in that, unlike most other lysozymes, it binds calcium and supports a stable molten globule at low pH (Morozova-Roche *et al.*, 1997). The ROA spectrum of A-state equine lysozyme shows it to be more native-like than that of bovine α-lactalbumin, in agreement with other studies (Morozova-Roche *et al.*, 1997). It is particularly interesting that the strong positive ROA band at $\sim$1340 cm^{-1} exhibited by the native state is shown also by the A-state, unlike that in bovine α-lactalbumin, which indicates that the sequences of well-defined hydrated α-helix survive in the molten globule. This is reinforced by the amide I ROA couplet which is characteristic mainly of α-helix. Furthermore, the positive $\sim$1551 cm^{-1} ROA band shown by native state equine lysozyme and assigned to tryptophan W3 vibrations

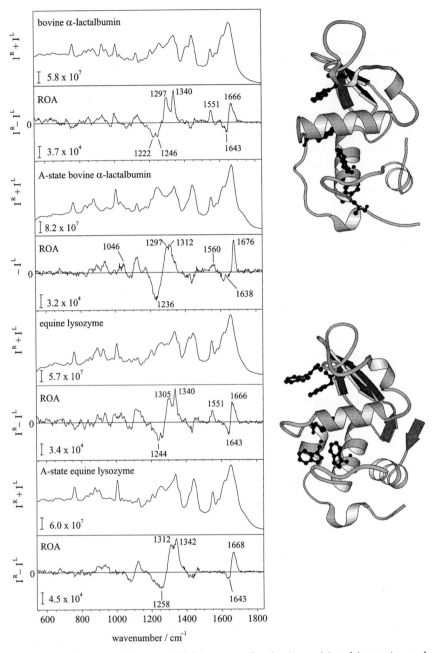

FIG. 6. Backscattered Raman and ROA spectra of native (top pair) and A-state (second pair) bovine α-lactalbumin, and of native (third pair) and A-state (bottom pair) equine lysozyme, together with MOLSCRIPT diagrams of the crystal structures (PDB codes 1hfz and 2eql) showing the tryptophans. The native proteins were in acetate buffer at pH ∼4.6 and 5.6, respectively, and the A-states in glycine buffer at pH 1.9. The native-state and A-state spectra were recorded at ∼20°C and ∼2°C, respectively.

completely disappears in the A-state. As discussed in Section III,B, this suggests conformational heterogeneity among the five tryptophans in the A-state. ROA has also demonstrated that the structure of calcium-free equine lysozyme is almost identical to that of the calcium-bound protein, and that the structure of the molten globule state adopted by the apo protein at pH 4.5 and 48°C is similar to that of the A-state (Blanch *et al.*, 1999c).

From temperature-dependent ROA measurements, Wilson *et al.* (1996c) suggested that, on reducing the temperature of A-state α-lactalbumin at pH 2.0, the tertiary fold develops and becomes native-like at 2°C. Since DSC revealed no latent heat, it was suggested that the tertiary fold developed under the control of a continuous (higher order) phase transition with a midpoint at ~20°C. No such changes were observed in a later study when the pH was ~2.0, but they were observed on raising the pH to ~3.2 (Blanch *et al.*, 1999c). At pH ~3.2, bovine α-lactalbumin is known to support a molten globule-to-native transition via a first-order phase transition showing a very small latent heat with a transition midpoint of ~30°C (Griko *et al.*, 1994), so it appears that Wilson *et al.* (1996c) were in fact monitoring this transition using ROA.

3. The Amyloidogenic Prefibrillar Intermediate of Human Lysozyme

Although having a stability similar to that of hen lysozyme with respect to temperature and reduced pH, being greater than that of equine lysozyme, the thermal denaturation behavior of human lysozyme is subtly different from that of hen lysozyme. Below ~pH 3.0, thermal denaturation of human lysozyme is not a two-state process: Unlike hen lysozyme, it supports a partially folded molten globule–like state at elevated temperatures (Haezebrouck *et al.*, 1995). Furthermore, incubation of human lysozyme at ~57°C at pH 2.0, under which conditions the partially folded state is most highly populated, induces the formation of amyloid fibrils; but if incubated at 70°C at pH 2.0, under which conditions the fully denatured state is most highly populated, only amorphous aggregates are formed (Morozova-Roche *et al.*, 2000). Since the prefibrillar intermediate of human lysozyme survives for up to ~24 h, it proved possible to measure its ROA spectrum, which revealed the conformation of the critical sequences that may be important in the formation of amyloid fibrils (Blanch *et al.*, 2000).

The ROA spectra of native and prefibrillar amyloidogenic human lysozyme are displayed in Figure 7, together with a MOLSCRIPT diagram of the native structure. The ROA spectrum of the native protein is very similar to that of hen lysozyme (Fig. 5). However, large changes have occurred in the ROA spectrum of the prefibrillar intermediate. In particular, the positive ~1340 cm^{-1} ROA band assigned to hydrated

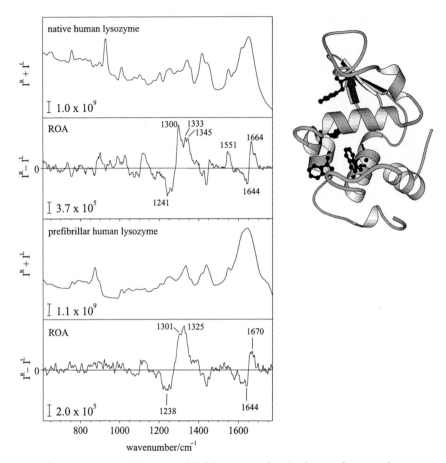

FIG. 7. Backscattered Raman and ROA spectra of native human lysozyme in acetate buffer at pH 5.4 measured at 20°C (top pair), and of the prefibrillar intermediate in glycine buffer at pH 2.0 measured at 57°C (bottom pair), together with a MOLSCRIPT diagram of the crystal structure (PDB code 1jsf) showing the tryptophans.

α-helix has disappeared, and a new strong positive ROA band centered at $\sim$1318 cm^{-1} and assigned to PPII helix (*vide infra*) has appeared. This suggests that hydrated α-helix has undergone a conformational change to PPII structure. Furthermore, the disappearance of the positive $\sim$1551 cm^{-1} ROA band assigned to tryptophan W3 vibrations indicates that, as discussed in Section III,B, major conformational changes have occurred among the five tryptophans present in human lysozyme, four of which are located in the α-domain (Fig. 7). From the appearance of the amide I couplet, and of other ROA bands, there does not appear to be any increase in β-sheet in the prefibrillar intermediate. All these ROA data suggest that a major destabilization and conformational change have occurred in the α-domain (and perhaps also in the β-domain), and that PPII helix may be the critical conformational element involved in amyloid fibril formation. This strikes a chord with prion disease in which the structured α-domain of the prion protein undergoes a conformational change to produce most of the β-sheet found in the associated amyloid fibrils (see Section V,D). The ROA spectrum of hen lysozyme, which shows less propensity for amyloid fibril formation (Krebs *et al.*, 2000), looks much more native-like at 57°C and pH 2.0 than does that of human lysozyme under the same conditions (Blanch *et al.*, 2000).

A recent NMR study of the structure and dynamics of two amyloidogenic variants of human lysozyme (Chamberlain *et al.*, 2001) showed that, although one variant destabilized the β-domain much more than the other, it had no greater propensity to form amyloid fibrils. It was concluded that the increased ability of the variants to access substantially unfolded conformations of the protein is the origin of their amyloidogenicity. This appears to reinforce the conclusions from ROA that a destabilized α-domain is involved in fibril formation.

B. *Polyproline II Helix*

Polyproline II helix appears to be a particularly important structural element in some unfolded proteins. Furthermore, as mentioned above, its observation in the prefibrillar partially unfolded intermediate of human lysozyme provided the first suggestion that it may be implicated in amyloid fibril formation. Largely on the basis of evidence from UVCD and VCD, PPII helix is thought to be the main conformational element present in disordered polypeptides such as poly-L-lysine and poly-L-glutamic acid, perhaps in the form of short segments interspersed with residues having other conformations (Tiffany and Krimm, 1968; Dukor and Keiderling, 1991; Woody, 1992; Keiderling *et al.*, 1999; Keiderling, 2000). We therefore consider the ROA spectra of disordered poly-L-lysine and poly-L-glutamic acid shown in Figure 3 to contain prominent bands characteristic of PPII structure, especially the strong positive ROA band at $\sim$1320 cm^{-1} (Smyth *et al.*, 2001). Further evidence is the presence of positive ROA bands in the range $\sim$1315–1325 cm^{-1} in some β-sheet and other proteins, assigned to PPII-helical elements known from X-ray crystal structures to be present in some of the longer loops (see Section III,B). To date, no reliable *ab initio* computations of the ROA spectrum of PPII helix have been performed, so our assignment of strong

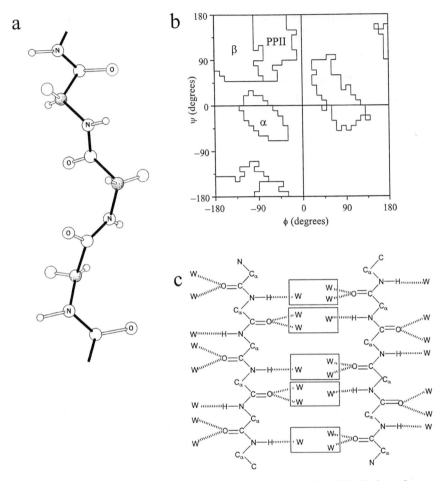

FIG. 8. (a) Sketch of the extended left-handed poly-L-proline II helical conformation. (b) Schematic diagram of a Ramachandran potential energy surface. (c) Extended
polypeptide chains with fully hydrated backbone C=O and N—H groups. In certain circumstances, chain association may occur via formation of β-sheet hydrogen bonds, each
involving elimination of three water molecules (W) between the chains.

positive ROA intensity at $\sim$1320 cm^{-1} to PPII structure relies mainly on
the evidence outlined here.

PPII helix is illustrated in Figure 8a. It consists of a left-handed extended helical conformation in which the ϕ,ψ angles of the constituent
residues are restricted to values around $-78°$, $+146°$ corresponding to a
region of the Ramachandran surface adjacent to the β-region (Fig. 8b).
This imparts a perfect threefold rotational symmetry to the structure

($n = -3$). The majority of proteins contain at least one short segment of PPII helix with a residue composition including not only proline, but also nearly all other amino acids (Adzhubei and Sternberg, 1993; Stapley and Creamer, 1999). PPII helix is most often found in longer loops connecting two different or the same types of secondary structure. The long sequences linking the separate β-domains in immunoglobulins are especially rich in PPII-helical structure (Adzhubei and Sternberg, 1994). Of particular relevance here is the fact that the extended nature of PPII helix precludes intrachain hydrogen bonds. However, the orientation and high accessibility of the peptide NH and C=O groups makes them favorable partners for hydrogen bonding with side chains and with water molecules, both from the protein hydration shell and from water molecules incorporated into the protein structure. Indeed, main-chain hydrogen bonding with water molecules is the most probable stabilizing factor for PPII-helical segments lying on the protein surface (Adzhubei and Sternberg, 1993).

The absence of intrachain hydrogen bonds, together with its inherent flexibility, means that PPII helix is not readily amenable to traditional methods of structure determination, and this has hindered the recognition of the widespread occurrence of PPII helices in globular proteins. For free peptides in solution, the PPII conformation is indistinguishable from an irregular backbone structure by ^{1}H NMR spectroscopy (Siligardi and Drake, 1995). However, it can be recognized for peptides in solution by means of UVCD (Tiffany and Krimm, 1968; Woody, 1992; Makarov et al., 1993; Siligardi and Drake, 1995) and VCD (Dukor and Keiderling, 1991; Keiderling et al., 1999; Keiderling, 2000). In globular proteins the UVCD spectrum of PPII helix has been deconvoluted from the UVCD spectra of a set of proteins and is very similar to that of PPII helix in model peptides (Sreerama and Woody, 1994).

The special characteristics of PPII helix allow the polypeptide chain to progress easily from this conformation to other types, including α-helix and β-sheet, which could explain its prevalence in longer loops connecting secondary structure elements in proteins (Adzhubei and Sternberg, 1993). Furthermore in alanine-based peptides, which possess a strong tendency to form α-helices, even in aqueous solution, it has been suggested on the basis of UVCD studies that the PPII conformation contributes significantly to the ensemble of unfolded states (Park et al., 1997). This concurs with ROA results on AK21 which reveal increasing conversion of hydrated α-helix to PPII helix with increasing temperature (Blanch et al., 2000). In addition, computational studies on the model dipeptide N-acetyl-L-alanine N'-methylamide, $CH_3CONHCH(CH_3)CONH(CH_3)$, have indicated that hydration stabilizes both α-helical and PPII conformations, which are therefore expected to dominate in aqueous solution (Han *et al.*, 1998). All this, together with the fact that PPII helix is highly hydrated, leads to the idea that the interconversion between hydrated α-helix and PPII helix is a facile process. This idea harmonizes with the suggestion that water molecules promote the unfolding of α-helices since hydrogen bonding of water molecules to the peptide backbone lowers the potential energy barrier separating the α from the adjacent β and PPII regions of the Ramachandran surface (Sundaralingam and Sekharudu, 1989).

C. Natively Unfolded (Rheomorphic) Proteins

Raman optical activity has proved valuable in recent studies of the structure and behavior of a number of natively unfolded proteins found in quite different biological situations. Although completely unfolded, the structures appear to be more stable than those generated by complete unfolding of proteins having a tertiary fold in their native states. Such natively unfolded proteins clearly have special characteristics built into their amino acid sequences that prevent aggregation under normal physiological conditions.

In one study, the bovine milk proteins β- and κ-casein were investigated, together with the human recombinant brain proteins α-, β-, and γ-synuclein plus the A30P and A53T mutants of α-synuclein associated with familial cases of Parkinson's disease and recombinant human tau 46 together with the tau 46 P301L mutant associated with inherited frontotemporal dementia (Syme *et al.*, 2002). The ROA spectra of all these proteins were found to be very similar, being dominated by a strong positive band centered at $\sim$1318 cm^{-1} and assigned to the PPII-helical conformation. DSC measurements on these proteins revealed no evidence for a high-temperature thermal transition associated with cooperative unfolding.

The ROA spectra of β-casein, γ-synuclein, and the tau 46 P301L mutant are displayed in Figure 9 as typical examples. Although very similar overall, there are differences of detail between the three ROA spectra which reflect the different residue compositions and also differences in minor structural elements. The absence of a well-defined amide I ROA couplet in any of the spectra indicates that none of the proteins contains a significant quantity of well-defined secondary structure.

These studies suggest that the caseins, synucleins, and tau have natively unfolded structures in which the sequences are based largely on the PPII conformation and are held together in a loose noncooperative fashion. However, rather than describing them as "random coil",

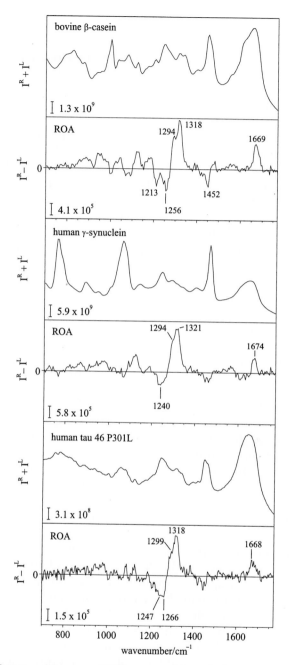

FIG. 9. Backscattered Raman and ROA spectra of the natively unfolded proteins bovine β-casein (top pair) and human γ-synuclein (middle pair) in aqueous buffer at pH ∼7.0, and of the human tau 46 P301L mutant at pH ∼4.3.

the term *rheomorphic* would seem to apply to the synucleins and tau as it does to the caseins for which it was originally coined (Holt and Sawyer, 1993). Syme *et al.* (2002) envisage a rheomorphic protein to have the following general properties. The radius of gyration and hydrodynamic radius are ~2–4 times larger than for a globular protein containing a similar number of residues. Over extensive lengths of its sequence the polypeptide chain is expected to be rather stiff, having a persistence length of ~5–10 residues. In other parts of the molecule there may be local interactions and small amounts of regular secondary structure, but interactions between remote parts of the sequence are expected to be minimal and many of the side chains are expected to have conformational flexibility. We do not consider the rheomorphic state of a protein to be the same as the molten globule state since the latter is almost as compact as the folded state (radius of gyration and hydrodynamic radius ~10–30% larger), has a hydrophobic core, and contains a large amount of secondary structure (Ptitsyn, 1995; Arai and Kuwajima, 2000).

In another study, several wheat gluten proteins were investigated. Specifically, the prolamins A- and ω-gliadin were investigated, together with a peptide called T-A-1 which constitutes fragment 147–440 of the high-molecular-weight glutenin subunit Dx5 (D. D. Kasarda, L. Hecht, K. Nielsen, L. D. Barron, and E. W. Blanch, unpublished work). The ROA spectra of these three samples are displayed in Figure 10. The spectra of the T-A-1 peptide and ω-gliadin are dominated by a strong sharp positive ROA band at ~1321 cm^{-1}, very similar to that seen in the ROA spectrum of disordered poly-L-glutamic acid displayed in Figure 3. This band, together with the absence of an ROA couplet in the amide I region, suggests that the two structures are very similar, being based on large amounts of PPII helix, perhaps with some turns but no secondary structure at all. Although the T-A-1 peptide and ω-gliadin may also be classed as rheomorphic like the milk and brain proteins discussed above, the fact that their ROA bands, especially the positive ~1321 cm^{-1} band, are generally sharper than those of the milk and brain proteins (Fig. 9) suggests that the PPII-helical structure is better defined. This could be due to the high proportion of proline and glutamine residues. The ROA spectrum of A-gliadin is rather different from those of the T-A-1 peptide and ω-gliadin, being dominated by a strong positive band at ~1342 cm^{-1} characteristic of hydrated α-helix. The clear amide I couplet is also characteristic of α-helix, as are several features in the backbone skeletal stretch region. The structure of A-gliadin is therefore probably based on a large amount of well-defined hydrated α-helix together with a smaller proportion of PPII structure, as evidenced by the weak but clear positive ROA band at ~1316 cm^{-1}, plus perhaps some turns. This

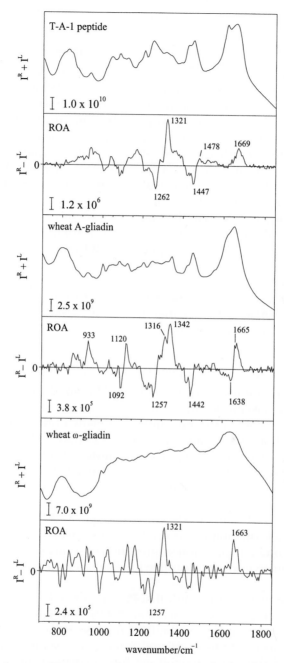

FIG. 10. Backscattered Raman and ROA spectra in H_2O of wheat glutenin T-A-1 peptide (top pair) at pH 3.5, wheat A-gliadin (middle pair) at pH 3.5, and wheat ω-gliadin (bottom pair) at pH 2.6.

conclusion is consistent with evidence that A-gliadin comprises two distinct domains: a largely α-helical C-terminal domain stabilized by three disulfide bonds and a largely disordered N-terminal domain where most of the proline residues are located (Purcell *et al.*, 1988).

The reason that the caseins, which constitute nearly 80% of bovine milk, are unfolded in their native states appears to be to facilitate digestion, since the open rheomorphic structures allow rapid and extensive degradation to smaller peptides by proteolytic enzymes. The natively unfolded structures of many cereal proteins may serve an analogous purpose since they provide nutrition for seedlings. The physiological function of the synucleins in the brain is as yet unclear, but tau is known to promote and stabilize the assembly of microtubules.

In another rather different application, ROA data indicated that the coat protein subunits of intact tobacco rattle virus contain a significant amount of PPII structure, which is possibly associated with sequences previously suggested to be mobile and to be exposed externally in the intact virus particle and which may be associated with its transmission by nematodes (Blanch *et al.*, 2001b).

D. *Protein Misfolding and Disease: Amyloid Fibril Formation*

The conformational plasticity supported by mobile regions within native proteins, partially denatured protein states such as molten globules, and natively unfolded proteins underlies many of the conformational (protein misfolding) diseases (Carrell and Lomas, 1997; Dobson *et al.*, 2001). Many of these diseases involve amyloid fibril formation, as in amyloidosis from mutant human lysozymes, neurodegenerative diseases such as Parkinson's and Alzheimer's due to the fibrillogenic propensities of α-synuclein and tau, and the prion encephalopathies such as scrapie, BSE, and new variant Creutzfeldt-Jacob disease (CJD) where amyloid fibril formation is triggered by exposure to the amyloid form of the prion protein. In addition, aggregation of serine protease inhibitors such as α_1-antitrypsin is responsible for diseases such as emphysema and cirrhosis.

It has been suggested recently that PPII helix may be the "killer conformation" in such diseases (Blanch *et al.*, 2000). This was prompted by the observation, described in Section III,B, of a positive band at $\sim$1318 cm^{-1}, not present in the ROA spectrum of the native state, that dominates the ROA spectrum of a destabilized intermediate of human lysozyme (produced by heating to 57°C at pH 2.0) that forms prior to amyloid fibril formation. Elimination of water molecules between extended polypeptide chains with fully hydrated C=O and N—H groups to form

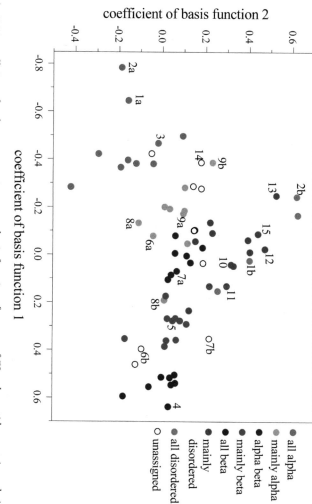

Fig. 11. Plot of the PCA coefficients for the two most important basic functions for a set of 73 polypeptide, protein and virus ROA spectra. The proteins discussed in this chapter are labeled as follows: α-helical (1a) and disordered (1b) poly-L-lysine; α-helical (2a) and disordered (2b) poly-L-glutamic acid; human serum albumin (3); jack bean concanavalin A (4); bovine β-lactoglobulin at pH 6.8 (5); native (6a) and A-state (6b) hen lysozyme; native (7a) and reduced (7b) ribonuclease A; native (8a) and A-state (8b) bovine α-lactalbumin; native (9a) and reduced (9b) equine lysozyme; bovine β-casein (10); human γ-synuclein (11); human tau 46 P301L mutant (12); wheat glutenin T-A-1 peptide (13); wheat A-gliadin (14); wheat ω-gliadin (15). More complete definitions of the structural types are as follows: all alpha, >~60% α-helix with the rest mainly disordered and little or no β-sheet; mainly alpha, significant amounts of α-helix and a small amount of β-sheet (~5–15%); alpha beta, significant amounts of α-helix and β-sheet; mainly beta, >~35% β-sheet and a small amount of α-helix (~5–15%); all beta, >~45% β-sheet with the rest mainly disordered and little or no α-helix; mainly disordered, little secondary structure; all disordered, no secondary structure.

β-sheet hydrogen bonds (Fig. 8c) is a highly favorable process entropically (Sekharudu and Sundaralingam, 1993); and since strands of PPII helix are close in conformation to β-strands, they would be expected to readily undergo this type of aggregation with one another and with the edges of established β-sheet to form the typical cross-β structure of amyloid fibrils. The more dynamic type of disorder associated with the true random coil is expected to lead to amorphous aggregates rather than ordered fibrils, as is observed in most examples of protein aggregation. However, although the presence of significant amounts of PPII structure may be necessary for the formation of regular fibrils, other factors must be important since, of the natively unfolded milk and brain proteins studied by Syme et al. (2002), only α-synuclein is known to readily form typical amyloid cross-β fibrils (the presence or otherwise of a cross-β substructure in filamentous aggregates of tau is currently unclear). The finding that a combination of low mean hydrophobicity and high net charge is an important prerequisite for proteins to remain natively unfolded (Uversky et al., 2000) may be especially pertinent. One example of the significance of charge is the observation that removal of the highly charged anionic C-terminal of α-synuclein results in more rapid fibril formation (Crowther et al., 1998).

Vigorous shaking is required to induce rapid amyloid fibril formation from full-length α-synuclein (Serpell et al., 2000). Shaking may lead to the shearing of α-synuclein assemblies, which then function as seeds, resulting in a marked acceleration of filament formation. On the other hand, Serio et al. (2000) found that only modest rotation of the yeast prion protein Sup35 was effective in inducing amyloid fibril formation. These observations could be consistent with the presence of large amounts of PPII structure, since any agitation that produces fluid flow, as in a circular motion, would tend to align the PPII-helical sequences, thereby making it more favorable for them to aggregate into ordered β-sheet. These two mechanisms (generation of new seeds plus alignment of PPII sequences) could strongly reinforce each other.

Prion disease is especially interesting because an understanding of some key aspects remains elusive, particularly the existence of an infectious form along with genetic and sporadic forms (Prusiner, 1998), something that has not been recognized in any other class of neurodegenerative disease. An especially subtle aspect is how differences between separate mammalian species variants of prion protein, and of genetic variants within species, can lead to barriers to infection between some species but not others. The basic current model involves conversion of a ubiquitous cellular form of prion protein PrP^c into a scrapie (amyloid fibril) form PrP^{Sc}, the prion protein being both target and infectious agent. PrP^c has a predominantly α-helical structured domain together with a long disordered tail, whereas PrP^{Sc} has a high β-sheet content. A recent ROA study of full-length and truncated recombinant sheep prion proteins in our laboratory has revealed that the disordered N-terminal tail of PrP^c is composed mainly of well-defined PPII helical elements (E. W. Blanch, A. C. Gill, A. G. O. Rhie, J. Hope, L. Hecht, and L. D. Barron, unpublished work). However, although containing a large amount of PPII structure, the disordered tail does not itself appear to be intrinsically fibrillogenic since most of the β-sheet in PrP^{Sc} appears to originate in sequences within the structured α-domain of PrP^c (Prusiner, 1998). Indeed, the N-terminal end carries high net charge and has a low net hydrophobicity which, as mentioned above, inhibits association of separate polypeptide chains; and the disordered tail has a high proportion of proline and glycine residues which have very low β-sheet forming propensities (Smith and Regan, 1997). Furthermore, NMR studies have revealed that the disordered tail has a stabilizing effect on the structured domain (Prusiner, 1998). Thus, although not participating directly in the β-fibril formation associated with prion disease, there is scope for the disordered tail to modulate the fibrillogenesis itself together with the infectivity and species barriers that characterize prion disease. Evidence is growing to support a functional role for PrP^c in copper metabolism: Cu(II) ions appear to bind to the protein in a highly conserved octarepeat region of the N-terminal tail (Viles *et al.*, 1999). Presumably the rheomorphic character of the PPII structure in the N-terminal tail of the apoprotein enables the appropriate sequences to readily adapt to the fixed Cu(II)-bound conformation in the holoprotein.

V. PRINCIPAL COMPONENT ANALYSIS

Because protein ROA spectra contain bands characteristic of loops and turns in addition to bands characteristic of secondary structure, they should provide information on the overall three-dimensional solution structure. We are developing a pattern recognition program, based on principal component analysis (PCA), to identify protein folds from ROA spectral band patterns (Blanch *et al.*, 2002b). The method is similar to one developed for the determination of the structure of proteins from VCD (Pancoska *et al.*, 1991) and UVCD (Venyaminov and Yang, 1996) spectra, but is expected to provide enhanced discrimination between different structural types since protein ROA spectra contain many more structure-sensitive bands than do either VCD or UVCD. From the ROA spectral data, the PCA program calculates a set of subspectra that serve as basis functions, the algebraic combination of which with appropriate expansion coefficients can be used to reconstruct any member of the original set of experimental ROA spectra. Our current set, which includes most of the proteins discussed in this chapter, contains 73 entries comprising ROA spectra of several poypeptides in model α-helical and disordered states, many proteins with well-defined tertiary folds known mostly from X-ray crystallography with a few from multidimensional solution NMR, and several viruses with coat protein folds known from X-ray crystallography or fiber diffraction. The method appears to be useful for unfolded and partially folded as well as completely folded protein structures: thus, a first preliminary level of PCA analysis is applied here to most of the proteins discussed above since this provides a convenient summary of their structural relationships.

Figure 11 (see color insert) shows a plot of the coefficients for the whole set of ROA spectra for the two most important raw basis functions. This serves to separate the spectra into clusters corresponding to different types of protein structure, allowing structural similarities between proteins of unknown structure with those of known structure to be identified. The protein positions are color-coded with respect to the seven different structural types listed on the figure, which provide a useful initial classification that will be refined in later work. The plot reveals increasing α-helix content to the left, increasing β-sheet content to the right, and increasing disorder from bottom to top. The proteins discussed in this chapter are numbered according to the list provided in the figure caption.

All of the native proteins appear in the correct regions of Figure 11 corresponding to their known X-ray crystal structures. Bovine β-casein, human γ-synuclein, human tau 46 P301L, the wheat glutenin T-A-1 peptide, and wheat ω-gliadin all appear toward the top of the plot in a cluster containing disordered poly-L-lysine and poly-L-glutamic acid consistent with their assignment to natively unfolded structures based on PPII-helix as found in the two disordered polypeptides. The casein, synuclein, tau, and ω-gliadin cluster closely with disordered poly-L-lysine lying a little way down from the top, suggesting that they are mainly but not completely disordered, unlike poly-L-glutamic acid which appears to be completely disordered since it lies close to the top. We have classified the T-A-1 peptide as completely disordered since it lies closer to disordered poly-L-glutamic acid. The reduced and A-state proteins appear outside of the mainly disordered and all disordered regions, suggesting that some secondary or other well-defined structural elements may be present. Reduced hen lysozyme is especially interesting in this respect since its position in the mainly β-region, well to the right of that of the native protein, suggests that it contains significantly more β-sheet than the native protein. The position of reduced ribonuclease A is again displaced to the right of the native protein, but also significantly higher than reduced hen lysozyme, which indicates that it contains rather more disordered structure. A-state bovine α-lactalbumin is placed well to the right and slightly higher than the native protein, suggesting some increase in β-sheet, whereas A-state equine lysozyme is placed significantly to the left of the native protein and higher, which suggests an increase in α-helix and disordered structure. This reinforces the earlier conclusion that the A-states of bovine α-lactalbumin and equine lysozyme are rather different (Blanch *et al.*, 1999c).

VI. Concluding Remarks

This chapter has reviewed the application of ROA to studies of unfolded proteins, an area of much current interest central to fundamental protein science and also to practical problems in areas as diverse as medicine and food science. Because the many discrete structure-sensitive bands present in protein ROA spectra, the technique provides a fresh perspective on the structure and behavior of unfolded proteins, and of unfolded sequences in proteins such as A-gliadin and prions which contain distinct structured and unstructured domains. It also provides new insight into the complexity of order in molten globule and reduced protein states, and of the more mobile sequences in fully folded proteins such as β-lactoglobulin. With the promise of commercial ROA instruments becoming available in the near future, ROA should find many applications in protein science. Since many gene sequences code for natively unfolded proteins in addition to those coding for proteins with well-defined tertiary folds, both of which are equally accessible to ROA studies, ROA should find wide application in structural proteomics.

Acknowledgments

We thank the EPSRC and BBSRC for research grants. We are also grateful to Dr. C. D. Syme for measuring some of the protein ROA spectra shown in this chapter, and to the other students and collaborators who have contributed to the Glasgow biomolecular ROA work. We are especially grateful to Dr. K. Nielsen for developing the PCA program, and to the following for their critical input in the form of samples and expertise into the ROA studies of natively unfolded proteins reported here: Drs. M. Goedert (synucleins and tau), C. Holt (caseins), A. C. Gill and J. Hope (prions), and D. D. Kasarda (glutens).

References

Adzhubei, A. A., Eisenmenger, F., Tumanyan, V. G., Zinke, M., Brodzinski, S., and Esipova, N. G. (1987). *J. Biomol. Struct. Dyn.* **5**, 689–704.
Adzhubei, A. A., and Sternberg, M. J. E. (1993). *J. Mol. Biol.* **229**, 472–493.
Adzhubei, A. A., and Sternberg, M. J. E. (1994). *Protein Sci.* **3**, 2395–2410.

Arai, M., and Kuwajima, K. (2000). *Adv. Protein Chem.* **53,** 209–282.

Atkins, P. W., and Barron, L. D. (1969). *Mol. Phys.* **16,** 453–466.

Barron, L. D. (1982). *Molecular Light Scattering and Optical Activity.* Cambridge University Press, Cambridge.

Barron, L. D., and Buckingham, A. D. (1971). *Mol. Phys.* **20,** 1111–1119.

Barron, L. D., and Buckingham, A. D. (1974). *J. Am. Chem. Soc.* **96,** 4769–4773.

Barron, L. D., and Hecht, L. (2000). In *Circular Dichroism: Principles and Applications,* 2nd Ed. (N. Berova, K. Nakanishi, and R. W. Woody, eds.), pp. 667–701. Wiley, New York.

Barron, L. D., Bogaard, M. P., and Buckingham, A. D. (1973). *J. Am. Chem. Soc.* **95,** 603–605.

Barron, L. D., Hecht, L., Bell, A. F., and Wilson, G. (1996). *Appl. Spectrosc.* **50,** 619–629.

Barron, L. D., Hecht, L., and Wilson, G. (1997). *Biochemistry* **36,** 13143–13147.

Barron, L. D., Hecht, L., Blanch, E. W., and Bell, A. F. (2000). *Prog. Biophys. Mol. Biol.* **73,** 1–49.

Bergethon, P. R. (1998). *The Physical Basis of Biochemistry.* Springer-Verlag, New York.

Blanch, E. W., Bell, A. F., Hecht, L., Day, L. A., and Barron, L. D. (1999a). *J. Mol. Biol.* **290,** 1–7.

Blanch, E. W., Hecht, L., and Barron, L. D. (1999b). *Protein Sci.* **8,** 1362–1367.

Blanch, E. W., Morozova-Roche, L. A., Hecht, L., Noppe, W., and Barron, L. D. (1999c). *Biopolymers (Biospectroscopy)* **57,** 235–248.

Blanch, E. W., Morozova-Roche, L. A., Cochran, D. A. E., Doig, A. J., Hecht, L., and Barron, L. D. (2000). *J. Mol. Biol.* **301,** 553–563.

Blanch, E. W., Hecht, L., Day, L. A., Pederson, D. M., and Barron, L. D. (2001a). *J. Am. Chem. Soc.* **123,** 4863–4864.

Blanch, E. W., Robinson, D. J., Hecht, L., and Barron, L. D. (2001b). *J. Gen. Virol.* **82,** 1499–1502.

Blanch, E. W., Hecht, L., Syme, C. D., Nielsen, K., and Barron, L. D. (2002a). *J. Gen. Virol.* In press.

Blanch, E. W., Robinson, D. J., Hecht, L., and Barron, L. D. (2002b). *J. Gen. Virol.* **83,** 241–246.

Bolin, K. A., and Millhauser, G. L. (1999). *Accs. Chem. Res.* **32,** 1027–1033.

Bour, P. (2001). *J. Comput. Chem.* **22,** 426–435.

Braun, W., Vašák, M., Robbins, A. H., Stout, C. D., Wagner, G., Kägi, J. H. R., and Wüthrich, K. (1992). *Proc. Natl. Acad. Sci. USA* **89,** 10124–10128.

Buckingham, A. D. (1967). *Adv. Chem. Phys.* **12,** 107–142.

Bychkova, V. E., and Ptitsyn, O. B. (1995). *FEBS Lett.* **359,** 6–8.

Bychkova, V. E., Pain, R. H., and Ptitsyn, O. B. (1988). *FEBS Lett.* **238,** 231–234.

Carey, P. R. (1982). *Biochemical Applications of Raman and Resonance Raman Spectroscopies.* Academic Press, New York.

Carrell, R. W., and Lomas, D. A. (1997). *Lancet* **350,** 134–138.

Chamberlain, A. K., Receveur, V., Spencer, A., Redfield, C., and Dobson, C. M. (2001). *Protein Sci.* **10,** 2525–2530.

Creighton, T. E., ed. (1992). Protein Folding. W. H. Freeman, New York.

Crowther, R. A., Jakes, R., Spillantini, M. G., and Goedert, M. (1998). *FEBS Lett.* **436,** 309–312.

Dalal, S., Balasubramanian, S., and Regan, L. (1997). *Fold. Des.* **2,** R71–R79.

Diem, M. (1993). *Modern Vibrational Spectroscopy.* Wiley, New York.

Dobson, C. M. (1994). *Curr. Biol.* **4,** 636–640.

Dobson, C. M., Ellis, R. J., and Fersht, A. R. (2001). Protein misfolding and disease. *Phil. Trans. Roy. Soc. Lond. B* **356,** 127–227.

Donate, L. E., Rufino, S. D., Canard, L. H. J., and Blundell, T. L. (1996). *Protein Sci.* **5,** 2600–2616.

Dukor, R. K., and Keiderling, T. A. (1991). *Biopolymers* **31,** 1747–1761.

Fasman, G. D. (1987). *Biopolymers* **26,** S59–S79.

Fiebig, K. M., Schwalbe, H., Buck, M., Smith, L. J., and Dobson, C. M. (1996). *J. Phys. Chem.* **100,** 2661–2666.

Ford, S. J., Wen, Z. Q., Hecht, L., and Barron, L. D. (1994). *Biopolymers* **34,** 303–313.

Ford, S. J., Cooper, A., Hecht, L., Wilson, G., and Barron, L. D. (1995). *J. Chem. Soc., Faraday Trans.* **91,** 2087–2093.

Geetha, V., and Munson, P. J. (1997). *Protein Sci.* **6,** 2538–2547.

Goedert, M., Spillantini, M. G., Serpell, L. C., Berriman, J., Smith, M. J., Jakes, R., and Crowther, R. A. (2001). *Phil. Trans. Roy. Soc. B* **356,** 213–227.

Griko, Y. V., Freire, E., and Privalov, P. L. (1994). *Biochemistry* **33,** 1889–1899.

Haezebrouck, P., Joniau, M., van Dael, H., Hooke, S. D., Woodruff, N. D., and Dobson, C. M. (1995). *J. Mol. Biol.* **246,** 382–387.

Han, W.-G., Jalkanen, K. J., Elster, M., and Suhai, S. (1998). *J. Phys. Chem. B* **102,** 2587–2602.

Hanson, P., Anderson, D. J., Martinez, G., Millhauser, G., Formaggio, F., Crisma, M., Toniolo, C., and Vita, C. (1998). *Mol. Phys.* **95,** 957–966.

Hecht, L., and Barron, L. D. (1990). *Appl. Spectrosc.* **44,** 483–491.

Hecht, L., and Barron, L. D. (1994). *Faraday Discuss.* **99,** 35–47.

Hecht, L., Barron, L. D., and Hug, W. (1989). *Chem. Phys. Lett.* **158,** 341–344.

Hecht, L., Barron, L. D., Blanch, E. W., Bell, A. F., and Day, L. A. (1999). *J. Raman Spectrosc.* **30,** 815–825.

Helgaker, T., Rudd, K., Bak, K. L., Jørgensen, P., and Olsen, J. (1994). *Faraday Discuss.* **99,** 165–180.

Holt, C., and Sawyer, L. (1993). *J. Chem. Soc., Faraday Trans.* **89,** 2683–2692.

Hug, W. (1994). *Chimia* **48,** 386–390.

Hug, W., and Hangartner, G. (1999). *J. Raman Spectrosc.* **30,** 841–852.

Karplus, M., and Shakhnovich, E. (1992). In *Protein Folding* (T. E. Creighton, ed.), pp. 127–195. W. H. Freeman, New York.

Keiderling, T. A. (2000). In *Circular Dichroism: Principles and Applications,* 2nd Ed. (N. Berova, K. Nakanishi, and R. W. Woody, eds.), pp. 621–666. Wiley, New York.

Keiderling, T. A., Silva, R. A. G. D., Yoder, G., and Dukor, R. K. (1999). *Bioorg. Med. Chem.* **7,** 133–141.

Kraulis, P. J. (1991). *J. Appl. Crystallog.* **24,** 946–950.

Krebs, R. H., Wilkins, D. K., Chung, E. W., Pitkeathly, M. C., Chamberlain, A. K., Zurdo, J., Robinson, C. V., and Dobson, C. M. (2000). *J. Mol. Biol.* **300,** 541–549.

Lee, S. H., and Krimm, S. (1998a). *J. Raman Spectrosc.* **29,** 73–80.

Lee, S. H., and Krimm, S. (1998b). *Biopolymers* **46,** 283–317.

Makarov, A. A., Adzhubei, I. A., Prostasevich, I. I., Lobachov, V. M., and Esipova, N. G. (1993). *J. Protein Chem.* **12,** 85–91.

Miura, T., and Thomas, G. J., Jr. (1995). In *Subcellular Biochemistry,* Vol. 24. *Proteins: Structure, Function and Engineering* (B. B. Biswas and S. Roy, eds.), pp. 55–99. Plenum Press, New York.

Morozova-Roche, L. A., Arico-Muendel, C. C., Haynie, D. T., Emelyanenko, V. I., Van Dael, H., and Dobson, C. M. (1997). *J. Mol. Biol.* **268,** 903–921.

Morozova-Roche, L. A., Zurdo, J., Spencer, A., Noppe, W., Receveur, V., Archer, D. B., Joniau, M., and Dobson, C. M. (2000). *J. Struct. Biol.* **130,** 339–351.

Nafie, L. A. (1996). *Appl. Spectrosc.* **50,** 14A–26A.

Nafie, L. A., and Che, D. (1994). *Adv. Chem. Phys.* **85**, 105–222.
Nafie, L. A., and Freedman, T. B. (2000). In *Circular Dichroism: Principles and Applications,* 2nd Ed. (N. Berova, K. Nakanishi, and R. W. Woody eds.), pp. 97–131. Wiley, New York.
Overman, S. A., and Thomas, G. J., Jr. (1999). *Biochemistry* **38**, 4018–4027.
Pain, R. H., ed. (1994). *Mechanisms of Protein Folding.* IRL Press, Oxford.
Pancoska, P., Yasui, S. C., and Keiderling, T. A. (1991). *Biochemistry* **30**, 5089–5103.
Park, S.-H., Shalongo, W., and Stellwagen, E. (1997). *Protein Sci.* **6**, 1694–1700.
Poland, D., and Scheraga, H. A. (1970). *Theory of Helix-Coil Transitions in Biopolymers.* Academic Press, New York.
Polavarapu, P. L. (1990). *J. Phys. Chem.* **94**, 8106–8112.
Polavarapu, P. L. (1998). *Vibrational Spectra: Principles and Applications with Emphasis on Optical Activity.* Elsevier, Amsterdam.
Polavarapu, P. L., and Deng, Z. (1994). *Faraday Discuss.* **99**, 151–163.
Prusiner, S. B. (1998). *Proc. Natl. Acad. Sci. USA* **95**, 13363–13383.
Ptitsyn, O. B. (1995). *Adv. Protein Chem.* **47**, 83–229.
Purcell, J. M., Kasarda, D. D., and Wu, C.-S. C. (1988). *J. Cereal Sci.* **7**, 21–32.
Qin, B. Y., Bewley, M. C., Creamer, L. K., Baker, H. M., Baker, E. N., and Jameson, G. B. (1998). *Biochemistry* **37**, 14014–14023.
Rader, S. D., and Agard, D. A. (1997). *Protein Sci.* **6**, 1375–1386.
Schweitzer- Stenner, R. (2001). *J. Raman Spectrosc.* **32**, 711–732.
Sekharudu, C. Y., and Sundaralingam, M. (1993). In *Water and Biological Macromolecules* (E. Westhof, ed.), pp. 148–162. CRC Press, Boca Raton, FL.
Serio, T. R., Cashikar, A. G., Kowal, A. S., Sawacki, G. J., Moslehi, J. J., Serpell, L., Arnsdorf, M. F., and Lindquist, S. L. (2000). *Science* **289**, 1317–1321.
Serpell, L. C., Berriman, J., Jakes, R., Goedert, M., and Crowther, R. A. (2000). *Proc. Natl. Acad. Sci. USA* **97**, 4897–4902.
Shewry, P. R., and Tatham, S. (1990). *Biochem. J.* **267**, 1–12.
de la Sierra, I. L., Quillien, L., Flecker, P., Gueguen, J., and Brunie, S. (1999). *J. Mol. Biol.* **285**, 1195–1207.
Siligardi, G., and Drake, A. F. (1995). *Biopolymers* **37**, 281–292.
Smith, L. J., Fiebig, K. M., Schwalbe, H., and Dobson, C. M. (1996). *Fold. Des.* **1**, R95–R106.
Smith, C. K., and Regan, L. (1997). *Accs. Chem. Res.* **30**, 153–161.
Smyth, E., Syme, C. D., Blanch, E. W., Hecht, L., Vašák, M., and Barron, L. D. (2001). *Biopolymers* **58**, 138–151.
Sreerama, N., and Woody, R. W. (1994). *Biochemistry* **33**, 10022–10025.
Stapley, B. J., and Creamer, T. P. (1999). *Protein Sci.* **8**, 587–595.
Sundaralingam, M., and Sekharudu, Y. C. (1989). *Science* **244**, 1333–1337.
Swindells, M. B., MacArthur, M. W., and Thornton, J. M. (1995). *Nature Struct. Biol.* **2**, 596–603.
Syme, C. D., Blanch, E. W., Holt, C., Jakes, R., Goedert, M., Hecht, L., and Barron, L. D. (2002). *Eur. J. Biochem.* **269**, 148–156.
Tanford, C., Bunville, L. G., and Nozaki, Y. (1959). *J. Am. Chem. Soc.* **81**, 4032–4036.
Teraoka, J., Bell, A. F., Hecht, L., and Barron, L. D. (1998). *J. Raman Spectrosc.* **29**, 67–71.
Thomas, G. J., Jr. (1999). *Ann. Rev. Biophys. Biomol. Struct.* **28**, 1–27.
Tiffany, M. L., and Krimm, S. (1968). *Biopolymers* **6**, 1379–1382.
Tsuboi, M., Suzuki, M., Overman, S. A., and Thomas, G. J., Jr. (2000). *Biochemistry* **39**, 2677–2684.
Tu, A. T. (1986). *Adv. Spectrosc.* **13**, 47–112.
Uversky, V. N., Gillespie, J. R., and Fink, A. L. (2000). *Proteins* **41**, 415–427.
Vargek, M., Freedman, T. B., and Nafie, L. A. (1997). *J. Raman Spectrosc.* **28**, 627–633.

Venyaminov, S. Y., and Yang, J. T. (1996). In *Circular Dichroism and Conformational Analysis of Biomolecules* (G. D. Fasman, ed.), pp. 69–107. Plenum Press, New York.

Viles, J. H., Cohen, F. E., Prusiner, S. B., Goodin, D. B., Wright, P. E., and Dyson, H. J. (1999). *Proc. Natl. Acad. Sci. USA* **96,** 2042–2047.

Wilson, G., Hecht, L., and Barron, L. D. (1996a). *J. Chem. Soc., Faraday Trans.* **92,** 1503–1510.

Wilson, G., Hecht, L., and Barron, L. D. (1996b). *Biochemistry* **35,** 12518–12525.

Wilson, G., Hecht, L., and Barron, L. D. (1996c). *J. Mol. Biol.* **261,** 341–347.

Wilson, G., Hecht, L., and Barron, L. D. (1997). *J. Phys. Chem.* **101,** 694–698.

Woody, R. W. (1992). *Adv. Biophys. Chem.* **2,** 37–79.

Woody, R. W., and Dunker, A. K. (1996). In *Circular Dichroism and the Conformational Analysis of Biomolecules* (G. D. Fasman, ed.), pp. 109–157. Plenum Press, New York.

Wright, P. E., and Dyson, H. J. (1999). *J. Mol. Biol.* **293,** 321–331.

WHAT FLUORESCENCE CORRELATION SPECTROSCOPY CAN TELL US ABOUT UNFOLDED PROTEINS

By CARL FRIEDEN, KRISHNANANDA CHATTOPADHYAY, and ELLIOT L. ELSON

Department of Biochemistry and Molecular Biophysics, Washington University
School of Medicine, St. Louis, Missouri 63110

I. INTRODUCTION

Fluorescence correlation spectroscopy (FCS) measures rates of diffusion, chemical reaction, and other dynamic processes of fluorescent molecules. These rates are deduced from measurements of fluorescence fluctuations that arise as molecules with specific fluorescence properties enter or leave an open sample volume by diffusion, by undergoing a chemical reaction, or by other transport or reaction processes. Studies of unfolded proteins benefit from the fact that FCS can provide information about rates of protein conformational change both by a direct readout from conformation-dependent fluorescence changes and by changes in diffusion coefficient.

Although the diffusion coefficient is relatively insensitive to protein conformation changes, it can be measured with high accuracy by FCS and so can be useful for monitoring structural changes that occur as the protein passes among partially unfolded states. As the protein takes on more compact or more extended conformations, the hydrodynamic radius and consequently the diffusion coefficient change correspondingly; if these changes are large enough, they can be observed in FCS measurements. Should the fluorescent probe be sensitive to formation of these residual structures, it may be possible to extract information about the kinetics of their interconversion. As will be discussed later, however, with respect to the specific example of the intestinal fatty acid binding protein (IFABP), it is possible to detect different conformational states at different denaturant concentrations.

Among kinetics approaches used to study dynamics within the unfolded state, FCS is closely related to relaxation methods in which small free energy perturbations (e.g., by temperature or pressure jump) displace the equilibrium point of a reaction system. The kinetic parameters are deduced from the rate of relaxation to the new equilibrium (Eigen and De Maeyer, 1963). There are, however, two important distinctions in principle between FCS and conventional relaxation kinetic measurements. First, conventional methods measure kinetic properties of macroscopic systems containing vast numbers of reactant molecules. Measurement of a single relaxation transient can yield estimates of kinetic parameters with an accuracy determined entirely by the accuracy of the single measurement. In contrast, FCS measurements focus on microscopic systems containing relatively few, perhaps only one, of the reactant molecules. The molecular fluctuations seen in these systems are therefore stochastic. Even if the time course of an individual fluctuation could be measured with infinite accuracy, this would not yield an accurate estimate of kinetic parameters. Rather, it is essential to measure many fluctuations and to analyze them statistically (Elson and Webb, 1975). Second, the perturbations of state in a conventional relaxation kinetics measurement, while small, are nevertheless macroscopic. In contrast, FCS measurements are carried out using no external perturbation. In principle, this could be useful in analyzing a complex process such as protein folding that involves many intermediates closely spaced in free energy. A macroscopic perturbation, even if small, could drive the system through many intermediates. The zero perturbation conditions of FCS, however, could sample the transition kinetics among a minimum number of neighboring states. The value of this favorable possibility is, however, often diminished in actual studies if the process is highly cooperative. In this case folding intermediates are not exposed to a substantial extent even when the system is poised in the center of the folding transition equilibrium. Rather, the system comprises mainly fully folded and fully unfolded molecules. If, however, the molecules were somehow constrained (e.g., mechanically) to a specific domain of intermediate conformations, FCS could be used to characterize the transition kinetics among the nearby intermediate states (Qian and Elson, 2001). Alternatively, if the cooperativity of the transition were reduced, similar measurements could be accomplished without external constraints. Transitions among substates within the unfolded state should be much less cooperative than the folding transition overall. Hence, the microscopic character of FCS measurements may prove especially useful in exploring closely spaced disordered states.

II. FLUORESCENCE CORRELATION SPECTROSCOPY TECHNIQUE AND THEORY

A laser beam highly focused by a microscope into a solution of fluorescent molecules defines the open illuminated sample volume in a typical FCS experiment. The microscope collects the fluorescence emitted by the molecules in the small illuminated region and transmits it to a sensitive detector such as a photomultiplier or an avalanche photodiode. The detected intensity fluctuates as molecules diffuse into or out of the illuminated volume or as the molecules within the volume undergo chemical reactions that enhance or diminish their fluorescence (Fig. 1). The measured fluorescence at time t, $F(t)$, is proportional to the number of molecules in the illuminated volume weighted by the

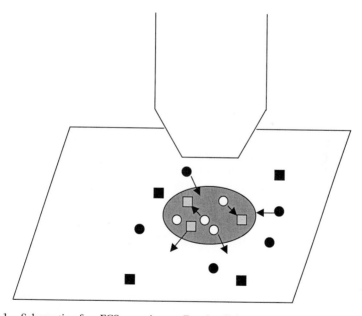

FIG. 1. Schematic of an FCS experiment. For simplicity we consider an FCS measurement on a chemical reaction system confined to a plane, e.g., a membrane. The reaction is a two-state isomerization: A (circles) ↔ B (squares). In the region of the plane illuminated by a laser beam (dark gray), A and B molecules appear white and light gray, respectively. Fluorescence fluctuations arise from interconversion of A and B and by A and B molecules diffusing into or out of the illuminated region. Molecules outside the illuminated region (black) are not detected.

excitation intensity profile, $I(r):F(t) = Q \int I(r)c(r, t) \, d^3r$, where $c(r, t)$ is the concentration of the fluorophore at position $\mathbf{r}$ and time t, and Q accounts for the spectroscopic properties of the fluorophore and the characteristics of the microscope. The fluorescence fluctuation is $\delta F(t) = Q \int I(r) \, \delta c(r,t) \, d^3r$, where $\delta F(t) = F(t) - \langle F \rangle$ and arguments in angle brackets indicate an ensemble or time average and therefore equilibrium value. Although the duration of each individual fluctuation depends on the rates both of contributing chemical reactions and of the diffusion of fluorescent molecules, because of the stochastic character of the fluctuations, conventional diffusion coefficients and chemical kinetic rate constants can be derived only from a statistical analysis of many fluctuations. The required analysis is typically carried out by computation of a fluorescence fluctuation autocorrelation function (Fig. 2):

$$G(\tau) = \int I(r)I(r') \langle \delta c(r', t) \delta c(r', t + \tau) \rangle \, d^3r \, d^3r'$$

In operational terms, with fluorescence measurements made at intervals t apart,

$$G(\tau) = \lim_{N \to \infty} \frac{1}{N} \sum_{i=1}^{N} \delta F(it) \delta F(it + \tau)$$

As for all chemical kinetic studies, to relate this measured correlation function to the diffusion coefficients and chemical rate constants that characterize the system, it is necessary to specify a specific chemical reaction mechanism. The rate of change of the jth chemical reactant can be derived from an equation that couples diffusion and chemical reaction of the form (Elson and Magde, 1974):

$$\partial \delta c_j(r, t) / \partial t = D_j \nabla^2 \delta c_j + \sum_i T_{ji} \delta c_i$$

The chemical kinetic mechanism is embodied in the rate coefficient matrix, T_{ij}, and D_j is the diffusion coefficient of the jth component.

The interpretation of FCS measurements of chemical kinetics is complicated by the coupling of diffusion and reaction in the open reaction system. Consider the simple bimolecular reaction:

$$A + B \underset{k_b}{\overset{k_f}{\rightleftharpoons}} C$$

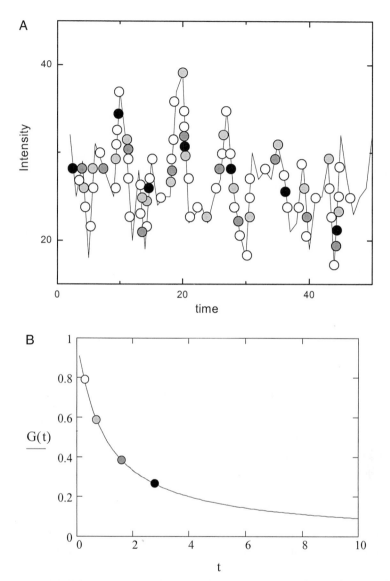

FIG. 2. Schematic analysis of fluorescence fluctuation autocorrelation function: (A) Intensity fluctuation record; (B) corresponding autocorrelation function. Equally spaced points define a record of intensity versus time in panel A. The most closely spaced points are white. Products of the intensity values associated with adjacent pairs of white points define the first (white) point in the correlation function in panel B at the shortest time lag. Light gray points are separated by a longer time lag, and so the sum of products of pairs of light gray points defines a lower value of the correlation function. Similarly, the greater time lags between dark gray points and still greater lags between black points in panel A define progressively lower values of the correlation function at longer times in panel B.

In a conventional relaxation kinetics experiment in a closed reaction system, because of mass conservation, the system can be described in a single equation, e.g., $\delta C_c(t) = \delta C_c(0)e^{-Rt}$ where $R = k_f(\langle C_A \rangle + \langle C_B \rangle) + k_b$. The forward and reverse rate constants are k_f and k_b, respectively. In an open system A, B, and C, can change independently and so three equations, one each for A, B, and C, are required, each equation having contributions from both diffusion and reaction. Consequently, three "normal modes" rather than one will be required to describe the fluctuation dynamics. Despite this complexity, some general comments about FCS measurements of reaction kinetics are useful.

The character of an FCS autocorrelation function for a chemical reaction system depends on the relative rates of reaction and diffusion. It is useful to illustrate this dependence by calculating the autocorrelation functions to be expected for a simple one-step reaction system (Elson and Magde, 1974). We take as an example the simplest possible isomerization within the unfolded state, a single-step isomerization:

$$A \underset{k_b}{\overset{k_f}{\rightleftarrows}} B$$

where A might be a molten globule conformation while B represents a more extended disordered form. The progress of the isomerization could be detected by a direct change in fluorescence or by a change in the diffusion coefficient. Two characteristic times are important: the relaxation time for the chemical reaction, $\tau_C = (k_f + k_b)^{-1}$ and the characteristic times for diffusion of A and B, $\tau_D(x) = w^2/4D_x$, where $x = A$ or B; D_x is the diffusion coefficient of component x; and w is the radius of the illuminated volume. When $\tau_C \ll \tau_D(x)$, any molecule will typically isomerize many times during the time required for it to diffuse across the illuminated volume. Then, the chemical kinetics can make an important contribution to the observed fluctuation dynamics that can be detected either in terms of an effect of the reaction on the fluorescence properties of the reactants or on their diffusion coefficients.

Figures 3 and 4 illustrate how $G(\tau)$ changes with K_{eq} in the fast reaction regime ($\tau_C \ll \tau_D$). For both figures, k_b and the equilibrium constant $K_{eq} = k_f/k_b$ are taken to vary as $k_b = 10^4/K_{eq}$ s^{-1} so that $k_f = 10^4$ s^{-1}. Hence, the chemical relaxation time, τ_C, varies from $\sim 0.9 \times 10^{-5}$ to $\sim 8 \times 10^{-5}$ s as K_{eq} varies from 0.1 to 5. If there is a sufficiently large difference in fluorescence between A and B, a term in $G(\tau)$ that varies as $\exp(-\tau/\tau_C)$ can provide a direct readout of the kinetics of

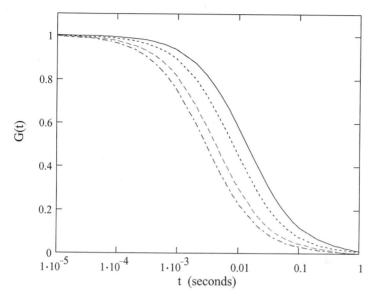

FIG. 3. Variation of autocorrelation function with changes in the equilibrium constant in the fast reaction limit. A and B have different diffusion coefficients but the same optical (fluorescence) properties. This figure illustrates how, for the simple isomerization process, A ↔ B, a change in the diffusion coefficient is sufficient to indicate the progress of the reaction. This example is calculated for a two-dimensional (planar) system in the fast reaction limit: $(k_f + k_b) \gg 4D_A/w^2$. Therefore, only a single diffusion process is seen because any molecule passes rapidly back and forth between the A and B states during the time required for it to diffuse across the illuminated volume; thus its effective diffusion coefficient is an average of the diffusion coefficients of A and B weighted by the relative fractions of A and B in the solution or, equivalently, the relative time any molecule spends as A or B. For this calculation $D_A = 10^{-7}$ and $D_B = 10^{-6}$ cm^2 s^{-1}; $k_b = 10^4/K_{eq}$ s^{-1}; $k_f = 10^4$ s^{-1}. If $C_0 = C_A + C_B$, then for solid, dotted, dashed, and dot-dashed curves, $C_B/C_0 = 0.091, 0.231, 0.545,$ and 0.833, respectively. For this example a tenfold difference in diffusion coefficients was chosen for illustrative purposes. Under most practical situations the change in diffusion coefficient due to a conformational change would be much smaller.

isomerization. This is shown in Figure 4. In this example, for purposes of illustration, the fluorescence of B is taken to be 10-fold greater than that of A ($Q_B/Q_A = 10$) while the diffusion coefficients of A and B are the same ($D_A = D_B = 10^{-7}$ cm^2 s^{-1}). We take $w = 10^{-4}$ cm, and so the characteristic diffusion time is $\tau_D = 0.025$ s. As seen in Figure 4, there is a substantial contribution to the correlation function from chemical relaxation at short times and one from diffusion at longer times. If, however, there is no effect of the chemical reaction on the fluorescence properties ($Q_A = Q_B$), the reaction can still be detected through an effect

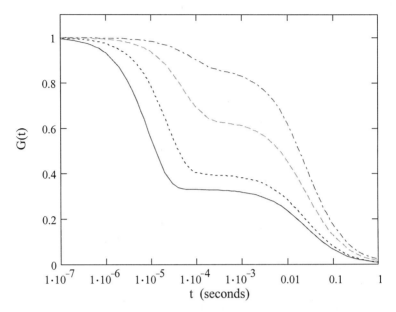

FIG. 4. Variation of autocorrelation function with changes in the equilibrium constant in the fast reaction limit. A and B have the same diffusion coefficients but different optical (fluorescence) properties. A difference in the fluorescence of A and B serves to indicate the progress of the isomerization reaction; the diffusion coefficients of A and B are the same. The characteristic chemical reaction time is in the range of 10^{-4}–10^{-5} s, depending on the value of the chemical relaxation rate; that for diffusion is 0.025 s. For this calculation parameter values are the same as those for Figure 3 except that $D_A = D_B = 10^{-7}$ cm^2 s^{-1} and $Q_A = 0.1$ and $Q_B = 1.0$. The relation of C_B/C_0 to the different curves is as in Figure 3.

on the diffusion coefficient. This is illustrated in Figure 3 for which $D_A = 10^{-7}$ cm^2 s^{-1} and $D_B = 10 D_A$. Although there is no direct indication of the chemical relaxation, the correlation function moves from longer to shorter times as the relative concentration of the faster moving species (B) increases. Because $\tau_C \ll \tau_D$, there is only a single effective diffusion process the rate of which varies as the relative fractions of A and B vary. If, however, $\tau_C \gg \tau_D$, then any molecule will be either in state A or state B but is unlikely to isomerize as it crosses the illuminated volume. Then the experiment will yield the equilibrium fraction of A and B but will not be sensitive to the kinetics of isomerization. This is shown in Figure 5 for which $D_A = 10^{-7}$ cm^2 s^{-1} and $D_B = 10 D_A$, as in Figure 3, but the rates of the chemical reaction have been decreased to 10^{-5} of their values in Figures 3 and 4. In the slow reaction limit A and B diffuse essentially independently. Hence, if the difference in diffusion coefficients between

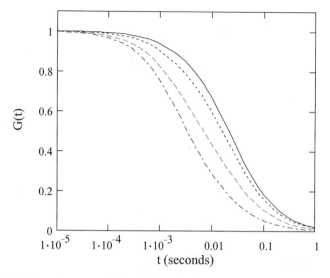

FIG. 5. Variation of autocorrelation function with changes in the equilibrium constant in the slow reaction limit. A and B have different diffusion coefficients but the same optical (fluorescence) properties. In the slow reaction limit molecules in the A and B states rarely interconvert as they diffuse across the illuminated volume. They are, therefore, seen as independently diffusing components. In this example $D_A = 10^{-7}\,\mathrm{cm}^2\,\mathrm{s}^{-1}$ while $D_B = 10D_A$. The optical properties of A and B are the same: $Q_A = Q_B = 1$. The rate of the chemical reaction is reduced to 10^{-5} of its values in Figures 3 and 4: $k_b(j) = 0.1/K_{eq}\,\mathrm{s}^{-1}$ and so $k_f = 0.1\,\mathrm{s}^{-1}$. Other conditions, including the relation of C_B/C_0 to the different curves, is as described for the preceding figures. Even though the D_B/D_A is large compared to what one might expect for a protein conformational change, it is not so large that the contributions of the two molecular species are resolved in the correlation function. This is because of the functional character of the diffusion autocorrelation function for a two-dimensional Gaussian beam shape. That is, the function $1/(1 + \tau/\tau_D)$ approaches 0 slowly as τ increases. A correlation function of the form $G(\tau) = f_A/(1 + \tau/\tau_A) + f_B/(1 + \tau/\tau_B)$, where $\tau_{A,B} = w^2/4D_{A,B}$ and $f_{A,B} = C_{A,B}/C_0$, the mole fractions of A and B, and the total concentration is $C_0 = C_A + C_B$. As is readily shown, distinguishable contributions from A and B to $G(\tau)$ are seen when D_B/D_A is large enough, e.g., $D_B/D_A = 100$.

A and B is very large (e.g. ≥ 100), two separate diffusion coefficients will be detected. If, however, the difference is smaller, a single broadened correlation relaxation will be observed as in Figure 5. Although a direct readout of the chemical kinetics cannot be seen in this limit, the effect of the reaction via its effect on diffusion is readily observed. Even a 10-fold change in diffusion coefficient is very large for protein conformational changes. Nevertheless, the accuracy of FCS measurements is sufficiently great that much smaller changes in diffusion coefficient can be reliably detected.

Thus for a direct measurement of the isomerization kinetics in $G(\tau)$, it is necessary that the reaction be fast compared to diffusion and that the fluorescence be a substantially changed by the reaction (Fig. 4). The progress of the reaction can be seen via its effect on diffusion even when there is no change in the fluorescence (Figs. 3 and 5). Therefore, it may be possible to measure the kinetics of the reaction if it is very slow compared to the time required for measurement of $G(\tau)$. In this situation the reaction system could be displaced from equilibrium, and the fractions of A and B measured by FCS as the system slowly re-equilibrates. It may also be possible to increase τ_D relative to τ_C to provide more favorable conditions for measuring the chemical kinetics. This might be accomplished by increasing w or by slowing the diffusion of A and B, for example, by increasing the viscosity of the reaction system or by confining the reactants within a matrix that hindered their diffusion.

III. APPLICATION TO CONFORMATIONAL CHANGES WITHIN THE UNFOLDED STATE

To study conformation changes of proteins by FCS, it is essential that the protein be strongly fluorescent and preferably that the fluorescence be sensitive to the conformation of the protein molecule. Tryptophan fluorescence is generally not suitable for FCS experiments because it must be excited in the UV, which requires UV microscope optics that are not generally available. Moreover, the brightness of tryptophan—the product of its extinction coefficient and quantum yield—is relatively low. Generally, it will be necessary to attach a fluorescent probe to the protein under study. The advantage of this approach is that the spectroscopic properties of the probe can be selected to match the requirements of the system, including the brightness, excitation and emission spectrum, and sensitivity to conformation. The fluorophore could be either covalently or noncovalently bound to the protein. In either case it is important to ascertain that the properties of the conformation change under study are not strongly perturbed by the presence of the fluorophore. If the fluorophore is bound noncovalently to the protein, its binding affinity might be selective for a specific protein conformation and its fluorescence might be strongly modulated by binding to the protein. For example, a fluorophore might bind selectively to the hydrophobic interior of a condensed disordered form of the protein in preference to a more fully solvated extended form. Then, providing that the kinetics of binding are fast compared to the rates of the conformational changes under study, the binding of the fluorophore could provide a good indicator of the kinetics of conformation change in FCS experiments.

IV. ADVANTAGES AND DISADVANTAGES OF USING FLUORESCENCE CORRELATION SPECTROSCOPY TO STUDY PROTEIN CONFORMATIONAL CHANGES

Perhaps the biggest advantage of FCS for studies of protein conformational changes is the wide range of accessible time scales, spanning from microseconds to seconds. FCS readily measures processes that occur in the microsecond time range and is limited in measuring fast processes mainly by the need to obtain a sufficient number of photons during the characteristic duration of the process. Hence, characterization of faster processes requires brighter fluorophores. Other advantages include sensitivity to conformation change (with an appropriate choice of fluorophore), the absence of externally applied perturbations of state, and the simultaneous determination of diffusion and chemical kinetic properties. It is also advantageous, especially for scarce proteins, that very small amounts are needed for FCS measurements: nanoliter concentrations in microliter volumes.

In principle, FCS can also measure very slow processes. In this limit the measurements are constrained by the stability of the system and the patience of the investigator. Because FCS requires the statistical analysis of many fluctuations to yield an accurate estimation of rate parameters, the slower the typical fluctuation, the longer the time required for the measurement. The fractional error of an FCS measurement, expressed as the root mean square of fluorescence fluctuations divided by the mean fluorescence, varies as $N^{-1/2}$, where N is the number of fluctuations that are measured. If the characteristic lifetime of a fluctuation is τ, the duration of a measurement to achieve a fractional error of $E = N^{-1/2}$ is $T = N\tau$. Suppose, for example, that $\tau = 1$ s. If 1% accuracy is desired, $N = 10^4$ and so $T = 10^4$ s.

Disadvantages include the need to apply a somewhat complicated theoretical analysis to measurements on systems that have many components or many states and the requirement for long measurement times for systems with slow processes, as mentioned above. For some fluorophores triplet kinetics could complicate measurements of fast conformational changes in the microsecond time range. The contribution of triplet decay to the correlation function can, however, be disentangled from changes due to conformational changes by measurements carried out at different light intensities that will cause triplet population levels to vary but that should have no effect on conformational equilibria or kinetics. Effects of conformational changes on the correlation function might be quite small, especially if the effect of the change on the diffusion coefficient and/or the fluorescence is small.

To measure these small changes reliably, it is essential to have both an adequate signal-to-noise ratio and a correct formal expression for the correlation function. Assessment of the signal-to-noise properties of FCS correlation functions has been intensively studied (Kask *et al.,* 1997; Koppel, 1974; Meseth *et al.,* 1999; Qian, 1990). Obtaining the correct formal expression for the correlation function depends both on correctly identifying and characterizing the dynamic processes and kinetic mechanisms under study and on properly defining the shape of the illuminated volume. As in all kinetic studies, the former task requires constructing an appropriate kinetic mechanism; the major purpose of experiment is to test this mechanism and to provide quantitative estimates of rate parameters. Obtaining the correct representation of the shape of the illuminated volume presents a subtler problem. An ellipsoidal Gaussian approximation is frequently used (Qian and Elson, 1991; Rigler *et al.,* 1993). But the validity of this approximation depends on the way in which the laser beam is focused within the sample. In principle, if the actual beam shape deviates sufficiently from the ellipsoidal representation, spurious components could arise in attempts to fit the experimental data. These artifactual components might have small amplitudes, but the contributions of conformational changes to the correlation function might also be expected to be small. Then, there is the danger of mistaking a fitting artifact as the contribution of a conformational change. When dealing with components of the correlation function that represent only a small fraction of its total amplitude, it is therefore essential to vary conditions of the experiment to cause the amplitude and/or rate of the component to vary in a predictable way (e.g., by changing the concentration of a denaturant) and to carry out appropriate controls (e.g., by careful measurement of the diffusion of a well defined single-component diffusant to test or calibrate the shape of the illuminated volume). Still other potential problems can arise when using denaturants such as urea or guanidine to vary the equilibrium levels of various conformations in the unfolded state. Using these denaturants at the high concentrations needed to influence protein conformations can cause substantial changes of both the refractive index of the solution and its viscosity. Changes of the refractive index change the size and shape of the illuminated volume, while changes of solution viscosity will change diffusion rates and possibly also chemical reaction rates. Careful calibration of both these effects is essential for correct interpretation of experimental results. The issue of how to correct experimentally for changes in refractive index and viscosity at different denaturant concentrations is dealt with in the next section.

V. Experimental Studies

A. Properties of the Intestinal Fatty Acid Binding Protein

The intestinal fatty acid binding protein (IFABP) belongs to a family of proteins that bind a variety of ligands (fatty acids, bile salts, and retinoids) into a large cavity located in the interior of the protein. The protein consists of two β-sheets, each containing five β-strands. The structure has been determined both by X-ray (Scapin et al., 1992) and by NMR (Hodsdon and Cistola, 1997) and the two structures are generally in good agreement. IFABP has been an excellent model system for folding studies of β-sheet proteins because it is small (15 kDa), monomeric, quite stable, and contains no proline or cysteine residues. In previous work a number of mutants had been made that contained a single cysteine residue placed at specific positions around the molecule (Jiang and Frieden, 1993). In later studies the properties of these mutants with fluorescein covalently attached to the cysteine residue were examined (Frieden et al., 1995). For two mutants there was a large change in fluorescein fluorescence between the folded and unfolded forms; it was concluded that, in these cases, the fluorescein was buried in the interior cavity. We have used one of these mutants (Val60Cys) for FCS experiments. Because fluorescein is somewhat photolabile, the cysteine residue was modified with the more stable probe, Alexa488. An autocorrelation function of Alexa488-modified IFABP in 20 mM phosphate buffer at pH 7.3 is shown in Figure 6. These data can be satisfactorily described by a model containing a single diffusing species with a diffusion time of 100 μs. The fit is improved with the addition of an exponential component of about 20 μs with an amplitude of 20–25%. As mentioned above, this rapid component could be artifactual owing to a non-Gaussian beam shape or triplet kinetics. Until this is clarified, we have chosen not to include this component in the discussion here. As pointed out in Section IV, however, there are ways to determine whether this is an artifact. For example, the diffusion rates could be differentially slowed by changing the beam width, increasing the viscosity, enclosing the IFABP within a matrix, or attaching it to a slow-moving particle such as a bead. Using different light intensity may uncover the contribution of triplet kinetics to the measurement.

B. Results as a Function of Urea Concentration

As noted earlier, there are refractive index and viscosity issues to be dealt with. Figure 7 shows the change of the diffusion coefficient

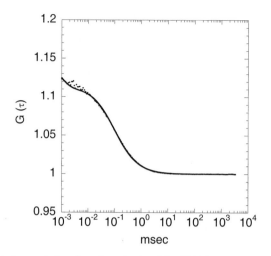

FIG. 6.   A typical correlation function obtained for IFABP in 20 mM phosphate buffer at pH 7.3 at room temperature. The experiment was performed using a ConfoCor 2 LSM combination instrument (Carl Zeiss-Evotec, Jena, Germany) and the correlation function data ($G(t)$) were fitted to the form $G(\tau) = G(0)/(1 + \tau/\tau_D)$, where τ_D is the characteristic diffusion time. An additional exponential component improves the fit.

(D, Fig. 7A) and apparent number of particles in the confocal volume (N, Fig. 7B) with urea concentration. The measured value of D decreases gradually but substantially as the urea concentration increases (Fig. 7A), which is inconsistent with any equilibrium denaturation curve. Moreover, a systematic increase in the number of particles (N) was observed with increasing urea concentration (Fig. 7B) that could not be accounted for by a physical phenomenon such as protein aggregation or related processes. These data show that changes in the viscosity and/or refractive index have a significant effect on τ_D and N. Viscosity, which increases linearly with urea concentration, is a straightforward effect on τ_D since τ_D should be directly proportional to the viscosity of the solution. Changes in the refractive index, as in urea solutions, can have a significant effect on the size of the illuminated volume and therefore on τ_D (and hence on D) as well as on N. If the size of the illuminated volume increases owing to an increase in refractive index, the value of N and the diffusion correlation time will increase correspondingly. Because an exact theoretical solution of the effects of refractive index on τ_D and N might be difficult, we have corrected our data in an empirical manner. To do this, we have measured τ_D and N of the free dye (Alexa488) at different urea concentrations. Using the linear correlation of changes in τ_D of the free dye and N with urea concentrations, the diffusion coefficient values (D) of

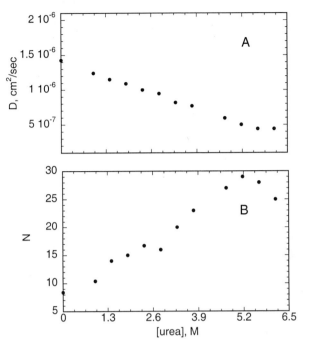

FIG. 7. Dependence of uncorrected (A) diffusion coefficient (*D*) and (B) number of particles in the observation volume (*N*) of Alexa488-coupled IFABP with urea concentration. The data shown here are not corrected for the effect of viscosity and refractive indices of the urea solutions. Experimental condition is the same as in Figure 6.

Alexa488-coupled protein have been corrected. Figure 8A and C shows corrected values of *D* and *N* of the Alexa488-bound protein as a function of urea concentration. After the viscosity and refractive index correction, *N* remains almost constant with urea concentration (Fig. 8C). The urea-induced unfolding transition of the Alexa488-bound protein monitored by steady-state measurements of Alexa488 fluorescence is shown in Figure 8B. The steady-state fluorescence data of Figure 8B can be fitted to a typical two-state reversible unfolding transition, although there is a steady increase in fluorescence intensity below 3 M urea before the unfolding transition. The large decrease in the diffusion coefficient, *D*, observed between 3 and 6 M urea concentration (Fig. 8A) matches the unfolding transition measured by steady-state fluorescence (Fig. 8B). There is, however, a significant and steady increase in *D* observed between 0 and 3 M urea prior to the actual unfolding event. A similar observation is made with guanidine as the denaturant and as a function of pH (data not shown). The urea-induced unfolding transition of

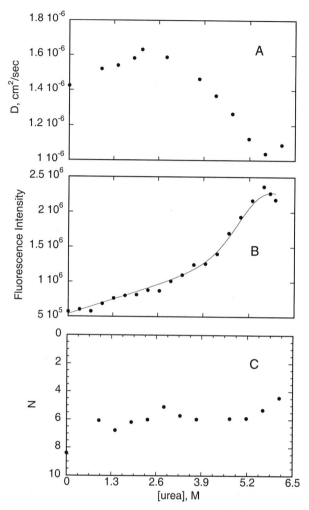

FIG. 8. Dependence of (A) corrected diffusion coefficient (D), (B) steady-state fluorescence intensity, and (C) corrected number of particles in the observation volume (N)
of Alexa488-coupled IFABP with urea concentration. The diffusion coefficient and number of particles data shown here are corrected for the effect of viscosity and refractive
indices of the urea solutions as described in text. For steady-state fluorescence data the
protein was excited at 488 nm using a PTI Alphascan fluorometer (Photon Technology
International, South Brunswick, New Jersey). Emission spectra at different urea concentrations were recorded between 500 and 600 nm. A baseline control containing only
buffer was subtracted from each spectrum. The area of the corrected spectrum was then
plotted against denaturant concentrations to obtain the unfolding transition of the protein. Urea data monitored by steady-state fluorescence were fitted to a simple two-state
model. Other experimental conditions are the same as in Figure 6.

Alexa488 bound to IFABP monitored by steady-state fluorescence was fitted to a two-state reversible unfolding model. This modified protein is slightly less stable (midpoint of 4.5 M compared to 4.7 M for wild-type IFABP).

C.   The Intermediate and the Unfolded State

The radius of a protein can be calculated from the value of D using the following equation:

$$r = kT/6\pi\eta D$$

where k is Boltzmann's constant, T is the absolute temperature, and r is the radius of the molecule assuming that the protein is spherical. Using the observed value of D, the calculated radius of the protein in the absence of denaturant is 15 Å. From the crystal structure (Scapin *et al.*, 1992), IFABP has a volume of 18,479 Å and the radius obtained from this volume (16 Å) is thus comparable to that observed from diffusion coefficient data. Using the same assumption—that the protein molecule is spherical—the radius of an intermediate state at low urea concentrations can also be calculated. In 2 M urea, well before any denaturation or secondary structure change as measured by far-UV circular dichroism, this value is 13.4 Å. This result suggests that the protein is more compact in 2 M urea than in the absence of urea. Although one could postulate that the large interior cavity collapses in 2 M urea, the reason for the increase in diffusion coefficient is not yet clearly understood.

For purposes of this discussion we have supposed that the protein undergoes a linear sequence of changes from the native to the intermediate to the unfolded state. We have further supposed that, in the lower range of denaturant concentration, the protein will be primarily in the native and intermediate states, whereas in the higher denaturant concentrations the intermediate and unfolded states will dominate. As indicated above, under conditions in which the conformational transformation is fast compared to the diffusion of the protein, the FCS autocorrelation functions can be analyzed as a simple diffusion process characterized by a single diffusion coefficient. This diffusion coefficient would vary as the relative fractions of fast and slow diffusing components (cf. Fig. 3). In both the concentration ranges the fast component would be the intermediate, the slow, either the native or the unfolded conformation. In contrast, if the conformational transformation is slow compared to diffusion, the autocorrelation function can be represented in general as the superimposition of the simple diffusion correlation functions of the fast and slow components (Fig. 5). These two components are not distinguishable by inspection, even if the differences in diffusion coefficient are as large as 10-fold (Fig. 6), but are readily demonstrated by fitting the measured correlation function to a theoretical model. The changes in Q and D in Figures 3–5 were taken to be large for illustrative purposes. In experimental systems the changes would be expected to be much smaller. At present we have no indication of a fluorescence difference between the intermediate and unfolded states that would be large enough to be detected in the FCS measurements. Therefore, our correlation functions do not provide a direct indication of the dynamics of the transformation between these two states. On the other hand, although the measured differences in the diffusion coefficients of the different conformations of IFABP are small, their presence in different ranges of the unfolding process is readily observed owing to the high accuracy of the correlation function measurements.

Assuming the unfolded state of IFABP is an extended random coil, the radius of the unfolded state can be calculated to be 46 Å (Tanford, 1968). However, the hydrodynamic radii of the unfolded states of IFABP obtained at high urea concentration is significantly smaller (around 20 Å), indicating that the unfolded state of IFABP is compact and not as extended as expected for a random coil. Although the unfolded state of the protein was earlier envisioned as unstructured, recent experimental and theoretical studies suggest the importance of structured unfolded states for efficient and rapid folding. For IFABP, for example, residual NMR peaks in HSQC spectra can be observed in 6 M urea (Hodsdon and Frieden, 2001). And there are several other examples. For staphylococcal nuclease native-like spatial positioning and orientation of chain segments were found to persist at 8 M urea (Shortle and Ackerman, 2001). In an another study, preorganization of one or more secondary structure elements in the B-domain of staphylococcal protein A was predicted in the unfolded state (Myers and Oas, 2001). Submillisecond small-angle X-ray scattering experiments recently suggested formation of compact denatured structures of cytochrome c (Pollack *et al.*, 1999) and β-lactoglobulin (Pollack *et al.*, 2001) before the formation of the native state (Pollack *et al.*, 2001). The present study also indicates that the hydrodynamic radius of IFABP in the unfolded state is only 33% more than that of the native state.

VI. CONCLUDING REMARKS

The material presented in this chapter demonstrates the utility of fluorescence correlation spectroscopy in the study of unfolded proteins.

In theory, both equilibrium and dynamic measurements can be made, the latter over a wide range of time regimes. Although the focus of this chapter is protein, the theory applies to any fluorescent macromolecule. Additionally, the theory is applicable to native, intermediate, or unfolded states of the macromolecule. Based on the theory presented here, experimental results for a single protein, the intestinal fatty acid binding protein, are presented. These data provide evidence for a compact intermediate state at low urea concentrations and a state of the unfolded protein more compact than expected for a random coil. The experimental results also revealed the need to take account of changes in viscosity or refractive index that can occur at high urea or guanidine concentrations.

ACKNOWLEDGMENTS

This work is supported by DK13332 to CF and GM 38838 to ELE.

REFERENCES

Eigen, M., and De Maeyer, L. C. (1963). In *Technique in Organic Chemistry,* 2nd ed. (S. L. Friess, E. S. Lewis, and A. Weissberger, eds.). Interscience, New York, pp. 895–1054.

Elson, E. L., and Magde, D. (1974). *Biopolymers* **13**, 1–27.

Elson, E. L., and Webb, W. W. (1975). *Ann. Rev. Biophys. Bioeng.* **4**, 311–334.

Frieden, C., Jiang, N., and Cistola, D. P. (1995). *Biochemistry* **34**, 2724–2730.

Hodsdon, M. E., and Cistola, D. P. (1997). *Biochemistry* **36**, 1450–1460.

Hodsdon, M. E., and Frieden, C. (2001). *Biochemistry* **40**, 732–742.

Jiang, N., and Frieden, C. (1993). *Biochemistry* **32**, 11015–11021.

Kask, P., Gunther, R., and Axhausen, P. (1997). *Eur. Biophys. J.* **25**, 163–169.

Koppel, D. (1974). *Phys. Rev. A* **10**, 1938–1945.

Meseth, U., Wohland, T., Rigler, R., and Vogel, H. (1999). *Biophys. J.* **76**, 1619–1631.

Myers, J. K., and Oas, T. G. (2001). *Nat. Struct. Biol.* **8**, 552–558.

Pollack, L., Tate, M. W., Darnton, N. C., Knight, J. B., Gruner, S. M., Eaton, W. A., and Austin, R. H. (1999). *Proc. Natl. Acad. Sci. USA* **96**, 10115–10117.

Pollack, L., Tate, M. W., Finnefrock, A. C., Kalidas, C., Trotter, S., Darnton, N. C., Lurio, L., Austin, R. H., Batt, C. A., Gruner, S. M., and Mochrie, S. G. (2001). *Phys. Rev. Lett.* **86**, 4962–4965.

Qian, H. (1990). *Biophys. Chem.* **38**, 49–57.

Qian, H., and Elson, E. L. (1991). *Appl. Optics* **30**, 1185–1195.

Qian, H., and Elson, E. L. (2001). In *Fluorescence Correlation Spectroscopy* (R. Rigler and E. L. Elson, eds.). Springer Berlin, pp. 65–83.

Rigler, R., Mets, U., Widengren, J., and Kask, P. (1993). *Eur. Biophys. J.* **22**, 169–175.

Scapin, G., Gordon, J. I., and Sacchettini, J. C. (1992). *J. Biol. Chem.* **267**, 4253–4269.

Shortle, D., and Ackerman, M. S. (2001). *Science* **293**, 487–489.

Tanford, C. (1968). *Adv. Protein Chem.* **23**, 121–282.

UNFOLDED PEPTIDES AND PROTEINS STUDIED WITH INFRARED ABSORPTION AND VIBRATIONAL CIRCULAR DICHROISM SPECTRA

By TIMOTHY A. KEIDERLING and QI XU

Department of Chemistry, University of Illinois at Chicago, Chicago, Illinois 60607

I. INTRODUCTION

Unfolded proteins and unstructured peptides form vital thermodynamic states critical to the overall protein folding problem. They represent one of a possible set of initial states in the folding process (or final state in the unfolding path). However, that state is intrinsically poorly defined. Most physical techniques commonly used for protein conformational studies develop a response that changes from one structure to another. If different residues are in a variety of local conformations but are the same in each molecule, and some technique can differentiate them from a different ensemble of conformations, much structural insight can be gained and folding pathways can potentially be followed. However, the unfolded state violates this basic premise in that any given residue in a protein or peptide may be in several different local states. This can cause the technique to have a completely nonsystematic response to conformational change, depending on the physical interactions it exploits. In this chapter, we address unfolded protein studies with optical spectroscopy, focusing on vibrational infrared (IR) absorption spectroscopy and its chiroptical variant, vibrational circular dichroism (VCD). Optical spectroscopy is low in resolution, so that site-specific information cannot

ADVANCES IN
PROTEIN CHEMISTRY, Vol. 62

be obtained without isotopic labeling, yet this failing may offer the best way to study the disordered unfolded state. Furthermore, these techniques are very fast, sampling conformations on the picosecond time scale. Thus, they can provide a representation of the equilibrium ensemble of structures, sensed as an average over all the residues, or can be used to follow fast kinetic changes in the structures, as might occur in the folding process. As such, optical techniques can be ideal for probing rapid fluctuations in a folding protein or peptide and exploring the distribution of local structures that can occur in an unfolded protein or peptide.

Determination of protein secondary structure has long been a major application of optical spectroscopic studies of biopolymers (Fasman, 1996; Havel, 1996; Mantsch and Chapman, 1996). These efforts have primarily sought to determine the average fractional amount of overall secondary structure, typically represented as helix and sheet contributions, which comprise the extended, coherent structural elements in well-structured proteins. In some cases further interpretations in terms of turns and specific helix and sheet segment types have developed. Only more limited applications of optical spectra to determination of tertiary structure have appeared, and these normally have used fluorescence or near-UV electronic circular dichroism (ECD) of aromatic residues to sense a change in the fold (Haas, 1995; Woody and Dunker, 1996).

ECD of far-UV $n-\pi^*$ and $\pi-\pi^*$ transitions of the amide linkage is a dominant optical technique for protein folding applications (Johnson, 1992; Sreerama and Woody, 1994; Greenfield, 1996; Venyaminov and Yang, 1996; Woody, 1996; Johnson, 1999). Being a differential absorption technique ($\Delta A = A_L - A_R$) CD's sensitivity to molecular conformation is enhanced by its signed character. The circular polarized absorption is due to interference of the magnetic and electric dipole moments, $\Delta A \sim Im\ (\mu \cdot m)$, which for a polymer of planar amides is dominated by transition dipole coupling. This results in an observed CD spectrum whose band shape, including sign, varies for different secondary structures. Because frequency changes are not significant with respect to the broad ECD bands, interpretations employ band shape–based methods that are often dependent on a statistical fit to a reference set of protein spectra (Sreerama and Woody, 1994; Venyaminov and Yang, 1996; Johnson, 1999). Given this basis, unfolded proteins are normally outliers to the statistical fits used to develop algorithms and thus are poor candidates for such structure prediction. Recently, data sets have been expanded to include unfolded structures, although defining their structures is problematic (Sreerama *et al.*, 2000). Whereas ECD data are critical for the best analyses of our data, they are covered in the chapter by Woody and Kallenbach, this volume (Woody and Kallenbach, 2002).

Infrared- and Raman-based vibrational spectral analyses of secondary structure took a different approach (Parker, 1983; Mantsch *et al.*, 1986; Haris and Chapman, 1995) owing to their natural improvement in resolution as compared to UV electronic transitions. These techniques offer the additional advantage that they are applicable to proteins in film or crystalline forms as well as solution samples. Amide-centered modes, particularly the C=O stretch (amide I) or the C—N—H deformation and C—N stretch (mixed amide II and III), are usually studied to get information about the fold of the chain, a polymer of amides. These local vibrations are coupled through the bonds (mechanically), as well as via dipolar and hydrogen bond–mediated interactions, into normal modes whose frequency distributions are characteristic of different secondary structure types in the molecule. Often the focus has been on assigning component frequencies to various secondary structural component types, as has been extensively reviewed (Tu, 1982; Krimm and Bandekar, 1986; Jackson and Mantsch, 1995; Mantsch and Chapman, 1996). This is a problem because a polymeric structure does not yield a single frequency for a given structure; rather, it has a large number of coupled modes, one for each local oscillator, which are observed as a distributed band shape whose components have varying intensities (Torii and Tasumi, 1992; Krimm, 2000). This gives rise to the characteristic patterns seen in IR. Most IR studies have focused on the amide I band, with some inclusion of the amide II; and Raman-based studies mostly used the amide III, with some reliance on the amide I (and amide II only in resonance). Other modes have had very little application, particularly for aqueous protein solutions (Masuda *et al.*, 1969; Krimm, 2000). Because IR and Raman techniques give rise to band shapes composed of contributions from all the component transitions which differ by relatively small frequency shifts, different proteins have similar IR band shapes. However, because IR and Raman have different intensity mechanisms, their intensity distributions over these components are different. As a consequence, IR and Raman offer complementary probes of secondary structure; however, this review deals exclusively with IR-based techniques, including VCD.

Larger frequency shifts lead to more sensitive structural discriminations. In the IR, the β-sheet amide I is distinctively lower in frequency than other structural types, particularly for aggregates, a form often seen in unfolding experiments. However, owing to the high signal-to-noise ratio (S/N) of Fourier transform IR (FTIR), one can detect components having smaller frequency separations. This effective resolution enhancement is accomplished using second-derivative or Fourier self-deconvolution (FSD) techniques (Kauppinen *et al.*, 1981; Byler and Susi, 1986; Surewicz and Mantsch, 1988). These methods partially compensate for the minimal band shape variation in IR and Raman spectra, but can be abused by the unwary user (Surewicz *et al.*, 1993; Jackson and Mantsch, 1995). Assigning the frequencies to specific secondary structural types and then assuming the assignments to be unique is a severe assumption, since any given conformation will shift in frequency response under the influence of different solvents and residues (Trewhella *et al.*, 1989; Jackson *et al.*, 1991; Pancoska *et al.*, 1993). Additionally, nonuniformity and end effects for segments of a given structural type will have an impact on the frequencies, resulting in some dispersion of that contribution over the spectrum (Dousseau and Pezolet, 1990; Kalnin *et al.*, 1990; Krimm and Reisdorf, 1994; Torii and Tasumi, 1996; Kubelka *et al.*, 2002). Finally, most methods assume that the dipole strengths (extinction coefficients) of all the residues are the same, regardless of their conformation; but, in fact, they vary (Venyaminov and Kalnin, 1990; Jackson and Mantsch, 1995; De Jongh *et al.*, 1996). Nonetheless, such FTIR methods have proved very useful for sorting out differences between proteins with similar IR band shapes.

An alternative way of enhancing IR resolution is by use of the polarization sensitivity of the transitions. The simplest of these is linear dichroism (LD) of oriented samples. By selecting the polarization (electric vector) parallel to the direction of dipolar excitation in a given mode, its absorbance is preferentially selected (Dluhy *et al.*, 1995). One also can use LD to obtain bond orientation information, provided something is known about the overall molecular orientation. These methods work best for extended polymers, fiber proteins, or membrane-bound systems (Goormaghtigh and Ruysschaert, 2000), but are not expected to have a large application to unfolded states themselves owing to their lack of order. Another method of abstracting more information from these IR transitions is to take advantage of their intrinsically rapid time response to structural change. IR experiments dealing with unfolding have been done on the millisecond scale with rapid mixing (White *et al.*, 1995), on the nanosecond scale with step-scan FTIR techniques (Siebert, 1996), and even faster with laser excitation (Cowen and Hochstrasser, 1996; Dyer *et al.*, 1998; Lim and Hochstrasser, 2001).

Band shape–based analyses similar to those used for ECD have also been applied to FTIR and Raman spectra with reasonable success (Williams and Dunker, 1981; Dousseau and Pezolet, 1990; Lee *et al.*, 1990; Sarver and Kruger, 1991; Pribic *et al.*, 1993; Baumruk *et al.*, 1996). Clearly, spectra–structure correlations based only on folded proteins would lead such analyses for unfolded proteins to the problem noted above for ECD. In principle, using the frequency assigned component approach frees the analyses from such constraints. The problem is to determine a characteristic spectrum for the unfolded state or states, which has been the subject of some studies (Byler and Susi, 1986; Fabian *et al.*, 1993; 1994, 1995; Byler *et al.*, 2000; Krimm, 2000).

A desire to couple the frequency resolution of FTIR with the chiroptical stereochemical sensitivity of ECD, or to obtain polarization sensitivity for IR studies of an unordered system, led to the development of vibrational CD (VCD), the differential absorbance for left and right circularly polarized light by vibrational transitions in the IR (Polavarapu, 1989; Diem, 1991; Nafie, 1994, 1997; Keiderling, 2000). Its counterpart, Raman optical activity (ROA), was developed in parallel, as reviewed in the separate chapter by Barron *et al.*, this volume (Barron *et al.*, 2002) and many times previously (Polavarapu, 1989; Barron *et al.*, 1996; Nafie, 1997). The vibrational region of the spectrum has many resolved transitions characteristic of localized parts of the molecule. VCD measurements have a distinct stereochemical sensitivity for each of these modes that arises specifically through the geometric nature of their coupling to other modes. The chromophores needed for VCD are the bonds themselves, as sampled by excitations of bond stretching and deformation. Chiral interaction of these bonds (necessary to establish optical activity) manifests itself in the resulting VCD spectral band shapes. Furthermore, these vibrational excitations are characteristic of the ground electronic state of the molecule, in which all thermally accessible conformational processes occur. In summary, VCD has the three-dimensional structural sensitivity natural to a chiroptical technique, but manifested in a response distributed over a number of localized probes of the structure.

This benefit comes at a cost, which arises from significantly reduced S/N and some interpretive difficulty as compared to IR. Developments on the latter front are bringing the theoretical prediction capability of VCD for small molecules to a level demonstrably superior to that for ECD (Freedman and Nafie, 1994; Stephens *et al.*, 1994; Stephens and Devlin, 2000), especially for peptide spectra (Kubelka *et al.*, 2002). Most previous protein applications of VCD used empirically based analyses (Keiderling, 1996, 2000). Theoretical methods are limited when applied to large molecules such as proteins; however, a hybrid approach using *ab initio* determination of spectral parameters with modest-sized molecules for transfer to large peptides has made simulation of spectra for large peptides possible (Bour *et al.*, 1997; Kubelka *et al.*, 2002). Theoretical techniques for simulation of small-molecule VCD are the focus of several previous reviews (Stephens and Lowe, 1985; Freedman and Nafie,

1994; Stephens *et al.,* 1994; Nafie, 1997; Polavarapu, 1998; Stephens and Devlin, 2000), and their extension to peptides is the topic of several recent papers (Bour *et al.,* 1997, 2000; Silva *et al.,* 2000a; Kubelka and Keiderling, 2001b; Kubelka *et al.,* 2002).

Experimentally, VCD instruments are now available to permit measurement of spectra for most molecular systems of interest under at least some sampling conditions (Keiderling, 1990; Diem, 1994; Nafie *et al.,* 1994; Freedman *et al.,* 1995; Keiderling *et al.,* 2002). In fact, stand-alone commercial VCD spectrometers (BioTools/Bomem) provide very high-quality spectra (Aamouche *et al.,* 2000; He *et al.,* 2001; Zhao and Polavarapu, 2001) Alternatively, modifications of standard FTIR spectrometers can be configured for VCD measurement (Urbanova *et al.,* 2000; Hilario *et al.,* 2001; Setnicka *et al.,* 2001). It is true that most VCD studies of proteins in aqueous solutions, the conditions of prime biological interest, are restricted to moderate concentration samples, much as for NMR. By comparison, ROA measurements demand even higher concentrations (Barron *et al.,* 1996). Experimental methods of biomolecular VCD are only briefly summarized in the next section and are more fully discussed in other reviews (Polavarapu, 1985; Keiderling, 1990; Diem, 1991, 1994; Nafie, 1997; Keiderling, 2000; Keiderling *et al.,* 2002).

Sampling conditions for obtaining experimental VCD spectra of protein and peptide samples are similar to those used in transmission FTIR studies. VCD has quite small amplitude signals (ΔA is of the order of 10^{-4}–10^{-5} of the sample absorbance, A) and thus lower S/N ($\sim \Delta A/A$). To obtain quality VCD spectra, much longer data collection times are required than for FTIR or ECD. These time scales have eliminated the possibility of obtaining VCD of rapid folding and unfolding events, restricting the technique to equilibrium studies. This is in contrast to time-dependent FTIR (Siebert, 1996). Nonetheless, the intrinsic time scale of the VCD measurement is still that of an IR absorption, so the spectrum may reflect even a short-lived species undergoing dynamic conformational change if it contributes to the equilibrium mixture.

The band shape variability of VCD coupled with its frequency leads to an enhanced sensitivity to secondary structure in folded proteins as compared to IR and ECD structure determinations alone (Pancoska *et al.,* 1989). However, by combining IR and ECD or VCD and ECD techniques, the best structural determinations are obtained, as the strengths of each compensate for the other (Pancoska *et al.,* 1995; Baumruk *et al.,* 1996; Keiderling, 2000). Differentiation of different unfolded structures is also possible, but is most useful when data from several techniques are combined. The distinction of IR and VCD from ECD for unfolded proteins lies in the short-range response that dominates the band shape–developing interactions in the vibrational techniques (Dukor and Keiderling, 1991; Keiderling and Pancoska, 1993; Keiderling *et al.*, 1999b). Because unfolded proteins and peptides lack long-range order, vibrational methods are better suited to analyses for their residual local structure effects.

Although theoretical interpretation of VCD and IR spectra, especially for unfolded proteins, remains a challenge, qualitative analyses can be done by utilizing the bandshapes and frequency positions to predict the dominant or any significant residual secondary structural types. This has become the standard approach for most peptide studies (Freedman *et al.*, 1995). In smaller fully solvated peptides, and presumably in fully solvated unfolded proteins, the frequency shifts due to solvent effects and the inhomogeneity of the peptide secondary structure (as well as end fraying) are severe problems for frequency-based analyses. In globular proteins, qualitative estimations of structure remain of interest for determining the dominant fold type, such as highly helical, highly sheet, or mixed helix and sheet type; but quantitative estimations of secondary structure content based on empirical spectral analyses are usually of more interest. Quantitative methods for analysis of protein VCD spectra in terms of structure follow the lead of most ECD analyses and employ band shape techniques referenced to a training set of protein spectra (Pancoska *et al.*, 1995; Baumruk *et al.*, 1996).

In summary, unlike ECD, VCD can be used to correlate data for several different spectrally resolved features; and, unlike the case of IR and Raman, each of these features will have a physical dependence on stereochemistry. The combination of these techniques allows one to compensate for the other, providing a balance between accuracy and reliability. Unfolded proteins provide a major challenge. The short-range dependence gives VCD (and IR, if somehow corrected for solvent and length effects) an advantage for identifying residual structure in long-range disordered systems. However, the VCD dependence on coupling of neighboring residues implies that some structure should remain if one wants to develop a significant signal. By contrast, ECD has residual signals arising through the chiral distortion of the π-electron system from the substitution on the residue C^{α} sites. Hence, by sensing local chirality, ECD can develop a response characteristic of a structure of low order; and through their short-range conformational sensitivity, VCD and IR can pick out the remains of the previous ordered conformation. It becomes obvious that, despite the fundamental physical differences leading to various advantages for any one technique over another in a given set of circumstances, progress in understanding complex structures, particularly those with little extended coherence, will come from synthesizing all the data gathered from various techniques.

II. Experimental Techniques

A. Spectrometers

1. IR Absorption Methods

IR and VCD instruments, both measuring absorption spectra, have much in common. Historically, IR spectrometers were dispersive in design; that is, they were based on dispersing light from a broadband (black-body) source with a monochromator that selected a narrow band pass for measurement. The monochromator was scanned through the spectral region of interest, and the response was recorded as either a single-beam intensity or, with alternating beams, double-beam transmittance. By contrast, virtually all ordinary IR absorption spectra are now measured with Fourier transform (FTIR) spectrometers because of their convenience, reproducibility, and optimal S/N. FTIR instruments send a light beam (typically from a heated ceramic rod, e.g., SiC) into a Michelson interferometer where it is separated (often with a Ge-coated KBr beam-splitter) into two paths, which are made different in length by means of a moving mirror. When recombined, the beams interfere with a variable optical retardation (depending on the mirror position). The resultant beam is encoded with respect to optical frequencies in the form of an interferogram (response versus retardation), which can be decoded to a spectrum (intensity versus wavenumber) by Fourier transformation. Spectral responses at all wavenumbers of interest are obtained simultaneously (multiplex advantage), and repetitive scans can be co-added, both improving S/N. Longer mirror displacements serve to improve the resolution of spectral features. For protein applications, the enhancement in precision by referencing wavenumber determination to a laser source (instrumentally through mirror and signal digitization control) is just as important.

These FTIR advantages have specific impacts on applications for peptide and protein secondary structure studies. The high S/N and digital character of the spectra allow one to enhance the apparent resolution using Fourier self-deconvolution (Kauppinen *et al.*, 1981; Byler and Susi, 1986; Surewicz and Mantsch, 1988) or second-derivative techniques (Byler *et al.*, 1995). Although the real resolution of the spectra cannot be changed by any computational method, these techniques bring out the underlying components in the overlapped multicomponent spectrum, typical of proteins, that are represented by very small fluctuations in the band shape. Because any feature sharper than the protein bands, such as noise or atmospheric water absorption, will be anomalously enhanced by these computational manipulations, one should obtain spectra with S/N approaching 3000. This is straightforward with FTIR, but requires removal of all residual vapor phase (typically H_2O) background absorption by careful subtraction. Well-ordered proteins give several distinct features, whereas truly denatured proteins and disordered peptides are expected to give heterogeneously broadened, featureless transitions. Most of this work has been done on the amide I modes whose distinct correlations with structural types have been developed.

Most protein conformational studies are done for samples dissolved in water, the most biologically relevant solvent system, which has a strong H—O—H bending mode at 1650 cm^{-1}, directly overlapping the amide I region. Many studies use D_2O as a substitute solvent to avoid this conflict for the amide I (which is now termed the amide I' mode, when N—H groups are exchanged to N—D). On the other hand, the amide II at ~1550 cm^{-1} and amide III at ~1300 cm^{-1} are studied best in H_2O. Use of D_2O for the amide I' in folded proteins involves a tradeoff because the less diagnostic amide II mode (now amide II') is partially shifted to about 1450 cm^{-1} and the protein has less well-defined, mixed amide I,I' modes due to partial exchange of N—H groups. Unfolded proteins and peptides can approach full exchange, as judged by loss of amide II intensity. The high S/N ratio of FTIR does allow one to measure the spectra in H_2O because small paths (~6 μm) and the bulk water background can be subtracted. However, the hydrogen-bonded water modes couple to the amide I modes, giving a contribution that effectively cannot (and should not be able to) be subtracted out. This probably poses an ultimate limit on interpretability of such spectra (Chen *et al.*, 1994b; Jackson and Mantsch, 1995; Sieler and Schweitzer-Stenner, 1997; Krimm, 2000; Manas *et al.*, 2000; Kubelka and Keiderling, 2001a).

Time-dependent FTIR data can be obtained in various ways, depending on the time resolution needed. For changes of the order of 100 ms or longer, collecting spectra from successive rapid scans is satisfactory, for example, following initiation with a stop-flow mixing (White *et al.*, 1995). However, folding experiments are not practical if one needs to dilute out a chemical denaturant, as is commonly done in CD and fluorescence experiments, since the final concentration would be too low. Thus, mixing experiments are better suited for rapid change of pH or solvent conditions and for monitoring the spectral response. For faster processes, alternative methods are used. For example, with a step-scan FTIR, a laser-induced temperature jump (T-jump) can initiate time decay relaxation data at each step. These data can be reassembled into a series of interferograms corresponding to successive time slices which can yield spectra sensitive to nanosecond changes (Siebert, 1996). Even faster time scales are accessible with various laser excitation and single-frequency detection techniques (Phillips *et al.*, 1995; Cowen and Hochstrasser, 1996; Dyer *et al.*, 1998; Lim and Hochstrasser, 2001).

2. VCD Methods

Development of a VCD instrument is normally accomplished by extending a dispersive IR or an FTIR spectrometer to accommodate, in terms of optics, time-varying modulation of the polarization state of the light and, in terms of electronics, detection of the modulated intensity that results from a sample with nonzero VCD. Our instruments and those of others are described in the literature in detail and in several reviews (Polavarapu, 1985; Keiderling, 1990; Diem, 1994; Nafie, 1997). Extended comparisons contrasting these methods, including a novel digital signal processing FT-VCD design (Hilario *et al.*, 2001; Keiderling *et al.*, 2002) and detailed discussion of components needed to construct or enhance either type of instrument, have already been published (Su *et al.*, 1981; Diem *et al.*, 1988; Malon and Keiderling, 1988; Keiderling, 1990; Yoo *et al.*, 1991; Chen *et al.*, 1994a; Wang and Keiderling, 1995; Long *et al.*, 1997a,b; Nafie, 2000). Linear dichroism can be measured with the same instrument by increasing the modulation level (and detecting at twice its frequency) or just by using an FTIR (without modulation) and subtracting the resulting steady-state spectra obtained at two different polarizer positions, separated by 90° (Marcott, 1984; Malon and Keiderling, 1997; Dluhy, 2000; Hilario *et al.*, 2001).

Both dispersive and FTIR-based VCD instruments provide for linear polarization of the light beam, normally with a wire grid polarizer, and its modulation between (elliptically) right- and left-hand polarization states with a photoelastic modulator (PEM). The beam then passes through the sample and onto the detector, typically a liquid N_2–cooled $Hg_{1-x}(Cd)_xTe$ (MCT) photoconducting diode. After preamplification, the electrical signal developed in the MCT is separated to measure the overall transmission spectrum (I_{trans}) of the instrument and sample via one channel and the polarization modulation intensity (I_{mod}), which is related to the VCD intensity, via the other. These signals are ratioed to yield the raw VCD signal, which normalizes out any dependence on the source intensity and instrument transmission characteristics. In the limit of small ΔA values,

$$I_{mod}/I_{trans} = g_I J_1(\alpha_0)(1.15\ \Delta A) \tag{1}$$

where $J_1(\alpha_0)$ is the first-order Bessel function at the maximum retardation of the modulator, α_0, and g_I is an instrument gain factor. Calibration of the VCD is done by measuring the VCD of a known sample or with a birefringent plate and a polarizer combined to create a signal of known magnitude (Nafie *et al.*, 1976; Polavarapu, 1985; Keiderling, 1990; Yoo *et al.*, 1991; Malon and Keiderling, 1996). Finally, one can employ baseline correction and spectral averaging or smoothing, or conversion to molar quantities, e.g. $\Delta\varepsilon = \Delta A/bc$, where b is the path length in centimeters and c is the concentration in moles per liter.

Although FT-VCD has many advantages, both the restriction to measurement only in the spectral windows of water and the relatively broad bands seen in biopolymer IR spectra can nullify the multiplex and throughput advantages of FTIR. Other things being equal, biopolymers can actually favor use of dispersive VCD (Pancoska *et al.*, 1989; Wang and Keiderling, 1995). A VCD spectral comparison of only the amide I′ band for exactly the same random coil poly-L-lysine sample in D_2O measured on our dispersive instrument (at UIC) and on the Chiral IR commercial, FTIR-based instrument (installed at Vanderbilt) is shown in Figure 1. These spectra, which used the same total data collection time and resolution, have less baseline and signal distortion (noise) in the dispersive case if only one band is measured (Keiderling *et al.*, 2002).

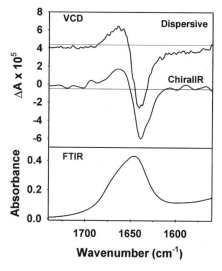

FIG. 1. Comparison of amide I′ VCD for an identical sample of poly-L-lysine in D_2O as measured on the UIC dispersive instrument (top) and on the ChiralIR FT-VCD instrument (at Vanderbilt University, kindly made available by Prof. Prasad Polavarapu). Sample spectra were run at the same resolution for the same total time ($\sim$1 h) in each case. The FTIR absorbance spectrum of the sample is shown below. VCD spectra are offset for sake of comparison. Each ideal baseline is indicated by a thin line, the scale providing a measure of amplitude. Noise can be estimated as the fluctuation in the baseline before and after the amide I′, which indicates the S/N advantage of the single band dispersive measurement.

VCD spectral collection is quite slow, eliminating the realistic possibility of fast kinetic measurements. Dispersive VCD spectra of a typical IR band can take about half an hour for a single scan at a 10-s time constant and must be averaged for several scans to obtain reliable S/N. FTIR-measured VCD spectra can sample a much wider spectral region but often require extensive averaging over longer times to attain desirable S/N. If, in the end, only one or two adjacent bands are needed for the analysis (such as for Fig. 1), much data would be collected for no purpose with the FTIR-based technique. On the other hand, if multiple bands are to be studied, FTIR-VCD retains its advantage, even for biological samples (Malon *et al.*, 1988; Wang and Keiderling, 1995; Tam *et al.*, 1996). In both cases, the VCD is collected single-beam, though corrected for light-beam intensity by the ratio I_{mod}/I_{trans}. Because the protein and peptide VCD signals are weak, baseline correction is critical, requiring equivalent collections of blank (solvent, buffer, and cell) spectra.

B. Sampling Techniques

For typical transmission experiments, protein samples in D_2O can be prepared at concentrations in the range of 10–50 mg/ml for VCD and FTIR. An aliquot of the solution (typically 20–30 μl) is placed in a standard demountable cell consisting of two BaF_2 windows separated by a 25–50 μm spacer. For systems with larger signal amplitudes or for FTIR alone, more dilute samples can be studied with 100 μm path cells. Similarly, for peptides or perturbation studies in which alternative or mixed solvents (e.g., TFE) are used, longer paths and more dilute solutions are practical. For proteins in H_2O, study of the amide I is possible for concentrations of $\sim$100 mg/ml (but <20 μl in volume) and path lengths of 6 μm, under which conditions the water alone gives $A \sim 0.9$ at 1650 cm^{-1} (Baumruk and Keiderling, 1993). This interference causes loss of S/N in VCD, but no major artifacts are found in our instruments when properly aligned. Lower concentrations can be used for FTIR, due to higher S/N.

VCD and FTIR spectra should always be obtained on the same samples, the FTIR at higher resolution and optimized S/N to permit computation of deconvolved (resolution-enhanced) spectra (Kauppinen *et al.*, 1981). VCD spectra of biomolecules are often normalized to the absorbance, since concentration and path lengths are rarely known to good accuracy. Because the absorbance coefficients for different molecules will vary, this is only an approximate correction for concentration variation.

For solid-phase samples or for membrane interaction studies, it can be very useful to use the alternative sampling technique of attenuated total reflectance (ATR). Films or solutions can be placed on a specially cut, long, thin ZnSe or Ge ATR plates. The light beam is passed through the length of the plate by repeated reflection off the surfaces, each time entering the sample coating. These measurements are capable of detecting of high-quality absorbance spectra for very small quantities of proteins (2–20 μg) (Singh, 2000). The major problem of ATR sampling for proteins, and especially for unfolded peptides and proteins, is their binding to the surface of the ATR cell and consequent distortion of the conformation. Because ATR preferentially senses the portion of sample on or near the surface, a modest distortion on the surface can have a large impact on the spectrum, particularly for solution samples. However, the same argument means that coating the surface with a membrane and exposing it to protein solution can enhance the detection of the membrane-bound protein molecules. Protein crystals have been studied with FTIR using a microscope accessory and a special sealed cell, allowing control over atmosphere (Hadden *et al.*, 1995). If the mother liquid is used to control the atmosphere, spectra virtually identical to solution transmission data are obtained. Linear polarization (LD) measurements (but not VCD, owing to partial loss of circular polarization on reflection) are also possible with ATR sampling and are especially useful for membrane orientation studies (Flach *et al.*, 1999; Goormaghtigh and Ruysschaert, 2000). Recently, a modified VCD instrument design using two polarization modulators has been reported that permits measurements of for solid-state samples in transmission (Nafie, 2000; Nafie and Dukor, 2000). This may broaden biosystem sampling possibilities and permit avoidance of solvent interference problems.

In our laboratory, ECD spectra provide important auxiliary data for the proteins and peptides we study. ECD spectra are usually obtained for more dilute samples using strain-free quartz cells having various sample path lengths from 0.2 to 10 mm for concentrations of 0.1–1 mg/ml. To test if concentration effects cause a difference in the interpretation of data from the two techniques, which can be very important for study of unfolded proteins and peptides, we also can use IR cells and samples directly in the ECD spectrometer (Baumruk *et al.*, 1994; Yoder, 1997; Yoder *et al.*, 1997b; Silva *et al.*, 2000b).

III. Theoretical Simulation of IR and VCD Spectra

Infrared spectra are straightforward to predict theoretically, demanding development of a force field (FF) to determine frequencies and dipole derivatives for intensities. These parameters were initially obtained using empirically fitted force constants and simple models for transition dipoles (Krimm and Bandekar, 1986; Torii and Tasumi, 1996).

With quantum-mechanical methods, the second derivatives of the energy could be used directly for the FF and atomic polar tensors (APT) for the dipole derivatives. Both are standardly computed in most quantum-chemistry programs; but for accurate results, moderately large basis sets and/or some accommodation for correlation interaction is needed. Until recently, this has restricted most *ab initio* studies to modest-sized molecules.

The theory of VCD has long been a challenge because there is no electronic contribution to the rotational strength within the Born–Oppenheimer approximation. This led to a series of approximate models, which, though offering useful physical insight early on, have been abandoned for an *ab initio* quantum-mechanically based approach. This approach, developed by Stephens and co-workers, has had significant impact on the study of small chiral molecules (Rauk, 1991; Freedman and Nafie, 1994; Stephens *et al.,* 1994; Stephens and Devlin, 2000). We have extended these models for peptide VCD computations and, with the help of a property transfer mechanism developed by Bour, have simulated spectra for a number of conformations (Bour *et al.,* 1997; Kubelka and Keiderling, 2001c; Kubelka *et al.,* 2002).

For larger molecules, the most accurate quantum-mechanical force fields, obtainable with practical computational requirements, are computed with density functional theory (DFT) methods (e.g., BPW91 or B3LYP) using medium-sized split-valence basis sets (e.g., 6-31G*). To add IR and VCD intensities, *ab initio* calculations of the electric and magnetic transition dipole moments are needed, usually in the form of the APT and atomic axial tensors (AAT). The magnetic field perturbation method (MFP) of Stephens and co-workers (Stephens, 1985; Stephens and Lowe, 1985; Stephens, 1987; Stephens *et al.,* 1994, 2002) has proved to be the most useful method of obtaining VCD spectral simulations. Model calculations for dipeptides were reported early (Bour and Keiderling, 1993); others for tripeptides and much longer species are more recent (Kubelka, 2002). Those *ab initio* FF, APT, and AAT tensor values have been transferred onto larger molecules with good success, resulting in useful simulations of spectra for relatively large oligopeptides (Bour *et al.,* 1997, 2000; Silva *et al.,* 2000a; Kubelka and Keiderling, 2001c; Kubelka, 2002). Because peptide theoretical calculations require use of molecular geometry, they might appear to have little relevance to unfolded proteins. However, as will be shown, the 3_1-helix provides a good spectroscopic model for many unfolded peptides and denatured proteins, and it can be modeled just as successfully as the standard secondary structure types of prime interest in folded proteins. The spectral differences between simulations for folded and unfolded species can then be compared to the differences seen experimentally.

IV. Peptide Studies

A. Empirical IR and VCD of Peptides

To qualitatively interpret VCD and IR data for peptides and proteins, one must establish how the spectra change between known structures having well-defined conformational types. Polypeptides provide the usual set of molecules for doing this. The initial method for characterizing of polypeptide conformation of solid-phase samples was IR spectra, which in turn could be directly correlated to X-ray structural data (Elliott and Ambrose, 1950; Elliott, 1954; Blout et al., 1961). For VCD, only solution measurements were initially reliable. Several α-helical polypeptides, studied as nonaqueous solution samples, provided VCD of characteristic amide transitions (Singh and Keiderling, 1981; Lal and Nafie, 1982; Sen and Keiderling, 1984a). For β-sheets, polypeptide VCD solution studies were usually not possible owing to solubility problems (Sen and Keiderling, 1984b; Narayanan et al., 1985, 1986). Polyionic peptides (typically having charged side chains) are soluble in aqueous solution and often undergo coil to helix or sheet transitions under pH or salt perturbations, which can be studied with IR and VCD in D_2O (Paterlini et al., 1986; Yasui and Keiderling, 1986a; Malon et al., 1988; Baumruk et al., 1994). Thus, the earliest aqueous peptide VCD spectra were of unfolded (random coil) species.

Although amide I and II IR and VCD can be measured for samples in H_2O (Baumruk and Keiderling, 1993), the concentrations needed are not compatible with many peptides. Similarly, this can be a problem for unfolded proteins if one wants to avoid aggregation complications. IR can be obtained with about 10-fold lower concentration for stable systems (long signal averaging and careful subtraction of baselines can reveal very weak absorbances). VCD studies using D_2O have focused on the amide I' (denoting H/D exchange); this is because the amide II' is shifted so that HOD (usually present owing to residual exchange and atmospheric water contamination) interferes with it, and the amide A' (N−D stretch) and III' are not detectable (owing to overlap with D_2O modes). In IR, these interferences can cause real uncertainty in what one measures, since the subtractions of the baselines must be very reliable; but in VCD, the interference results in just a loss of S/N, since the achiral solvent has no VCD.

The IR and VCD for α-helix, β-sheet, and coil forms have been shown to have distinctly different amide I and II spectra (see Fig. 2 for typical spectra of charged peptides of different conformation in H_2O). Similar band shapes have been measured for a variety of polypeptides (Paterlini et al., 1986; Yasui and Keiderling, 1986a,b; Baumruk et al.,

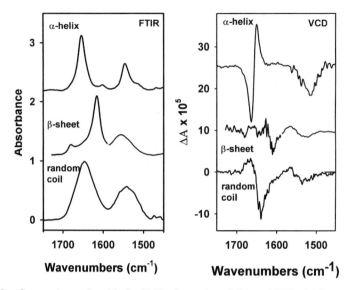

FIG. 2. Comparison of amide I + II IR absorption (left) and VCD (right) spectra of polypeptides dissolved in H_2O obtained for α-helical poly-(LKKL) in high salt (top), β-sheet poly-(LK) with high salt (middle; amide I' measured only in D_2O), and random coil poly-K at pH $\sim$7 (bottom), where L is L-Leu and K is L-Lys. All spectra were rescaled to have an amide I absorbance maximum of $A = 1.0$, but are offset to avoid overlap.

1994). IR differences key on frequencies of band maxima and band widths, whereas VCD varies in terms of band shapes associated with those IR transitions, thus being secondarily influenced by frequency and width as well. Two strong bands, the amide I and II, dominate the mid-IR, whereas the amide III is weak and mixed with local CαH deformation modes, spreading its small intensity over a broad region (Diem *et al.*, 1992; Baello *et al.*, 1997; Asher *et al.*, 2001).

For right-handed α-helices (Fig. 2 top), these IR bands are associated with VCD features that include strong a positive amide I (centered at $\sim$1655 cm^{-1}) couplet (+ then −, with increasing frequency) and a broad negative amide II (to low frequency of the IR maximum at $\sim$1550 cm^{-1}) band (Singh and Keiderling, 1981; Lal and Nafie, 1982; Sen and Keiderling, 1984a; Yasui and Keiderling, 1986b; Malon *et al.*, 1988; Yoder *et al.*, 1997a). A very broad and weak, net positive VCD is often seen in the general amide III regions (Roberts *et al.*, 1988; Birke *et al.*, 1991; Baello *et al.*, 1997). Deuteration of the amide N–H yields an amide I' IR peak to $\sim$1650 cm^{-1} and amide II' at $\sim$1450 cm^{-1}. Amide I' VCD changes to a three-peaked (−,+,−) pattern, but the amide II' retains the negative sign with a loss of intensity (Sen and Keiderling, 1984a; Baumruk and

Keiderling, 1993). Other structures have similar H/D shifts, but, as yet, have not shown a significant amide I' VCD shape change.

The β-sheet amide I spectra have been characterized by the two widely split IR absorbance features at ~1620 cm^{-1} and 1690 cm^{-1} seen for antiparallel β-sheets and fibrous proteins, most probably indicating appreciable degrees of aggregation (Clark *et al.*, 1981; Byler and Purcell, 1989; Surewicz *et al.*, 1993; Kubelka and Keiderling, 2001c). The polypeptide models yield very weak negative amide I' VCD corresponding to the absorbance peaks (Fig. 2 middle). Studies of oligopeptides that form β-aggregates in solution reflect the polypeptide results (Brauner *et al.*, 2000; Kubelka and Keiderling, 2001b; Kubelka, 2002), but those that apparently adopt a β-structure in nonaqueous solution or those with β-hairpin structures evidenced a higher frequency (~1635 cm^{-1}) couplet VCD whose detailed shape and sign were very influenced by aggregation and solvent (Narayanan *et al.*, 1986; Silva *et al.*, 2000b; Hilario *et al.*, 2002). On the other hand, even aggregate polypeptide β-sheet structures give rise to medium-intensity negative couplet amide II VCD coincident with its IR at ~1530 cm^{-1} and a negative amide III VCD (from protein correlations) (Gupta and Keiderling, 1992; Baumruk and Keiderling, 1993; Baello *et al.*, 1997).

The coil form is most relevant to unfolded proteins in this review, but its characteristics are meaningful only by contrast with the above-noted results for structured systems. On an empirical basis, random coil peptides have a broadened amide I (~1645 cm^{-1}) that develops a surprisingly intense, negative couplet VCD (Fig. 2 bottom). On H/D exchange, the frequency shifts down only a few wavenumbers, a shoulder to high frequency (~1655–1660 cm^{-1}) becomes apparent in the IR, but the VCD band shape is unchanged. This pattern is also characteristic of the final state for many peptides undergoing a helix–coil transition under various perturbations. However, at high temperatures, the IR band shifts even higher and broadens, obscuring the high-frequency shoulder, and its VCD magnitude is reduced (Dukor, 1991; Keiderling *et al.*, 1999b; Silva, 2000). The amide II is less well-defined, being broad, having a IR maximum at ~1540 cm^{-1} with a typically weak, negative VCD. No characteristic amide III VCD is established for random coils or unfolded proteins, but the IR has a feature at ~1280 cm^{-1}, lying between the helix and sheet frequencies, assigned to coil (Singh, 2000).

The random coil amide I VCD pattern is exactly the same shape, but smaller in amplitude and shifted in frequency from the pattern characteristic of poly-L-proline II (PLP II) which is a left-handed 3$_1$helix of *trans* peptides (Kobrinskaya *et al.*, 1988; Dukor and Keiderling, 1991; Dukor *et al.*, 1991; Dukor and Keiderling, 1996; Keiderling *et al.*, 1999b). This coincidence of band shape, coupled with extensive oligomer studies and with previous ECD (Tiffany and Krimm, 1972) and VCD studies (Paterlini *et al.*, 1986), has indicated that the typical "coil" VCD is characteristic of a local left-handed twist. Such a pattern implies that substantial local structure similar to that of left-hand helical PLP II exists in such polypeptide "random coils" which otherwise lack long-range order. If one imagines extending a peptide, such as that present in a β-strand, then twisting the strand in a left-handed sense would lead to a left-handed "extended helix" (Tiffany and Krimm, 1972) and, if twisted more ($\sim 60°$), would yield a 3_1-helix like that of PLP II. It should be realized that, in this situation, VCD data could not determine both the degree of twist and its extent (fractional contribution to the structure). The shape of the VCD or ECD would reflect the sense of the twist, and the intensity would depend on the degree coupled to the extent. The IR would be most affected by solvent interactions, so that hydrogen bonding to the solvent, if water, would lead to significant shifts of the amide I. VCD has the advantage of being sensitive to shorter range interactions than ECD, which can highlight such local structures in long-range disordered systems, and the band shape can move with the IR shift. All of these observations are consistent with the "random coil" having locally ordered regions of a left-handed helical twist sense.

Tests of this property have shown that it is possible to disorder such "coil" molecules by heating or by addition of various salts (Dukor, 1991; Keiderling *et al.*, 1999b). This point is consistent with the sort of spectra found for very short peptides, which in solution are expected to have little structure, sampling much of conformational space. Di- and tripeptides have quite weak and nondescript VCD, often negative for the amide I', see, for example, spectra for K_2 and K_3 (Fig. 3). This combination of shape and intensity is approached by the broadened, weak, primarily negative VCD seen for modest-length random coil peptides which are further unfolded at high temperatures (Fig. 4).

The short-range length dependence of peptide VCD, which was established by oligomer studies, is perhaps its most important property and is shaped to some extent by IR. At short lengths the VCD patterns found for oligomers with stable helical structures [e.g., Pro_n (Dukor and Keiderling, 1991); Aib_n (Yasui *et al.*, 1986a); $(\alpha MeVal)_n$ (Yoder *et al.*, 1997b); $(Pro\text{-}Aib)_n$ (Yoder *et al.*, 1995b)] closely parallel those of longer chains, whereas ECD magnitude continues to increase for longer oligomers, indicating that a shorter range interaction is dominant in VCD. Oligopeptide ECD primarily arises from through-space, electric dipole coupling, which yields a length dependence capable of discriminating long versus short helices, but prevents it from having high sensitivity to very short segments of structure. IR similarly has

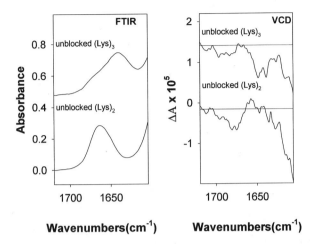

FIG. 3.   Amide I′ FTIR (left) and VCD (right) of effectively fully disordered systems, unblocked (L-Lys)$_2$ (bottom) and (L-Lys)$_3$ (top) in D$_2$O, having one and two amide groups, respectively. The large deflection to low wavenumber is due to the terminal COO$^-$ group. The breadth of the K$_3$ spectrum is due to different environments for the two C=O groups. Spectra are offset for clarity, and the actual signals measured are roughly at our detection limits for reasonable S/N ($\Delta A \sim 5 \times 10^{-6}$).

frequency shifts due to dipolar fields caused by long coherent structural units (Dousseau and Pezolet, 1990; Graff *et al.*, 1997); but, since all absorptions are positive and overlap, IR has a hard time detecting small fractions of structure. In addition to the through-space dipolar coupling, VCD has a through-bond mechanical coupling. This added vibrational interaction would normally be represented by force-field mixing of vibrational modes on adjacent subunits, which naturally interact over only short ranges, impacting mostly near-neighbor residues (aside from interresidue H-bonds). This behavior is supported by theoretical studies of oligopeptides (see below) (Bour and Keiderling, 1993; Kubelka *et al.*, 2002). Further support for this short-range dependence is provided by our VCD studies of helical peptides containing two adjacent isotopic labels (^{13}C on the amide C=O) which show a ^{13}C shifted band ($\sim$1600 cm^{-1}) with a similar α-helical VCD band pattern as seen (much more intensely) for the bulk of the oligopeptide (^{12}C at $\sim$1640 cm^{-1}) (Keiderling *et al.*, 1999a; Kubelka *et al.*, 2002).

Because of this length dependence, IR and VCD of β-turns may provide a means of discriminating and detecting them in a protein structure. Short Aib peptide results provide examples of type III β-turn VCD (Yasui *et al.*, 1986b), while cyclic peptides have been used to study type I and II turns, which are implied to have unique VCD band shapes (Wyssbrod and Diem, 1992; Xie *et al.*, 1995). We have identified distinct turn modes

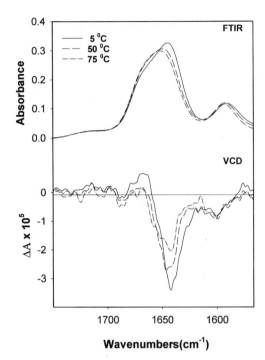

FIG. 4.   Amide I′ FTIR (top) and VCD (bottom) of thermally further unfolded random coil peptide, oligo-L-lysine, at 5°C (solid line), 50°C (dashed) and 75°C (dash-dot). Low temperature results reflect the polymer spectrum (Fig. 2, bottom), but with somewhat reduced intensity. Higher temperatures result in an IR frequency shift and loss of VCD amplitude, indicating a loss of structure. Measured amplitudes shown. Reprinted from Keiderling, T. A., Silva, R. A. G. D., Yoder, G., and Dukor, R. K. (1999b). *Bioorg. Med. Chem.* **7,** 133–141. © 1999, with permission from Elsevier Science.

in β-hairpin structures having an X-D-Pro-Gly-X-stabilized turn (Hilario *et al.*, 2002). Such identifications can be important for identifying residual structure in unfolded proteins. However, since short segments of sheet or helix can have component bands overlapping where one might expect turns, this application will most likely not move forward until labels are introduced to isotope-edit the turn or other structure elements one wishes to highlight.

B.   *Theoretically Simulated Spectra*

Before providing empirical examples of protein spectra and unfolding, it is useful to present some computational results that represent the state of the art in terms of theoretical analyses for the IR and VCD of idealized peptides. IR spectra had been theoretically computed using empirical force fields and idealized geometries for a number of years

(Krimm and Bandekar, 1986) following the pioneering efforts (Miyazawa *et al.*, 1958) that developed a normal mode description of the amide functional group. Very few of these empirical studies addressed intensities (Snir *et al.*, 1975; Chirgadze and Nevskava, 1976a,b; Nevskaya and Chirgadze, 1976; Krimm and Bandekar, 1986; Krimm and Reisdorf, 1994; Krimm, 2000), and even fewer tried to come to terms with the unfolded peptide or random coil. More recent efforts to incorporate *ab initio* force fields into peptide spectral analyses have appeared, along with intensity simulations (Qian and Krimm, 1994; Sieler and Schweitzer-Stenner, 1997; Kubelka and Keiderling, 2001b,c; Kubelka *et al.*, 2002).

Ab initio generated force fields (e.g., DFT/BPW91/6-31G* level) give computed IR amide I frequencies much too high (which is typical of DFT calculations) (Bour *et al.*, 2000; Kubelka and Keiderling, 2001a; Kubelka *et al.*, 2002). Since the amide II frequency is a bit low, this makes the splitting of the amide I–II too large by about 100 cm^{-1}. The intensity ratios are also a bit off, having too much amide II strength. Although the precise numbers vary with DFT method used and with basis set, the pattern is remarkably stable. Based on computational results for *N*-methyl acetamide and other small peptides, when solvent effects are considered, the problem is indeed seen to be due to solvation and can be corrected by including explicit waters along with a reaction field when computing the energy second derivatives (Knapp-Mohammady *et al.*, 1999; Kubelka and Keiderling, 2001a). Although such explicit water computations can, in principle, be done for some extended structures, published results are not yet available (Kubelka, 2002).

Conformational variation effects on the frequencies are difficult to judge, since one should compensate for systematic computation errors and for solvent effects; but for the amide I, the β-strand is predicted lower than the other helices, and the coil-like 3_1-helix is lower than the α-helix. When multiple strands are taken into account, the large amide I splitting in a β-sheet has been shown to be characteristic of the antiparallel form for extended, multistranded sheet structures that are necessarily flat (Kubelka and Keiderling, 2001c). Parallel structures, with more pucker, and twisted structures, as found in proteins and β-hairpin structures, have much less split amide I bands. The low frequency characteristic for β-sheets should then not be used as an indication of increased hydrogen bond strength in β-sheets, but rather primarily as a consequence of interstrand as well as intrachain coupling (Kubelka and Keiderling, 2001c). The width of various modes is strongly affected by the length of the coherent structure. Shorter segments give rise to more intensity in various components, thereby broadening the band. Since coil forms should have the shortest segments, this is consistent with the broader spectra observed for coils and unfolded (disordered) peptides.

The first simulations of peptide VCD used the MFP calculational approach at the SCF level (4-31G) for a glycine model system containing two peptide bonds whose relative φ,ψ torsional angles were varied to replicate those in α-helix, β-sheet, 3_{10}-helix, and left-handed 3_1-helix conformations (Bour and Keiderling, 1993). The computed results for the amide I and II bands of these constrained diamides qualitatively reflect the experimental VCD patterns found for proteins or peptides having those dominant forms. Calculations at the DFT level for pseudo-trialanines containing three peptide bonds linked by chiral (L-Ala) Cα centers(Bour *et al.*, 2000) gave consistent results. The simulated IR and VCD for an α-helix as a heptaamide (Ac-Ala$_6$-NHCH$_3$) and 3_1-helix as a pentaamide (Ac-Ala$_4$-NHCH$_3$) are shown in Figure 5 (top left and right). The computed VCD exhibits a positive couplet in the amide

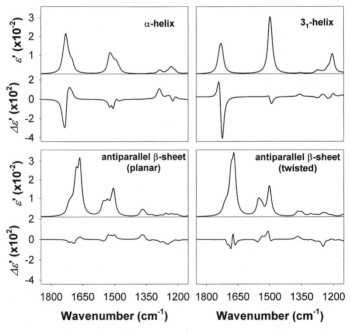

FIG. 5. Comparison of *ab initio*, DFT/BPW91/6-31G*-computed IR and VCD spectra over the amide I, II, and III regions for model peptides (of the generic sequence Ac-Ala$_n$-NHCH$_3$). These are designed to reproduce the major structural features of an α-helix (top left, $n = 6$, in which the center residue is fully H-bonded), a 3_1 helix (PLP II-like, top right, $n = 4$), and an antiparallel β-sheet ($n = 2$, 3 strands, central residue fully H-bonded) in planar (bottom left) and twisted (bottom right) conformations. The computations also encompass all the other vibrations in these molecules, but those from the CH$_3$ side chains were shifted by H/D exchange (CH$_3$) to reduce interference with the amide modes.

I and overall negative in the amide II as seen experimentally. The 3_1-helix is predicted to have a quite strong amide I couplet of the opposite sign and a weak negative amide II VCD, in excellent agreement with experimental results (see Fig. 2). By comparison, the β-sheet does give a split amide I. But the experimental splitting like that in Figure 2 (and potentially the low amide I' frequency $\sim$1620 cm^{-1}) can be computed only for flat, extended, multistrand antiparallel sheets (Kubelka and Keiderling, 2001c). Parallel and twisted sheets give smaller splitting, consistent with 1630–1640 cm^{-1} absorbance. VCD for sheets is predicted to be weak, primarily negative in the amide I with some increase in intensity for twisted forms (compare Fig. 5, bottom left and right). Because of interference from computed side-chain modes coupled with errors in the DFT frequencies for the amide modes, the most useful calculations are simulated with deuterated methyl groups. H/D exchange of the amides can also be simulated (Bour *et al.*, 2000; Kubelka *et al.*, 2002), which yields the expected frequency shifts, relatively small for the amide I' but large ($\sim$100 cm^{-1}) for the amide II', the latter mixing with side-chain modes and giving complex VCD. The amide I' α-helix VCD does not change shape; that is, the extra negative feature is not found, although it has been identified recently in some longer helical simulations (Kubelka, 2002). But the 3_1-helix VCD does change somewhat, becoming more positive, which may be important for coil simulations (Bour *et al.*, 2000).

That this method is so successful, based originally only on dipeptide (and later for tripeptide and longer) model calculations (but all with constrained conformations), supports the conclusions drawn above, based only on empirical observations, that short-range effects strongly influence VCD. Such constrained conformation methods do not at first seem appropriate for study of unfolded proteins, which must sample many conformations. However, the observations noted above, that the spectral character of random coil peptides reflects that of Pro-II helices, suggest that computation for peptide sequences with a 3_1-helical conformation will reasonably simulate random coil vibrational spectra. This approach has proved useful for simulating thermal denaturation and analyzing isotopic labeling studies (Silva *et al.*, 2000a). IR shifts in such a case are more affected by the change in solvation, which can be quite different for proteins and peptides and for sheets or helices versus coils. Such simulations will require inclusion of solvent, an effort whose results are just beginning to appear (Knapp-Mohammady *et al.*, 1999; Kubelka and Keiderling, 2001a). Finally, simulation of truly disordered systems will require some weighted averaging over an ensemble of structural types, a process not yet possible with currently available data sets.

C. Unfolded Peptide IR and VCD Examples

Some applications of IR and VCD for peptide conformational studies have been alluded to above in the course of describing the qualitative band shape patterns that can be related to secondary structure. An important aspect of vibrational spectra is their resolution of contributions from the amide group (the repeating aspect) from those originating in the side chains or other parts of the molecule. This allows one to focus on the conformation of the chain independent of interference from the variations in the side chain. In IR there is some residual overlap, for example, between the $-COO-$ modes and the amide I; but in VCD, these side chain modes rarely have any significant chirality, making their contribution minimal at most. In ECD, by contrast, the aromatic side chains gain chirality by long-range interactions and can distort the near-UV CD transitions. This has allowed VCD to be used to sort out conformations in aromatic polypeptides (Yasui and Keiderling, 1986b; Yasui et al., 1987; Yasui and Keiderling, 1988), to establish the 3_{10}-helical nature of several α-Me-substituted oligopeptides (Yasui et al., 1986a; Yoder et al., 1995a,b; 1997b), and to sort out the conformation and thermal unfolding pathways of Ala-rich peptides with various other residues incorporated for solubility or as concentration markers (Yoder et al., 1997a).

This latter problem provides an excellent example of use of VCD and IR to detect and differentiate unfolded peptide conformations. Alanine-rich peptides have a high propensity for helix formation at low temperatures, even in H_2O. Their amide I' frequencies are <1640 cm^{-1} in D_2O; this was eventually interpreted as being due to full solvation of the α-helix, which may or may not have some component of 3_{10}-helix as an intermediate in its unfolding (Williams et al., 1996; Millhauser et al., 1997; Yoder et al., 1997a; Keiderling et al., 1999b; Silva et al., 2000a). The confusion due to the low IR amide I' frequency is clarified by the VCD spectra in H_2O, which shows that the Ac-(AAKAA)$_{3,4}$-GY-NH$_2$ peptides are both dominantly α-helical (Yoder et al., 1997a). Temperature variation of their amide I' in D_2O shows the IR band shifting up to ~ 1645 cm^{-1} and the VCD shape changing to a negative couplet characteristic of a coil form (Fig. 6). Factor analysis (Malinowski, 1991) of the resulting band shapes demonstrated that both the IR and VCD, owing to short-range length dependence, were able to detect an intermediate structure attributable to growth and decay of the junctions between the central helical portion of the molecule and the steadily fraying ends, which were coil-like in VCD. In other words, the unfolding was not a simple two-state process, but maintained some local structure, an α-helix–coil junction, whose position in the sequence was unknown. ECD does not detect such an intermediate, owing to its dependence on long-range coupling, but the

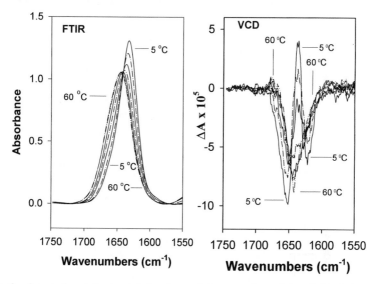

Fig. 6. Spectral monitoring of the thermal denaturation of the highly helical, Ala-rich peptide Ac-(AAAAK)$_3$AAAA-Y-NH$_2$ in D$_2$O from 5 to 60°C, as followed by changes in the amide I' IR (left) and VCD (right). IR show a clear shift to higher wavenumber from the dominant α-helical peak (here at an unusually low value, ~1637 cm^{-1}, due to full solvation of the helix) to a typical random coil value (~1645 cm^{-1}). VCD loses the (−,+,−) low-temperature helical pattern to yield a broad negative couplet, characteristic of a disordered coil, at high temperature. Spectra were normalized to $A = 1.0$ by ~45°C.

local structure in an unfolded state can be seen with the shorter range vibrational techniques.

Time-resolved IR spectra of similar peptides following a laser-excited temperature jump showed two relaxation times, unfolding ~160 ns and faster components <10 ns (Williams $et\ al.$, 1996). These times are very sensitive to the length, sequence, and environment of these peptides, but do show that the fundamental helix unfolding process is quite fast. These fast IR data have been contrasted with Raman and fluorescence-based T-jump experiments (Thompson $et\ al.$, 1997). Raman experiments at various temperatures have suggested a folding in ~1 μs, based on an equilibrium analysis (Lednev $et\ al.$, 2001). But all agree that the mechanism of helix formation is very fast.

To get site-specific information about this unfolding process, a similar series of peptides, Ac-(AAAAK)$_3$AAAA-Y-NH$_2$, was sequentially labeled with ^{13}C on the C=O of four Ala residues, first on those at the N terminus, and then sequentially for the three other tetra-Ala positions (Decatur and Antonic, 1999). IR results showed a band shifted down 40 cm^{-1} from the amide I', which formed a clearly resolved band

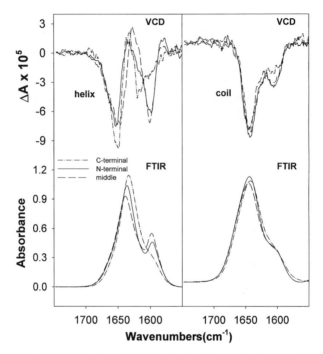

FIG. 7.   Amide I' VCD (top) and IR absorption (bottom) spectra of highly helical alanine-rich peptides in D_2O at low temperature (left, 5°C) and high (right, 50°C). Data for three isotopically labeled (^{13}C on the amide C=O of four Ala residues selected in sequence) peptides of the general formula Ac-(AAAAK)$_3$AAAA-Y-NH$_2$. The labels are on the N-terminal tetrad of Ala (4AL1, solid line), middle (4AL2, dashed), and C-terminal tetrad (4AL4, dash-dot). The ^{13}C feature is shifted ~40 cm^{-1} down from the ^{12}C feature and at low temperature yields a VCD shape mimicking that of the main chain for all except the C-terminal labels, indicating it to be unfolded. The high-temperature spectra of the three isotopomers are more uniform, indicating full unfolding. Reprinted from Silva, R. A. G. D., Kubelka, J., Decatur, S. M., Bour, P., and Keiderling, T. A. (2000a). *Proc. Natl. Acad. Sci. USA* **97**, 8318–8323. © 2000 National Academy of Science, U.S.A.

except for the C-terminally labeled residues, which were broad and different from the rest even at low temperatures. VCD then confirmed our unfolding hypothesis in which the frequency-shifted ^{13}C-labeled residue VCD indicates a helix–coil transition at a different temperature, depending on where the substitution is placed (Silva *et al.*, 2000a). Furthermore, even the lowest temperature data indicated the C-terminal residues to be unfolded whereas the N-terminal ones were helical (Fig. 7). The band shape in VCD moved with the IR transition, thereby providing a means of determining the conformation for both the ^{12}C and ^{13}C segments in the peptide. The characteristic random coil spectra could be detected for both segments after the transition. At high temperatures, the VCD band shape continued to change, getting weaker while the IR merely broadened (Fig. 7). This conclusion that the C-terminal residues were less stable than the middle or N-terminal residues was reaffirmed in a subsequent IR and thermal analysis study of a longer, isotopically labeled peptide with salt bridge stabilization proteins (Venyaminov *et al.,* 2001).

These low-temperature amide I′ IR and VCD isotope-edited results could be modeled with near-quantitative accuracy with DFT parameters by transferring *ab initio* FF, APT, and AAT parameters from computations for an α-helical heptapeptide model compound onto an α-helical 20-mer oligopeptide (Fig. 8 left). This simulation does not agree with data for the C-terminally labeled oligomer, because experimentally that end of

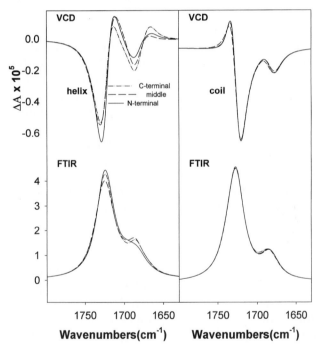

FIG. 8.   Theoretical simulation of VCD (top) and IR absorption (bottom) spectra of alanine dodecapeptides for the amide I′ bands for a fully α-helical conformation (left) and a fully left-handed 3_1-helical conformation (right). The simulations are for the same three isotopically labeled (^{13}C on the amide C=O for four Ala residues selected in sequence) peptides as in Figure 7: N-terminal tetrad (4AL1), middle (4AL2), and C-terminal (4AL4). The ^{13}C feature is the same for all sequences, confirming the experimentally found unfolding of the C-terminus. The agreement with the shapes in Figure 7 is near quantitative. Reprinted from Silva, R. A. G. D., Kubelka, J., Decatur, S. M., Bour, P., and Keiderling, T. A. (2000a). *Proc. Natl. Acad. Sci. USA* **97,** 8318–8323. © 2000 National Academy of Science, U.S.A.

the sequence was not helical where labeled. Of importance for application to unfolded peptides, the high-temperature random coil results could also be modeled well by transferring parameters from a 3_1-helical pentapeptide (Fig. 8 right). It might be noted that at these higher temperatures, the negative couplet shape, typical of low-temperature random coils, is obscured by broadening of the IR transition, resulting in a mostly negative VCD.

Another case of identifying an intermediate formation with local structure by use of IR and VCD comes from a study of the intermediate formed due to mutarotation of poly-L-proline (PLP) from the right-handed *cis*-amide PLP I form to the left-handed *trans*-amide PLP II form (Dukor and Keiderling, 1996). This intermediate formation could not be detected with ECD, whose analysis had erroneously implied a two-state transition. In this case, the peptide did not unfold, but went from one form to the other with a local conformational distortion that was too short range in extent to be seen in ECD. The key to interpreting the results was again the frequency resolution of IR and VCD coupled with their short-range conformational sensitivity. Although the PLP I and PLP II VCD band shapes are only marginally distinct, they and the associated IR undergo a significant frequency shift (which is quite solvent-dependent) that allows detection of the intermediate. The magnitude of the signal corresponding to the intermediate (again, a junction between the two forms) was quite significant, rising and falling during the transition.

Detection of this junction between left- and right-handed helical forms was presaged by a study of DL alternate proline oligomers whose helical preference in terms of handedness was determined by VCD to depend on the chirality of the C-terminal residue (Mastle *et al.,* 1995). In this case the IR was of less use. For very long DL peptides, both the VCD and ECD signals decay to zero, the sample becoming effectively racemic on an average scale. This pseudo-racemate spectral result is important because it is distinguishable from that of an unfolded peptide that is still quite chiral but lacks long-range order. The truly disordered peptide should sample much of Ramanchandran space and yield a VCD like the spectrum seen in Figure 3 for very short peptides or coils at very high temperatures.

V. Protein Studies

The above peptide results established general patterns that are apparent in protein spectra. However, most proteins differ from small peptides in terms of the degree of solvation and the uniformity of secondary structure segments. Although a peptide helix may terminate in a large number of conformations representing the "fraying" of that segment, in a protein this termination is likely to be a relatively well-defined or, at least, a conformationally constrained turn or loop sequence. Furthermore, the segment lengths in a protein are determined by the fold rather than the thermodynamics of a structure's stability. When unfolded, some similarity of dispersion of structures and lengths might be expected between proteins and peptides, assuming they both would achieve a fully disordered state. But they do not, since the degree of solvation remains different owing to residual (nonrandom) hydrophobic interactions arising from the heterogeneous sequence in proteins and residual specific disulfide and other cross-linking interactions. The propensity to form structure in each segment is enhanced by interaction with groups (e.g., peptide and hydrophobic side chains) other than solvent. Interpretation of IR and VCD data for proteins has largely been dependent on data obtained from native-state proteins of known structure and is based qualitatively on the background of peptide data described above. Explicit accounting for unfolded or denatured proteins has not been part of many quantitative IR and VCD studies. However, qualitative observations on unfolded proteins are available and the component-resolved, frequency-based quantitative approach of IR has been applied in some limited circumstances to study protein denaturation. In our laboratory, these interpretations are also constrained to be consistent with ECD data of the same system, thereby avoiding false conclusions.

A. Protein Spectral Qualitative and Quantitative Techniques

1. Qualitative Fold Patterns

Comparison of the spectra for selected proteins shows that their amide I or amide I' IR spectra heavily overlap, but that the amide I for high β-sheet proteins are significantly shifted down in frequency from helix, turn, and coil contributions (Surewicz *et al.*, 1993). If the structure is only part β-sheet with substantial α-helix—for example, in an α/β protein such as triosephosphate isomerase—this band is hard to distinguish from other conformations; but typically, on deconvolution, some low-frequency components can be identified (Wi *et al.*, 1998). Amide II and III spectra are even broader, and any shifts with conformation in these regions are complicated by overlap with local C—H and side chain modes. However, some progress on their interpretation has been made (Gupta and Keiderling, 1992; Singh, 2000).

Amide I VCD spectra for these examples of protein folds are indeed very different from one another, as would be expected from peptide

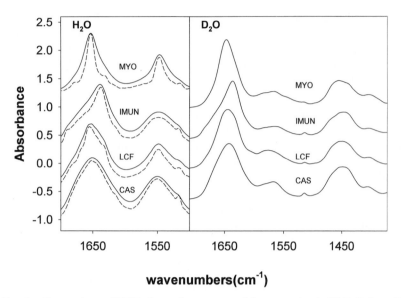

FIG. 9.   Comparison of FTIR absorption spectra of four proteins in H_2O (left, amide I + II) and D_2O (right, amide I′ + II′). Comparison between protein spectra for dominant secondary structure contributions from α-helix (myoglobin, MYO, top), β-sheet (immunoglobin, IMUN), from both helix and sheet (lactoferrin, LCF) and from no long-range order (α-casein, CAS, bottom). The comparisons emphasize the high similarity, differing mostly by small frequency shifts of the amide I with the changes in secondary structure.

results. Complete sign pattern inversions and peak frequency variations are found for a set of globular proteins (Pancoska *et al.*, 1989, 1991). However, the β-sheet amide I VCD is not the same as that seen for polypeptides, being higher in frequency and broader in shape, but is still often net negative, though typically with some couplet shape. This frequency variance from the polypeptide model is consistent with the predicted effects on VCD of the twist normally seen in protein β-sheets (Kubelka and Keiderling, 2001c). The amide II and III VCD also differ between fold types, but less so, owing to less resolution of components. The degree of variation seen in VCD is also not seen with ECD.

Figures 9 and 10 represent a selected comparison of amide I′ and I+II FTIR and VCD for four proteins in D_2O solution. Of these, myoglobin (MYO) has a very high fraction of α-helix, immunoglobulin (IMU) has substantial β-sheet component, lactoferrin (LAF) has both α and β contributions, and α-casein (CAS) supposedly has no extended structure. The FTIR spectra of these proteins change little, the primary difference

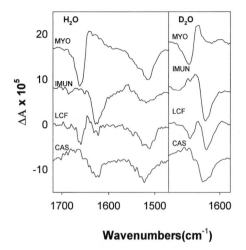

Wavenumbers(cm⁻¹)

FIG. 10. Comparison of VCD spectra of four proteins in H_2O (left, amide I + II) and D_2O (right, amide I′ + II′) with dominant secondary structure contributions from α-helix (myoglobin, MYO, top), β-sheet (immunoglobin, IMUN), both helix and sheet (lactoferrin, LCF) and "no structure" (α-casein, CAS, bottom). The comparisons emphasize the distinct band shapes developed in the amide I and I′ for each structural type. Note the reduced S/N in the H_2O-based spectra and the shape changes upon H/D exchange for helix and sheet (and mixed) structures, but relatively little for the unstructured CAS.

being a frequency shift of the peak amide I absorbance that roughly correlates with the amount of β-sheet in the protein. To enhance their variance, the Fourier self-deconvolution spectra for the proteins in H_2O are also presented (as dashed lines) in Figure 9. The amide III IR (not shown) are complex, with a broad peak at $\sim$1300–1320 cm^{-1} for helices and a loss of that intensity for other structures, exposing an underlying feature at $\sim$1240–1260 cm^{-1} in sheet and mixed structures. This contrasts with amide III Raman spectra, in which the characteristic frequencies appear to be opposite (Tu, 1982; Asher *et al.*, 2001). In VCD, the amide I frequency and the band shape contribute to the analysis. Amide I VCD in H_2O solution has only small differences from the D_2O results aside from the α-helix couplet changing to the $(-,+,-)$ band shape. The amide II band VCD sign pattern (and amplitude) is more important than frequency, since the bands are broader than the amide I or I′ and all contributions overlap. In summary, it is clear that VCD band shapes, owing to their signed nature, have more characteristics and variation than do the respective FTIR absorbance bands for each of the accessible protein amide modes, I, II, and III.

The gross fold type identification is established such that highly helical proteins have an amide I at $\sim$1650 cm^{-1} that is relatively sharp in the

IR with positive couplet, a strongly peaked amide II at ~1550 cm^{-1} with negative VCD to lower frequency, and an amide III IR at ~1310 cm^{-1} with a broader net positive VCD (Pancoska et $al.$, 1989; Baumruk and Keiderling, 1993). High β-sheet–content proteins have a broadly sloped amide I reaching a maximum at ~1630 cm^{-1} which has primarily negative VCD and often weak, dispersed positive features to higher frequency, a broader amide II (1550–1520 cm^{-1}) corresponding to a negative couplet VCD, and an amide III maximum at ~1240 cm^{-1} but a very broad net negative VCD at ~1300 cm^{-1}. The positive VCD at ~1230 cm^{-1} is structure-independent (Baello et $al.$, 1997). Mixed helix–sheet proteins are more easily identified in D_2O where they give rise to a distinctive, but weak, W-shaped $(-,+,-)$ predominantly negative amide I' VCD pattern (Pancoska et $al.$, 1991; Keiderling et $al.$, 1994). Their amide II and III patterns are just combinations of the limiting spectra.

So how does the unfolded protein fit in this structure–spectra pattern? As a first approximation we can consider the native–state spectra for the nominally unstructured protein α-casein. Its IR bands are broader and less characteristic than for the other proteins, but not hugely so, with the amide I frequency components spread above and below ~1650 cm^{-1} (Fig. 9). The negative couplet amide I VCD (Fig. 10) in this case may seem surprisingly strong for an "unordered protein" until the polypeptide results are recalled. The amide II band is broad and its VCD is very weak, with a negative component to low frequency, thus providing little discrimination for conformational studies. Amide III absorbances of coil conformations are assigned to lie between those of the sheet and helix forms (Singh, 2000). For truly disordered forms, the ECD does offer a significant tool for discrimination, since the unfolded peptide gives a strong negative far-UV (<200 nm) response that differs substantially from the helix and sheet spectra (Woody and Kallenbach, 2002). For proteins, the longer wavelength (n–π^*) transitions are often more confused, being negative in many cases yet positive in model peptides. These bands also have significant negative contributions close to 200 nm for various β-forms, which can cause confusion between these conformations, especially in a transition. Coupling IR or VCD, which have some distinction for coil systems, with these characteristic ECD shapes, which differentiate them on a different basis, leads to the safest structural interpretations.

Thermally denatured proteins have been studied for a variety of systems using FTIR and VCD. The resulting high-temperature spectra often reflect the characteristics seen earlier for random coil peptides as well as that seen for the "unstructured" casein. Particularly the amide I IR bands show a frequency shift to center on a broadened band at ~1645–50 cm^{-1}. The amide I VCD loses its distinctive character (Fig. 11) and tends toward

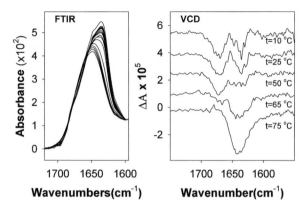

FIG. 11. Amide I' thermal denaturation spectra for ribonuclease A as followed by FTIR (left) and VCD (right), which show the IR peak shifting from the dominant β-sheet frequency (skewed with a maximum at $\sim$1635 cm^{-1}) to the random coil frequency ($\sim$1645–1650 cm^{-1}) and the VCD shape changing from the W-pattern characteristic of an $\alpha + \beta$ structure to a broadened negative couplet typical of a more disordered coil form. The process clearly indicates loss of one form and gain of another while encompassing recognition of an intermediate form. (This is seen here most easily as the decay and growth back of the $\sim$1630 cm^{-1} VCD feature, but is more obvious after factor analysis of the data set, Fig. 15).

a broadened negative couplet. An example of this behavior is provided by thermal unfolding of ribonuclease A, as will be discussed later.

Chemical denaturation poses an incompatibility problem for IR and similarly for VCD studies. Although some perturbants such as pH, salt, and reducing agents (such as dtt) can be added without too much interference, the typical full denaturation conditions using GuHCl or urea (typically >6 M) lead to a major interfering band in the amide I region (Byler *et al.*, 2000). However, one IR test showed that an apparently equivalent amide I' band shape is obtained for chemically denatured ribonuclease A and the thermally denatured protein (Fabian *et al.*, 1995). In this case denaturation was accomplished by addition of ^{13}C-labeled urea, whose isotope shift causes its interfering band (C=O stretch) to be shifted away from the amide I'.

2. Quantitative Secondary Structure Determinations

Quantitative analysis of protein IR and VCD spectra in terms of the fractional components (FC) of their secondary structure has taken different approaches, as noted earlier. The FTIR approach of assigning frequencies to specific components can, in principle, identify amounts of unordered structure in a protein fold. The viability of this approach is highly dependent on the reliability of the assignment. As shown repeatedly (Pancoska *et al.*, 1993; Jackson and Mantsch, 1995; Haris, 2000), there are problems with the nonuniqueness of such assignments. For example, solvated helices absorb in the same region assigned to the random coil; thus, after unfolding the tertiary structure and solvating helical components, the frequency shift on further unfolding those helices could be negligible. In such a case, an intermediate would be difficult to detect with IR alone. Nonetheless, several studies have used frequency characteristics (typically of amide I') to monitor structural change in an unfolding process and have correlated these changes with other data to interpret the mechanisms or thermodynamics of unfolding. The alternative approach, analysis of band shape variations, is subject to the limitations of the training set of proteins (Dousseau and Pezolet, 1990; Lee *et al.*, 1990; Sarver and Kruger, 1991; Pribic *et al.*, 1993; Baumruk *et al.*, 1996). Current IR (and VCD) band shape analyses have not included denatured proteins or random coil peptides in their training sets of "known" structures precisely because such structures are not known in the same way the normal training set protein structures are known. In ECD analyses, such additions have been made through the use of reasonable, but arbitrary assumptions as to their structures, and significant improvements in quantitative secondary structure prediction have resulted (Sreerama *et al.*, 1999, 2000). In the future, this should be implemented with IR and VCD methods as well to determine if similar improvements could obtained.

Our band shape methods have made use of the principal component method of factor analysis (Pancoska *et al.*, 1979; Malinowski, 1991) to characterize the protein spectra in terms of a relatively small number of coefficients (loadings) (Pancoska *et al.*, 1994; 1995; Baumruk *et al.*, 1996). This approach is similar, in its initial stages, to various methods (Selcon, Variselect, etc.) that have been used for determining protein secondary structure from ECD data (Hennessey and Johnson, 1981; Provencher and Glockner, 1981; Johnson, 1988; Pancoska and Keiderling, 1991; Sreerama and Woody, 1993, 1994; Venyaminov and Yang, 1996). At this point, one can say these traditional quantitative methods have had little impact upon structural studies of denatured proteins.

With regard to quantitative secondary structure spectral analyses, we should recognize that any training set, such as those based on X-ray crystal structures, would have strong correlations between the helix and sheet contents. This correlation is general and is independent of the algorithm used to determine the structural components (Pancoska *et al.*, 1992; 1995; Keiderling, 1996; Pancoska, 2000). In these native state structure analyses, ECD-based structural determinations are good for helix but are relatively insensitive to the spectral manifestation of the sheet, while the structures determined by VCD (and IR) are more reliable for sheet determination owing to their more direct spectral dependence on sheet content (Pancoska *et al.*, 1995). The helix–sheet correlation explains the apparent ability of ECD to "predict" sheet content while being qualitatively insensitive to it (with respect to helix). When a protein being analyzed is normal (i.e., like the others in the training set), the ECD does well by utilizing the underlying helix–sheet relationship from the training set, but the resulting prediction is really based on the excellent sensitivity of ECD to helix content. When a protein structure deviates from a typical α–β distribution, erratic sheet predictions can result from use of ECD alone. This is precisely the pattern to be guarded against in analysis of protein unfolding data where, even if one expected the final unfolded states to have similar characteristics, the intermediates are all likely to be different and to be different from typical native folds. Examples of strange predictions of secondary structure from ECD alone are common. (One hopes only few of these are published!) Certainly, similar erratic structure predictions could result from FTIR and VCD analyses if they were also used alone and were subject to some spectra interference. It is the combination of ECD and FTIR/VCD that provides some protection in this regard (Sarver and Kruger, 1991; Pribic *et al.*, 1993; Pancoska *et al.*, 1995; Baumruk *et al.*, 1996).

Our very best predictions of helix and sheet have come from a different approach, using only FTIR data (Baello *et al.*, 2000). In this case, however, we added another perturbation and used the spectral response to further discriminate structures. FTIR spectra were measured for proteins at various stages of H/D exchange and factor analysis was used to find the characteristic spectrum of each protein under exchange. These exchange-modified spectra for 19 proteins were then used as a training set to determine algorithms (FA-RMR) for predicting secondary structure. An extension of these data with a neural net work technique also predicted fold type from H/D exchange FTIR data (Baello, 2000). This perturbation method applied to unfolded proteins should be distinctive because the extent of exchange would differ radically, but it would be of limited use for unfolding experiments. An alternative approach used H/D exchange rates measured with ATR-FTIR to sort out coil components of folded proteins from helix contributions (Goormaghtigh and Ruysschaert, 2000). Presumably, this could be extended to partially folded system as well.

3. Factor Analyses of Unfolding

Left with a useful, but ultimately qualitative (or semiquantitative) appraisal of the spectra, one can normally establish the dominant type of native secondary structure in the fold, but cannot determine much detail. However, if a protein is systematically studied under the influence of some unfolding perturbation (e.g., pH, solvent, or temperature change), small structural variations are reliably detectable using difference spectra techniques (Urbanova *et al.,* 1991; Keiderling *et al.,* 1994; Urbanova *et al.,* 1996). The nature of the change caused by the perturbation can be rationalized from the character of the difference spectra, all other things being constant. To automate such analyses, one can do factor analysis of the spectra (using essentially the same mathematical techniques as above for quantitative studies), this time obtained for a single protein with varying degrees of perturbation. The loadings of the component spectra will reflect the shift in equilibrium between native and unfolded states (Yoder *et al.,* 1997a; Stelea *et al.,* 2001; Stelea *et al.,* 2002). In this analysis the variation of contribution to the training set spectra with perturbation is model-independent. This variance in contribution (loadings) of different factors will be an indicator of the relevant structural change. If a two-state transition, it will be simple; if multistate, all or just some of the loadings will reflect this. Often the component spectra derived from such an analysis (or those projected out by rotation of the component vectors onto target states) will mimic the characteristic spectral signatures of the components either lost or gained in the transition. The loadings of those components in the experimental spectra will map out the equilibrium transition as a function of perturbant. The result can be used for thermodynamic modeling, identifying intermediates, or postulating mechanisms of unfolding. Such sensitivity to relative change has always been the *forté* of optical spectral analyses.

In such an analysis, the first spectral component (or factor) represents the most common elements of all the experimental spectra used. The second component represents the major variance from that average. Each successive component then becomes less significant, eventually being just noise contributions. The components are constructed to be orthogonal, but their orientation is arbitrary and is dependent on the set of spectra used, since the first component is the average. The original spectra, θ_i, can be represented as a linear combination of factors, Φ_j, and loadings α_{ij} as

$$\theta_i = \sum_j \alpha_{ij} \Phi_j \qquad (2)$$

The sum over j can be truncated by eliminating those Φ_j corresponding to noise.

Some of the lessons learned from quantitative predictions of secondary structure based on a training set of protein spectra are applicable to the use of such statistically based, band shape variance methods for unfolding experiments. The first is that a combination of IR-based measurements (FTIR absorption or VCD) with ECD leads to much more stable predictive results (fewer high error outliers when tested with known structures) (Pancoska et al., 1995; Baumruk et al., 1996). Because avoidance of singular predictions of high error is vital when trying to analyze an unknown structure, we developed a system for coupled analyses. The same can be said for denaturing studies, since the partially and terminally unfolded protein structure is unlikely to be understood, that being the point of the exercise. Similar coupled analyses have not yet been developed for unfolding, since the spectra are typically measured on different samples under different conditions peculiar to the needs of each experiment. However, the results from different spectroscopies are correlated to determine more structural detail of the process, taking advantage of the different length dependencies of IR and VCD as opposed to ECD.

The second lesson is that the spectra reflect many aspects of the structure, and if one is primarily interested in secondary structure, improved predictive results can be obtained by using just those selected loadings that have the most dependence on the structural changes being sought (Van Stokkum et al., 1990; Pribic et al., 1993; Pancoska et al., 1994; 1995; Baumruk et al., 1996). This has led to our development of a restricted multiple regression (RMR) method which uses only a few loadings that correspond to the optimal components for prediction of each secondary structure type (Pancoska et al., 1995). Alternatively, ECD-based analyses have taken a different restrictive approach, altering the training set or optimizing the method for consistent prediction behavior (Manavalan and Johnson, 1987; Sreerama and Woody, 1993). In unfolding analyses, different component spectra obtained with different techniques can provide insight into different types of structural transitions. We avoid taking the whole spectral change as the result, but decompose it as much as possible. Using more of a global fitting analysis to multiple spectral components for unfolding of six proteins in D_2O buffer as followed by FTIR, thermodynamic parameters were developed and problems of intermediate formation were addressed, interpretating band components in terms of the unfolded state (Van Stokkum et al., 1995).

4. Segment Determination

The short-range dependent techniques such as IR and VCD will sense the distortions characteristic of the conformation of residues at the ends of uniform segments of secondary structure in a different manner than does ECD with its longer-range sensitivity. Both spectral responses will still depend on segment length in some way. Our method of using this information led to a matrix form of protein secondary structure descriptor that accounted for both the number of segments of uniform structure as a value on the diagonal and the interconnectivity of segments of each type as off-diagonal elements(Pancoska et al., 1994, 1996, 1999a). This effectively encoded the segment length and distribution character. Using ECD data, other workers (Sreerama et al., 1999) have similarly determined segment length and numbers with a simpler method that disregards the interconnectivity aspects. These efforts moved spectra–structure analyses beyond conventional approaches to provide insight into the secondary structure distribution that might eventually be used as an experimental constraint for sequence-based structure prediction algorithms.

This segment method has been applied to thermal unfolding of ribonuclease T_1, which has one helical segment and seven sheet segments. RNase T_1 above 55°C was shown to have a loss of the helical segment and a fraying or shortening of the sheet segments but with some retention of a number of sheet components in a hydrophobic core (Pancoska et al., 1996). The thermal unfolding VCD data is summarized in Figure 12 for native and denatured state spectra, with the results of the analyses.

5. Two-Dimensional (2-D) Correlation Analysis

Alternatively, 2-D correlation methods can be used to show which spectral regions are changing with increase of a perturbation and, from their dependence on a given secondary structure type, can be used to assign the structural change in the process (Noda, 1990; Noda et al., 1996a; Pancoska et al., 1999b). Furthermore, heterocorrelation between different types of spectra (ECD with IR and Raman, IR with Raman, VCD with IR and Raman) can be used to identify unique bands and associate them with a secondary structure source through correlation to a more easily assigned method, such as ECD (Noda et al., 1996b; Kubelka et al., 1999a, 1999b). Such an approach can also be used to identify the spectral regions of highest sensitivity for later analyses using the methods described above. For example, the thermal partial unfolding of bovine serum albumin (BSA) as a function of temperature yields a 2-D correlation (Fig. 13) showing loss of amide I' at $\sim$1650 cm^{-1} anticorrelated with gain at $\sim$1620 cm^{-1} and 1680 cm^{-1} and similarly anticorrelated with amide II' gain at $\sim$1430 and 1380 cm^{-1}. This map was realized using a modification of the general 2-D method(Pancoska et al., 1999b), and the changes seen with temperature variation suggest a helix-to-sheet transition.

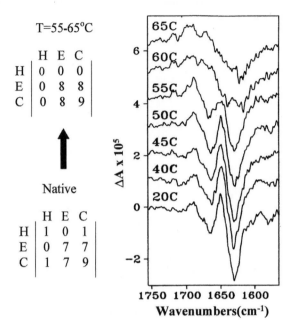

FIG. 12.   Thermal denaturation for ribonuclease T_1 as followed by VCD, from 20° to 65°C. The matrix descriptors determined for the native state and the unfolded high-temperature data are indicated. The values indicate a loss of the helix segment but maintenance of sheet segments. Also listed are the spectrally determined fractional contributions (FC) to the secondary structure. When combined with the segment analysis, this implies that the residual sheet segments must be very short. Reprinted with permission from Pancoska, P., *et al.* (1996). *Biochemistry* **35**(40), 13094–13106, the American Chemical Society.

B.   Unfolded Protein IR and VCD Examples

Unfolding data were described above in Figures 11 and 12 for RNase A and T_1. Analyses of the RNase A data (Stelea *et al.*, 2001) have shown an identifiable intermediate in the thermal unfolding pathway (also evident in the differential scanning calorimetry results) that is highlighted by factor analysis of the entire spectrum. It is made interpretable by the UV ECD showing a pretransition that can be associated primarily with a loss in intensity (α-helical change) and the FTIR having a corresponding change in both helical and sheet regions (Fig. 14). The main transition, seen in all techniques, correlates to the near-UV ECD intensity loss (tertiary structure unfolding). The pretransition unfolding step appears to be a loss of helical secondary structure which, owing to its correlation with some loss of β-sheet, has been postulated to be due to helix 2

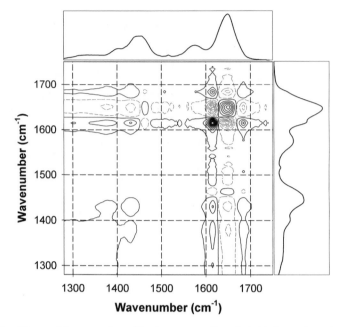

Fig. 13. Thermal denaturation of bovine serum albumin in D_2O monitored with FTIR and displayed as a 2-D correlation diagram. The normal spectrum at the top is the low-temperature native state spectrum, and that at the side is the high-temperature final state. Solid contour lines are positive correlations and dashed are negative. That the main amide I' at 1650 cm^{-1} negatively correlates with the amide II' features at 1430 and 1380 cm^{-1} suggests that, as the main feature disappears, the minor ones appear. In this case, use of several regions allows an interpretation. As native state helix is unfolded, a sheet-like structure is formed, which follows from the negative correlation of the helical 1650 cm^{-1} with the β-sheet-like 1620 and 1685 cm^{-1} features. By extension, the amide II' features become assignable to secondary structures as well.

unwinding where it couples to a sheet segment. This is supported by parallel studies of RNase S unfolding, which show almost the same pattern but with transitions shifted to lower temperatures, despite the break in the sequence at residue 20, at the end of the first helical segment (Stelea, 2002; Stelea *et al.,* 2002). The RNase S data further indicate that the pretransition is due to loss of secondary structure without dissociation of the peptide, a step that occurs along with the main denaturation transition.

Bovine α-lactalbumin (BLA) is a protein whose structure appears to be unusually malleable and, as such, has been the focus of many studies of what is termed the molten globule transition. At low pH, BLA expands and is said to lose tertiary structure, but it maintains substantial secondary structure in a partial unfolding transition (molten globule

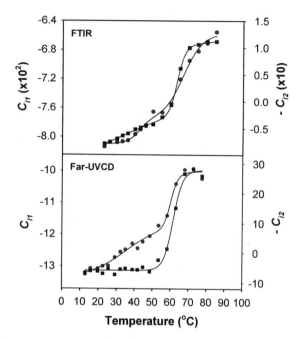

FIG. 14.   Factor analysis loadings (first and second spectral components) for thermal unfolding of RNase A as monitored with amide I′ FTIR and far-UV ECD. In each case a pretransition is evident in the curves before the main transition at ~55°C. This full band shape analysis can sense smaller variations and can be partitioned to give added insight. Since the main ECD change could be shown to be loss of intensity, the major structural change was unfolding of a helix. The frequency dispersion of the FTIR change showed that some β-sheet loss accompanied this pretransitional helix unfolding, but that most sheet loss was in the main transition.

state). VCD analysis shows that it actually gains secondary structure, in particular α-helix content, on unfolding or loosening tertiary structure (Keiderling *et al.*, 1994). Furthermore, although its crystal structure is nearly identical to that of hen egg white lysozyme (HEWL), its spectra (ECD, FTIR, and VCD) are noticeably different (Urbanova *et al.*, 1991), implying that a dynamic native state structure exists in solution for BLA, which could be viewed as fluxionally unfolded.

A major problem in unfolding studies of large proteins is irreversibility. In a study of elastase temperature-induced denaturation, second-derivative FTIR show a distinct loss of several sharp amide I′ features (dominant β-sheet components and growth in broadened bands at 1645 and 1668 cm^{-1} (Byler *et al.*, 2000). These features persisted on cooling, indicating lack of reversibility, a feature common to longer multidomain proteins. A graphic example of this is seen in the triosephosphate

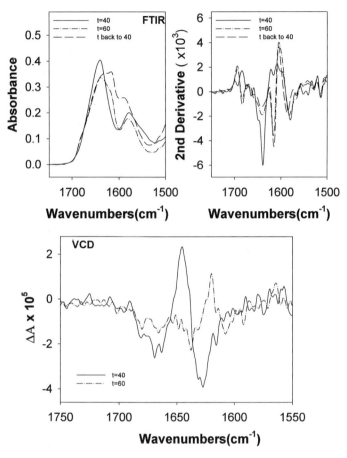

FIG. 15. Thermal denaturation of triosephosphate isomerase with FTIR (upper left), second-derivative FTIR (upper right), and VCD (bottom) showing irreversible aggregation effects. The IR shift from a simple maximum at ~1650–1640 cm^{-1} to a lower frequency distorted to low wavenumber is seen to be irreversible when the original spectrum is not recovered. The second-derivative result makes the changes more dramatic and shows the original native state spectrum to be more complex (negative second derivatives correspond to peak positions). Loss of structure is even more evident in the VCD, which loses most of its intensity at 60°C.

isomerase IR and VCD at 40 and 60°C (Fig. 15). The narrow amide I′ at ~1645 cm^{-1} broadens and shifts to ~1640 cm^{-1} and a strong shoulder grows at ~1620 cm^{-1} and a weak one at 1690 cm^{-1}. This shoulder increases in strength with time (several hours) and the band shape does not revert to the original form on recooling. These quite dominant second-derivative features corresponding to the low- and high-frequency

IR shoulders are an indicator of a structured aggregate as opposed to disordered coil, while the shift of the central frequency broaden band correlates with loss of order (unfolding). In the VCD, the intense (40°C) "W shape" amide I' corresponding to a mixed helix sheet structure collapses to a small residual negative intensity with a sharp positive feature at ~1620 cm^{-1}. VCD features correlated to aggregation are notoriously unstable and have been observed as couplets, evidencing components of both signs. The unfolding here has multiple effects: aggregation, loss of order and hence intensity, and (though hard to see in Fig. 15) some growth of spectral components characteristic of disorder (random coil).

Fink and co-workers have done extensive studies of these aggregation effects on protein spectra and structure (Fink, 1998; Fink *et al.*, 2000). In ATR-FTIR studies of films of disordered aggregates and inclusion bodies, they were able to identify both a shift down to ~1632 cm^{-1} and broadening of the β-sheet peak and a broad feature at ~1658 cm^{-1} attributed to disorder or loop structures. These aggregation peaks have been shown to vary in frequency from ~1630 to ~1620 cm^{-1}, depending on form of aggregation and sampling environment. This variation implies that the various fibril-forming denatured states are not all well-ordered and may have significant unfolded components. (Khurana and Fink, 2000)

We have similarly carried out a number of unfolding studies on cytochrome *c*, a helical bundle protein. At high pH, with increasing temperature, similar signs of aggregation as noted above are seen in addition to unfolding. VCD measured at 40, 65, 75, and 85°C are shown in Figure 16 (right) showing a steady progression from the mixed helix–sheet spectrum to a much weaker and broader coil form. In this case, aggregation is implied by failure of the spectra to revert to its original form on cooling, but the spectral interference was a major problem (Q. Xu, and T. A. Keiderling, unpublished). This irreversibility can sometimes be modified by addition of some chemical denaturants such as guanidine hydrochloride (GuHCl) (Byler *et al.*, 2000). We were able to measure spectra with 1 M GuHCl that shifted the transition down in temperature. However, in the case of cytochrome *c*, even more reversible spectra were obtained by lowering the pH to 1.7 where the tertiary structure is relaxed, yet significant secondary structure remains. The structure now melts between 30 and 40°C as indicated in the IR, second-derivative IR (left top), and VCD (middle) spectra in Figure 16. The second-derivative IR for the unfolded protein has fewer and less distinct components and the VCD becomes a broad negative with a potentially negative couplet component, as expected for a high-temperature unfolded structure.

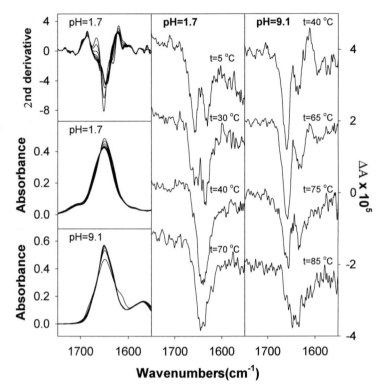

FIG. 16. Thermal denaturation monitored with FTIR and VCD of cytochrome c at pH 9.1 and 1.7 in D_2O. The process at high pH leads to aggregation, evident in the FTIR growth of a 1615 cm^{-1} peak (lower left). The transition to a disordered state occurred only above 75°C at high pH, as seen in the pH = 9.1 VCD data (far right). Lowering pH also lowered the stability, leading to unfolded state below 70°C, which was reversible; the original state could be recovered. The second-derivative representation (top left) illustrates the changes occurring and their complexity more clearly than do the FTIR data alone. The low pH state, though expanded, retained much secondary structure as seen by comparing the VCD in the two pHs (middle and right, top spectra).

VI. CONCLUSION

The lower resolution and lack of site specificity of IR and VCD spectroscopy are normally difficulties with which one deals in order to gain access to structural information on systems difficult to probe otherwise. Unfolded proteins, with their lack of homogeneity and long-range order, fit these techniques well owing to the short-range structure sensitivity and intrinsically short time scale of vibrational spectroscopy. Although applications of IR and VCD techniques to questions of unfolded protein structure are scattered, their potential suggests expansive future use.

ACKNOWLEDGMENT

This work was partially supported by a grant from the Research Corporation and by the Donors of the Petroleum Research Fund administered by the American Chemical Society. Early work was supported by the National Institutes of Health (GM 30147), while instrumentation and theoretical development have been supported by the National Science Foundation. Much work described stems from the research of a number of postdoctoral and graduate student co-workers, whose names are noted in the cited references. Many of the samples for special applications came to us as gifts or via collaborations with researchers from around the world, most of whom are also coauthors in the publications.

REFERENCES

Aamouche, A., Devlin, F. J., and Stephens, P. J. (2000). *J. Am. Chem. Soc.* **122,** 7358–7367.

Asher, S. A., Ianoul, A., Mix, G., Boyden, M. N., Karnoup, A., Diem, M., and Schweitzer-Stenner, R. (2001). *J. Am. Chem. Soc.* **123,** 11775–11781.

Baello, B. (2000). Ph.D. Thesis. Department of Chemistry, University of Illinois at Chicago.

Baello, B., Pancoska, P., and Keiderling, T. A. (2000). *Anal. Biochem.* **280,** 46–57.

Baello, B. I., Pancoska, P., and Keiderling, T. A. (1997). *Anal. Biochem.* **250,** 212–221.

Barron, L. D., Blanch, E. W., and Hecht, L. (2002). *Adv. Protein Chem.* **62,** 51–90.

Barron, L. D., Hecht, L., and Bell, A. D. (1996). In *Circular Dichroism and the Conformational Analysis of Biomolecules* (G. D. Fasman, ed.), pp. 653–695. Plenum, New York.

Baumruk, V., Huo, D. F., Dukor, R. K., Keiderling, T. A., Lelievre, D., and Brack, A. (1994). *Biopolymers* **34,** 1115–1121.

Baumruk, V., and Keiderling, T. A. (1993). *J. Am. Chem. Soc.* **115,** 6939–6942.

Baumruk, V., Pancoska, P., and Keiderling, T. A. (1996). *J. Mol. Biol.* **259,** 774–791.

Birke, S. S., Farrell, C., Lee, O., Agbaje, I., Roberts, G. M., and Diem, M. (1991). In *Spectroscopy of Biological Molecules* (R. E. Hester and R. B. Girling, eds.), pp. 131–132. Royal Society of Chemistry, Cambridge.

Blout, E. R., De Loze, C., and Asadourian, A. (1961). *J. Am. Chem. Soc.* **83,** 1895–1900.

Bour, P., and Keiderling, T. A. (1993). *J. Am. Chem. Soc.* **115,** 9602–9607.

Bour, P., Kubelka, J., and Keiderling, T. A. (2000). *Biopolymers* **53,** 380–395.

Bour, P., Sopkova, J., Bednarova, L., Malon, P., and Keiderling, T. A. (1997). *J. Comput. Chem.* **18,** 646–659.

Brauner, J. W., Dugan, C., and Mendelsohn, R. (2000). *J. Am. Chem. Soc.* **122,** 677–683.

Byler, D. M., Lee, D. L., and Randall, C. S. (2000). In *Infrared Analysis of Peptides and Proteins: Principles and Applications* (B. R. Singh, ed.), pp. 145–158. American Chemical Society, Washington D.C.

Byler, D. M., and Purcell, J. M. (1989). *SPIE Fourier Transform Spectrosc.* **1145,** 539–544.

Byler, D. M., and Susi, H. (1986). *Biopolymers* **25,** 469–487.

Byler, D. M., Wilson, R. M., Randall, C. S., and Sokoloski, T. D. (1995). *Pharmaceut. Res.* **12,** 446–450.

Chen, C. C., Polavarapu, P. L., and Weibel, S. (1994a). *Appl. Spectrosc.* **48,** 1218–1223.

Chen, X. G., Schweitzer-Stenner, R., Krimm, S., Mirkin, N. G., and Asher, S. A. (1994b). *J. Am. Chem. Soc.* **116,** 11141–11142.

Chirgadze, Y. N., and Nevskava, N. A. (1976a). *Bioploymers* **15,** 607–625.

Chirgadze, Y. N., and Nevskaya, N. A. (1976b). *Biopolymers* **15,** 627–636.

Clark, A. H., Saunderson, D. H. P., and Sugget, A. (1981). *Int. J. Peptide Res.* **17,** 353–364.

Cowen, B. R., and Hochstrasser, R. M. (1996). In *Infrared Spectroscopy of Biomolecules,* pp. 107–129. Wiley-Liss, Chichester, UK.

De Jongh, H. H. J., Goormaghtigh, E., and Ruysschaert, J. M. (1996). *Anal. Biochem.* **242,** 95–103.

Decatur, S. M., and Antonic, J. (1999). *J. Am. Chem. Soc.* **121,** 11914–11915.

Diem, M. (1991). *Vib. Spectra Struct.* **19,** 1–54.

Diem, M. (1994). In *Techniques and Instrumentation in Analytical Chemistry* (N. Purdie, and H. G. Brittain, eds.), Vol. 4, pp. 91–130. Elsevier, Amsterdam.

Diem, M., Lee, O., and Roberts, G. M. (1992). *J. Phys. Chem.* **96,** 548–554.

Diem, M., Roberts, G. M., Lee, O., and Barlow, O. (1988). *Appl. Spectrosc.* **42,** 20–27.

Dluhy, R. A. (2000). *Appl. Spectrosc.* **35,** 315–351.

Dluhy, R. A., Stephens, S. M., Widayati, S., and Williams, A. D. (1995). *Spectrochim. Acta, Part A* **51,** 1413–1447.

Dousseau, F., and Pezolet, M. (1990). *Biochemistry* **29,** 8771–8779.

Dukor, R. K. (1991). Ph.D. thesis. Department of Chemistry, University of Illinois at Chicago.

Dukor, R. K., and Keiderling, T. A. (1991). *Biopolymers* **31,** 1747–1761.

Dukor, R. K., and Keiderling, T. A. (1996). *Biospectroscopy* **2,** 83–100.

Dukor, R. K., Keiderling, T. A., and Gut, V. (1991). *Int. J. Peptide Protein Res.* **38,** 198–203.

Dyer, R. B., Gai, F., and Woodruff, W. H. (1998). *Acc. Chem. Res.* **31,** 709–716.

Elliott, A. (1954). *Proc. Royal Soc. Lond. A* **226,** 408–421.

Elliott, A., and Ambrose, E. J. (1950). *Nature* **165,** 921–922.

Fabian, H., Reinstadler, D., Zhang, M., Vogel, H., Naumann, D., and Mantsch, H. H. (1995). In *Spectroscopy of Biological Molecules* (J. C. Merlin, S. Turrell, and J. P. Huvenne, eds.), pp. 83–84. Kluwer Academic Publishers, Dordrecht.

Fabian, H., Schultz, C., Backmann, J., Hahn, U., Saenger, W., Mantsch, H. H., and Naumann, D. (1994). *Biochemistry* **33,** 10725–10730.

Fabian, H., Schultz, C., Naumann, D., Landt, O., Hahn, U., and Saenger, W. (1993). *J. Mol. Biol.* **232,** 967–981.

Fasman, G. D. (1996). *Circular Dichroism and the Conformational Analysis of Biomolecules.* Plenum, New York.

Fink, A. L. (1998). *Folding* & *Design* **3,** R9–R23.

Fink, A. L., Seshadri, S., Khurana, R., and Oberg, K. A. (2000). In *Infrared Analysis of Peptides and Proteins: Principles and Applications* (B. R. Singh ed.). pp. 132–144. American Chemical Society, Washington, D.C.

Flach, C. R., Gericke, A., Keough, K. M. W., and Mendelsohn, R. (1999). *Biochim. Biophys. Acta* **1416,** 11–20.

Freedman, T. B., and Nafie, L. A. (1994). In *Modern Nonlinear Optics* (M. Evans and S. Kielich eds.), Vol. 3, pp. 207–263. Wiley, New York.

Freedman, T. B., Nafie, L. A., and Keiderling, T. A. (1995). *Biopolymers* **37,** 265–279.

Goormaghtigh, E., and Ruysschaert, J. M. (2000). In *Infrared Analysis of Peptides and Proteins: Principles and Applications* (B. R. Singh, ed.). pp. 96–116. American Chemical Society, Washington, D.C.

Graff, D. K., Pastrana-Rios, B., Venyaminov, S. Y., and Prendergast, F. G. (1997). *J. Am. Chem. Soc.* **119,** 11282–11294.

Greenfield, N. J. (1996). *Anal. Biochem.* **235,** 1–10.

Gupta, V. P., and Keiderling, T. A. (1992). *Biopolymers* **32,** 239–248.

Haas, E. (1995). In *Spectroscopic Methods for Determining Protein Structure in Solution* (H. H. Havel, ed.), pp. 28–61. VCH, New York.

Hadden, J. M., Chapman, D., and Lee, D. C. (1995). *Biochim. Biophys. Acta* **1248**, 115–122.

Haris, P. I. (2000). In *Infrared Analysis of Peptides and Proteins: Principles and Applications* (B. R. Singh, ed.), pp. 54–95. American Chemical Society, Washington, D.C.

Haris, P. I., and Chapman, D. (1995). *Biopolymers* **37**, 251–263.

Havel, H. H. (1996). *Spectroscopic Methods for Determining Protein Structure in Solution.* VCH, New York.

He, Y. N., Cao, X. L., Nafie, L. A., and Freeman, T. B. (2001). *J. Am. Chem. Soc.* **123**, 11320–11321.

Hennessey, J. P., and Johnson, W. C. (1981). *Biochemistry* **20**, 1085–1094.

Hilario, J., Drapcho, D., Curbelo, R., and Keiderling, T. A. (2001). *Appl. Spectrosc.* **55**, 1435–1447.

Hilario, J., Kubelka, J., Syud, F. A., Gellman, S. H., and Keiderling, T. A. (2002). *Biospectroscopy.* **67**, 233–236.

Jackson, M., Haris, P. I., and Chapman, D. (1991). *Biochemistry* **30**, 9681–9686.

Jackson, M., and Mantsch, H. H. (1995). *Crit. Rev. Biochem. Mol. Biol.* **30**, 95–120.

Johnson, W. C. (1988). *Ann. Rev. Biophys. Chem.* **17**, 145–166.

Johnson, W. C. (1992). *Methods Enzymol.* **210**, 426–447.

Johnson, W. C. (1999). *Proteins* **35**, 307–312.

Kalnin, N. N., Baikalov, I. A., and Venyaminov, S. Y. (1990). *Biopolymers* **30**, 1273–1280.

Kauppinen, J. K., Moffatt, D. J., Mantsch, H. H., and Cameron, D. G. (1981). *Appl. Spectrosc.* **35**, 271–276.

Keiderling, T. A. (1990). In *Practical Fourier Transform Infrared Spectroscopy* (K. Krishnan and J. R. Ferraro, eds.), pp. 203–284. Academic Press, San Diego.

Keiderling, T. A. (1996). In *Circular Dichroism and the Conformational Analysis of Biomolecules* (G. D. Fasman, ed.), pp. 555–598. Plenum, New York.

Keiderling, T. A. (2000). In *Circular Dichroism: Principles and Applications* (N. Berova, K. Nakanishi, and R. A. Woody, eds.), pp. 621–666. Wiley-VCH, New York.

Keiderling, T. A., Hilario, J., and Kubelka, J. In *Vibrational Spectroscopy of Polymers and Biological Systems* (M. Braiman and V. Gregoriou, eds.) Marcel Dekker, Amsterdam–New York (in press).

Keiderling, T. A., and Pancoska, P. (1993). In *Biomolecular Spectroscopy, Part B* (R. E. Hester and R. J. H. Clarke eds.), Vol. 21, pp. 267–315. Wiley, Chichester.

Keiderling, T. A., Silva, R. A. G. D., Decatur, S. M., and Bour, P. (1999a). In *Spectrocopy of Biological Molecules: New Directions* (J. Greve, G. J. Puppels, and C. Otto, eds.), pp. 63–64. Kluwer Academic Publishers, Dordrecht.

Keiderling, T. A., Silva, R. A. G. D., Yoder, G., and Dukor, R. K. (1999b). *Bioorg. Med. Chem.* **7**, 133–141.

Keiderling, T. A., Wang, B., Urbanova, M., Pancoska, P., and Dukor, R. K. (1994). *Faraday Disc.* **99**, 263–286.

Khurana, R., and Fink, A. L. (2000). *Biophys. J.* **78**, 994–1000.

Knapp-Mohammady, M., Jalkanen, K. J., Nardi, F., Wade, R. C., and Suhai, S. (1999). *Chem. Phys.* **240**, 63–77.

Kobrinskaya, R., Yasui, S. C., and Keiderling, T. A. (1988). In *Peptides, Chemistry and Biology, Proceedings of the 10th American Peptide Symposium* (G. R. Marshall, ed.), pp. 65–66. ESCOM, Leiden.

Krimm, S. (2000). In *Infrared Analysis of Peptides and Proteins: Principles and Applications* (B. R. Singh, ed.), pp. 38–53. American Chemical Society, Washington, D.C.

Krimm, S., and Bandekar, J. (1986). *Adv. Protein Chem.* **38**, 181–364.

Krimm, S., and Reisdorf Jr., W. C. (1994). *Faraday Disc.* **99**, 181–194.

Kubelka, J. (2002). Ph.D Thesis. Department of Chemistry, University of Illinois at Chicago.

Kubelka, J., and Keiderling, T. A. (2001a). *J. Phys. Chem.* **105,** 10922–10928.

Kubelka, J., and Keiderling, T. A. (2001b). *J. Am. Chem. Soc.* **123,** 6142–6150.

Kubelka, J., and Keiderling, T. A. (2001c). *J. Am. Chem. Soc.* **123,** 12048–12058.

Kubelka, J., Pancoska, P., and Keiderling, T. A. (1999a). *Appl. Spectrosc.* **53,** 666–671.

Kubelka, J., Pancoska, P., and Keiderling, T. A. (1999b). In *Spectroscopy of Biological Molecules: New Directions* (J. Greve, G. J. Puppels, and C. Otto, eds.), pp. 67–68. Kluwer Academic Publishers, Dordrecht.

Kubelka, J., Silva, R. A. G. D., Bour, P., Decatur, S. M., and Keiderling, T. A. (2002). In *The Physical Chemistry of Chirality* (J. M. Hicks, ed.). pp. 50–64, American Chemical Society, Washington, D.C.

Lal, B. B., and Nafie, L. A. (1982). *Biopolymers* **21,** 2161–2183.

Lednev, I., K., Karnoup, A. S., Sparrow, M. C., and Asher, S. A. (2001). *J. Am. Chem. Soc.* **123,** 2388–2392.

Lee, D. C., Haris, P. I., Chapman, D., and Mitchell, R. C. (1990). *Biochemistry* **29,** 9185–9193.

Lim, M., and Hochstrasser, R. M. (2001). *Ultrafast Infrared Raman Spectrosc.* **26,** 273–374.

Long, F., Freedman, T. B., Hapanowicz, R., and Nafie, L. A. (1997a). *Appl. Spectrosc.* **51,** 504–507.

Long, F., Freedman, T. B., Tague, T. J., and Nafie, L. A. (1997b). *Appl. Spectrosc.* **51,** 508–511.

Malinowski, E. R. (1991). *Factor Analysis in Chemistry.* Wiley, New York.

Malon, P., and Keiderling, T. A. (1988). *Appl. Spectrosc.* **42,** 32–38.

Malon, P., and Keiderling, T. A. (1996). *Appl. Spectrosc.* **50,** 669–674.

Malon, P., and Keiderling, T. A. (1997). *Appl. Optics* **36,** 6141–6148.

Malon, P., Kobrinskaya, R., and Keiderling, T. A. (1988). *Biopolymers* **27,** 733–746.

Manas, E. S., Getahun, Z., Wright, W. W., Degrado, W. F., and Vanderkooi, J. M. (2000). *J. Am. Chem. Soc.* **122,** 9883–9890.

Manavalan, P., and Johnson, Jr., W. C. (1987). *Anal. Biochem.* **167,** 76–85.

Mantsch, H. H., Casel, H. L., and Jones, R. N. (1986). In *Spectroscopy of Biological Systems* R. J. H. Clark and R. E. Hester, eds.), Vol. 13, pp. 1–46. Wiley, London.

Mantsch, H. H., and Chapman, D. (1996). *Infrared Spectroscopy of Biomolecules.* Wiley-Liss, Chichester, UK.

Marcott, C. (1984). *Appl. Spectrosc.* **38,** 442–443.

Mastle, W., Dukor, R. K., Yoder, G., and Keiderling, T. A. (1995). *Biopolymers* **36,** 623–631.

Masuda, Y., Fukushima, K., Fujii, T., and Miyazawa, T. (1969). *Biopolymers* **8,** 91–99.

Millhauser, G. L., Stenland, C. J., Hanson, P., Bolin, K. A., and Vandeven, F. J. M. (1997). *J. Mol. Biol.* **267,** 963–974.

Miyazawa, T., Shimanouchi, T., and Mizushima, S. (1958). *J. Chem. Phys.* **29,** 611–616.

Nafie, L. A. (1994). In *Modern Nonlinear Optics, Part 3* (M. Evans and S. Kielich eds.), Vol. 85, pp. 105–206. Wiley, New York.

Nafie, L. A. (1997). *Ann. Rev. Phys. Chem.* **48,** 357–386.

Nafie, L. A. (2000). *App. Spectrosc.* **54,** 1634–1645.

Nafie, L. A., and Dukor, R. K. (2000). In *The Physical Chemistry of Chirality* (J. M. Hicks ed.). pp. 79–88, American Chemical Society, Washington, D.C.

Nafie, L. A., Keiderling, T. A., and Stephens, P. J. (1976). *J. Am. Chem. Soc.* **98,** 2715–2723.

Nafie, L. A., Yu, G. S., Qu, X., and Freedman, T. B. (1994). *Faraday Disc.* **99,** 13–34.

Narayanan, U., Keiderling, T. A., Bonora, G. M., and Toniolo, C. (1985). *Biopolymers* **24**, 1257–1263.
Narayanan, U., Keiderling, T. A., Bonora, G. M., and Toniolo, C. (1986). *J. Am. Chem. Soc.* **108**, 2431–2437.
Nevskaya, N. A., and Chirgadze, Y. N. (1976). *Biopolymers* **15**, 637–648.
Noda, I. (1990). *Appl. Spectrosc.* **44**, 550–561.
Noda, I., Liu, Y., and Ozaki, J. (1996a). *J. Phys. Chem.* **100**, 8665–8673.
Noda, I., Liu, Y., and Ozaki, Y. (1996b). *J. Phys. Chem.* **100**, 8674–8680.
Pancoska, P. (2000). In *Encyclopedia of Analytical Chemistry* (R. A. Meyers, ed.), pp. 244–383. John Wiley & Sons Ltd., Chichester.
Pancoska, P., Bitto, E., Janota, V., and Keiderling, T. A. (1994). *Faraday Disc.* **99**, 287–310.
Pancoska, P., Bitto, E., Janota, V., Urbanova, M., Gupta, V. P., and Keiderling, T. A. (1995). *Protein Sci.* **4**, 1384–1401.
Pancoska, P., Blazek, M., and Keiderling, T. A. (1992). *Biochemistry* **31**, 10250–10257.
Pancoska, P., Fabian, H., Yoder, G., Baumruk, V., and Keiderling, T. A. (1996). *Biochemistry* **35**, 13094–13106.
Pancoska, P., Fric, I., and Blaha, K. (1979). *Collect. Czech. Chem. Commun.* **44**, 1296–1312.
Pancoska, P., Janota, V., and Keiderling, T. A. (1999a). *Anal. Biochem.* **267**, 72–83.
Pancoska, P., and Keiderling, T. A. (1991). *Biochemistry* **30**, 6885–6895.
Pancoska, P., Kubelka, J., and Keiderling, T. A. (1999b). *Appl. Spectrosc.* **53**, 655–665.
Pancoska, P., Wang, L., and Keiderling, T. A. (1993). *Protein Sci.* **2**, 411–419.
Pancoska, P., Yasui, S. C., and Keiderling, T. A. (1989). *Biochemistry* **28**, 5917–5923.
Pancoska, P., Yasui, S. C., and Keiderling, T. A. (1991). *Biochemistry* **30**, 5089–5103.
Parker, F. S. (1983). *Applications of Infrared, Raman and Resonance Raman Spectroscopy.* Plenum, New York.
Paterlini, M. G., Freedman, T. B., and Nafie, L. A. (1986). *Biopolymers* **25**, 1751–1765.
Phillips, C. M., Mizutani, Y., and Hochstrasser, R. M. (1995). *Proc. Natl. Acad. Sci. USA* **92**, 7292–7296.
Polavarapu, P. L. (1985). In *Fourier Transform Infrared Spectroscopy* (J. R. Ferraro and L. Basile eds.), Vol. 4, pp. 61–96. Acadamic Press, New York.
Polavarapu, P. L. (1989). *Vib. Spectra Struct.* **17B**, 319–342.
Polavarapu, P. L. (1998). *Vibrational Spectra: Principles and Applications with Emphasis on Optical Activity.* Elsevier, New York.
Pribic, R., Van Stokkum, I. H. M., Chapman, D., Haris, P. I., and Bloemendal, M. (1993). *Anal. Biochem.* **214**, 366–378.
Provencher, S. W., and Glockner, J. (1981). *Biochemistry* **20**, 33–37.
Qian, W. L., and Krimm, S. (1994). *J. Phys. Chem.* **98**, 9992–10000.
Rauk, A. (1991). In *New Developments in Molecular Chirality* (P. B. Mezey, ed.), pp. 57–92. Kluwer Academic Publishers, Dordrecht.
Roberts, G. M., Lee, O., Calienni, J., and Diem, M. (1988). *J. Am. Chem. Soc.* **110**, 1749–1752.
Sarver, R. W., and Kruger, W. C. (1991). *Anal. Biochem.* **199**, 61–67.
Sen, A. C., and Keiderling, T. A. (1984a). *Biopolymers* **23**, 1519–1532.
Sen, A. C., and Keiderling, T. A. (1984b). *Biopolymers* **23**, 1533–1545.
Setnicka, V., Urbanova, M., Bour, P., and Volka, K. (2001). *J. Phys. Chem.* **105**, 8931–8938.
Siebert, F. (1996). In *Infrared Spectroscopy of Biomolecules* (H. H. Mantsch and D. Chapman, eds.), pp. 83–106. Wiley-Liss, Chichester, UK.
Sieler, G., and Schweitzer-Stenner, R. (1997). *J. Am. Chem. Soc.* **119**, 1720–1726.

Silva, R. A. G. D. (2000). Ph.D. Thesis Department of Chemistry, University of Illinois at Chicago.

Silva, R. A. G. D., Kubelka, J., Decatur, S. M., Bour, P., and Keiderling, T. A. (2000a). *Proc. Natl. Acad. Sci. USA* **97,** 8318–8323.

Silva, R. A. G. D., Sherman, S. A., Perini, F., Bedows, E., and Keiderling, T. A. (2000b). *J. Am. Chem. Soc.* **122,** 8623–8630.

Singh, B. R., ed. (2000). *Infrared Analysis of Peptides and Proteins: Principles and Applications,* ACS Symposium Series. American Chemical Society, Washington, D.C.

Singh, R. D., and Keiderling, T. A. (1981). *Biopolymers* **20,** 237–240.

Snir, J., Frankel, R. A., and Schellman, J. A. (1975). *Biopolymers* **14,** 173–196.

Sreerama, M., Venyaminov, S. Y., and Woody, R. W. (2000). *Anal. Biochem.* **287,** 243–251.

Sreerama, N., Venyaminov, S. Y., and Woody, R. W. (1999). *Protein Sci.* **8,** 370–380.

Sreerama, N., and Woody, R. W. (1993). *Anal. Biochem.* **209,** 32–44.

Sreerama, N., and Woody, R. W. (1994). *J. Mol. Biol.* **242,** 497–507.

Stelea, S. (2002). Ph.D. Thesis. Department of Chemistry, University of Illinois at Chicago.

Stelea, S., Keiderling, T. A., and Pancoska, P. (2002). *Biophys. J.* **83,** (in press).

Stelea, S. D., Pancoska, P., Benight, A. S., and Keiderling, T. A. (2001). *Protein Sci.* **10,** 970–978.

Stephens, P. J. (1985). *J. Phys. Chem.* **89,** 748.

Stephens, P. J. (1987). *J. Phys. Chem.* **91,** 1712.

Stephens, P. J., and Devlin, F. J. (2000). *Chirality* **12,** 172–179.

Stephens, P. J., Devlin, F. J., and Amouche, A. (2002). In *The Physical Chemistry of Chirality,* (M. Hicks ed.), pp. 18–33. American Chemical Society, Washington, D.C.

Stephens, P. J., Devlin, F. J., Ashvar, C. S., Chabalowski, C. F., and Frisch, M. J. (1994). *Faraday Disc.* **99,** 103–119.

Stephens, P. J., and Lowe, M. A. (1985). *Ann. Rev. Phys. Chem.* **36,** 213–241.

Su, C. N., Heintz, V., and Keiderling, T. A. (1981). *Chem. Phys. Lett.* **73,** 157–159.

Surewicz, W., Mantsch, H. H., and Chapman, D. (1993). *Biochemistry* **32,** 389–394.

Surewicz, W. K., and Mantsch, H. H. (1988). *Biochem. Biophys. Acta* **952,** 115–130.

Tam, C. N., Bour, P., and Keiderling, T. A. (1996). *J. Am. Chem. Soc.* **118,** 10285–10293.

Thompson, P. A., Eaton, W. A., and Hofrichter, J. (1997). *Biochemistry* **36,** 9200–9210.

Tiffany, M. L., and Krimm, S. (1972). *Biopolymers* **11,** 2309–2316.

Torii, H., and Tasumi, M. (1992). *J. Chem. Phys.* **96,** 3379–3387.

Torii, H., and Tasumi, M. (1996). In *Infrared Spectroscopy of Biomolecules* (H. Henry, H. H. Mantsch, and D. Chapman eds.), pp. 1–18. Wiley-Liss, Chichester.

Trewhella, J., Liddle, W. K., Heidorn, D. B., and Strynadka, N. (1989). *Biochemistry* **28,** 1294–1301.

Tu, A. T. (1982). *Raman Spectroscopy in Biology.* Wiley, New York.

Urbanova, M., Dukor, R. K., Pancoska, P., Gupta, V. P., and Keiderling, T. A. (1991). *Biochemistry* **30,** 10479–10485.

Urbanova, M., Keiderling, T. A., and Pancoska, P. (1996). *Bioelectrochem. Bioenerg.* **41,** 77–80.

Urbanova, M., Setnicka, V., and Volka, K. (2000). *Chirality* **12,** 199–203.

Van Stokkum, I. H. M., Linsdell, H., Hadden, J. M., Haris, P. I., Chapman, D., and Bloemendal, M. (1995). *Biochemistry* **34,** 10508–10518.

Van Stokkum, I. H. M., Spoelder, H. J. W., Bloemendal, M., Van Grondelle, R., and Groen, F. C. A. (1990). *Anal. Biochem.* **191,** 110–118.

Venyaminov, S. Y., Hedstrom, J. F., and Prendergast, F. G. (2001). *Proteins* **45,** 81–89.

Venyaminov, S. Y., and Kalnin, N. N. (1990). *Biopolymers* **30,** 1259–1271.

Venyaminov, S. Y., and Yang, J. T. (1996). In *Circular Dichroism and the Conformational Analysis of Biomolecules* (G. D. Fasman, ed.), pp. 69–107. Plenum, New York.

Wang, B., and Keiderling, T. A. (1995). *Appl. Spectrosc.* **49**, 1347–1355.
White, A. J., Drabble, K., and Wharton, C. W. (1995). *Biochem. J.* **306**, 843–849.
Wi, S., Pancoska, P., and Keiderling, T. A. (1998). *Biospectroscopy* **4**, 93–106.
Williams, R. W., and Dunker, A. K. (1981). *J. Mol. Biol.* **152**, 783–813.
Williams, S., Causgrove, T. P., Gilmanshin, R., Fang, K. S., Callender, R. H., Woodruff, W. H., and Dyer, R. B. (1996). *Biochemistry* **35**, 691–697.
Woody, R. W. (1996). In *Circular Dichroism and the Conformational Analysis of Biomolecules* (G. D. Fasman, ed.), pp. 25–67. Plenum, New York.
Woody, R. W., and Dunker, A. K. (1996). In *Circular Dichroism and the Conformational Analysis of Biomolecules* (G. D. Fasman, ed.), pp. 109–157. Plenum, New York.
Woody, R. W., and Kallenbach, N. R. (2002). *Adv. Protein Chem.* **62**, 163–240.
Wyssbrod, H. R., and Diem, M. (1992). *Biopolymers* **31**, 1237.
Xie, P., Zhou, Q., and Diem, M. (1995). *Faraday Disc.* **99**, 233–244.
Yasui, S. C., and Keiderling, T. A. (1986a). *J. Am. Chem. Soc.* **108**, 5576–5581.
Yasui, S. C., and Keiderling, T. A. (1986b). *Biopolymers* **25**, 5–15.
Yasui, S. C., and Keiderling, T. A. (1988). In *Peptides: Chemistry and Biology,* Proceedings of the 10th American Peptide Symposium (G. R. Marshall, ed.), pp. 90–92. ESCOM, Leiden.
Yasui, S. C., Keiderling, T. A., Bonora, G. M., and Toniolo, C. (1986a). *Biopolymers* **25**, 79–89.
Yasui, S. C., Keiderling, T. A., Formaggio, F., Bonora, G. M., and Toniolo, C. (1986b). *J. Am. Chem. Soc.* **108**, 4988–4993.
Yasui, S. C., Keiderling, T. A., and Sisido, M. (1987). *Macromolecules* **20**, 403–2406.
Yoder, G. (1997). Ph.D. Thesis Department of Chemistry, University of Illinois at Chicago.
Yoder, G., Keiderling, T. A., Formaggio, F., and Crisma, M. (1995a). *Tetrahedron Asymm.* **6**, 687–690.
Yoder, G., Keiderling, T. A., Formaggio, F., Crisma, M., and Toniolo, C. (1995b). *Biopolymers* **35**, 103–111.
Yoder, G., Pancoska, P., and Keiderling, T. A. (1997a). *Biochemistry* **36**, 15123–15133.
Yoder, G., Polese, A., Silva, R. A. G. D., Formaggio, F., Crisma, M., Broxterman, Q. B., Kamphuis, J., Toniolo, C., and Keiderling, T. A. (1997b). *J. Am. Chem. Soc.* **119**, 10278–10285.
Yoo, R. K., Wang, B., Croatto, P. V., and Keiderling, T. A. (1991). *Appl. Spectrosc.* **45**, 231–236.
Zhao, C. X., and Polavarapu, P. L. (2001). *Biopolymers* **62**, 336–340.

IS POLYPROLINE II A MAJOR BACKBONE CONFORMATION IN UNFOLDED PROTEINS?

By ZHENGSHUANG SHI,* ROBERT W. WOODY,† and NEVILLE R. KALLENBACH*

*Department of Chemistry, New York University, New York 10003; and †Department of Biochemistry and Molecular Biology, Colorado Sate University, Fort Collins, Colorado 80523

I. Introduction

Protein folding is a process by which a polypeptide chain acquires its native structure from an unfolded state through a transition state. This process is the subject of active investigation in many laboratories. Much is known about native structures of proteins by X-ray crystallography and NMR spectroscopy. The structures of the transition states for folding a number of proteins have also been mapped using a combination of protein engineering and kinetic Φ-value analysis (Matouschek *et al.*, 1989). Less is known about the structures of unfolded states. It is now widely appreciated that understanding the mechanism of protein folding requires a detailed structural characterization of the starting state (Arcus *et al.*, 1994, 1995; Bai *et al.*, 2001; Neri *et al.*, 1992a,b; Serrano, 1995; Shortle and Ackerman, 2001; Smith *et al.*, 1996; Yao *et al.*, 2001; Zhang *et al.*, 1997a). Tanford's pioneering experiments on denatured states of proteins in 6 M Gdm·HCl (Lapanje and Tanford, 1967; Nozaki and Tanford, 1967; Tanford, 1968; Tanford *et al.*, 1967a,b) were interpreted on the basis of a polymeric random coil model. These studies helped foster the general belief that the unfolded state represents a vast ensemble of random

ADVANCES IN
PROTEIN CHEMISTRY, Vol. 62

conformations, one idea that led Levinthal to frame his famous paradox (Levinthal, 1968).

Many recent studies of the unfolded states of proteins are actually based on a modification of the random coil model (Schwalbe *et al.*, 1997; Smith *et al.*, 1996), recognizing that in many cases some residual native or non-native structure persists (Bond *et al.*, 1997; Gillespie and Shortle, 1997; Penkett *et al.*, 1998; Schwalbe *et al.*, 1997; Wong *et al.*, 1996; Zhang and Forman-Kay, 1997; Zhang *et al.*, 1997b). An important question then is, are all allowed backbone conformations in unfolded proteins occupied in proportion to their probabilities in the Ramachandran map as implied by the random coil model? Obviously, the presence of residual secondary or tertiary structure argues against this possibility; but we can ask the question more precisely in the context of short peptides that are predominantly unfolded under native folding conditions. Combined evidence from the theoretical study of a blocked alanine peptide in aqueous solution (Han *et al.*, 1998) and a variety of spectroscopic studies, including ultraviolet circular dichroism (CD) (Rucker and Creamer, 2002), nuclear magnetic resonance (NMR) (Poon *et al.*, 2000), two-dimensional vibrational spectroscopy (Woutersen and Hamm, 2000, 2001), vibrational circular dichroism (VCD) (Keiderling *et al.*, 1999), and vibrational Raman optical activity (VROA) (Blanch *et al.*, 2000; Schweitzer-Stenner *et al.*, 2001), reveal that the polyproline II (P_{II}) conformation is the dominant conformation in a variety of short model peptides. This conclusion has important implications for the structure in the unfolded state of proteins in general. Several excellent reviews on CD (Woody, 1992), VCD (Keiderling *et al.*, 1999), and VROA (Barron *et al.*, 2000) discuss the conformations of the unfolded states of proteins. In Section II, we discuss the evidence from short peptides. In Section III, we review the circular dichroism of unfolded proteins and address the role of P_{II} in unfolded proteins. Before piecing together recent lines of evidence from short peptides, we first recapitulate some of the early CD and related studies.

II. POLYPROLINE II DOMINATES IN SHORT PEPTIDES

A. *Evidence from Circular Dichroism Studies*

Early CD Studies on Polypeptides

More than 36 years ago, the pH-induced cooperative transitions in CD spectra and other physical properties of poly-L-glutamic acid (PGA) and poly-L-lysine (PL) were reported and taken to correspond to a helix-to-coil transition (Carver *et al.*, 1966; Holzwarth and Doty, 1965; Legrand and Viennet, 1964; Velluz and Legrand, 1965). The CD spectra of PGA and PL at low degrees of ionization have characteristics of CD spectra assigned to the α-helix (negative bands at 222 and 208 nm, strong positive band at 192 nm). On the other hand, those of the charged forms have bands near 238 nm (negative), 218 nm (positive), and 198 nm (negative) (see Fig. 1B and 1C), similar at longer wavelengths to spectra observed for unfolded protein in 6 M Gdm·HCl. As a result, the CD spectra at the extremes of transitions were initially believed to represent those of α-helix and of putative "random coils." Although ample evidence supported the assignment of the α-helix spectrum from fiber diffraction and viscosity data, the assignment of the "random coil" spectrum was less secure and was subsequently challenged in a series of papers by Tiffany and Krimm (Krimm and Mark, 1968; Tiffany and Krimm, 1968a,c). They emphasized that the "random coil" spectrum resembled that of poly(Pro), and was too strong to correspond to a true random coil in any case. This last point had been noted earlier by Schellman and Schellman (1964) and Beychok (1967). This proposal initiated a controversy with far-reaching implications for the mechanism of protein folding that is still not fully resolved.

Before discussing details of their model and others, it is useful to review the two main techniques used to infer the characteristics of chain conformation in "unordered" polypeptides. One line of evidence came from hydrodynamic experiments—viscosity and sedimentation—from which a statistical end-to-end distance could be estimated and compared with values derived from calculations on polymer chain models (Flory, 1969). The second is based on spectroscopic experiments, in particular CD spectroscopy, from which information is obtained about the local chain conformation rather than global properties such as those derived from hydrodynamics. It is entirely possible for a polypeptide chain to adopt some particular local structure while retaining characteristics of "random coils" derived from hydrodynamic measurements; this was pointed out by Krimm and Tiffany (1974). In support of their proposal, Tiffany and Krimm noted the following points:

1. The P_{II} conformation of poly-L-proline (PP) or collagen in the solid state could be identified from X-ray fiber diffraction results (Cowan and McGavin, 1955). Persistence of this basic structure in solution was inferred from the resemblance between the CD spectra of solutions and films of the polypeptide. The CD spectra of the charged forms of PGA and PL closely resemble that of P_{II} (compare Fig. 1B, 1C, and 1D); however, these spectra differ significantly from those of PP peptides at high temperature or in the presence of high concentration of salts

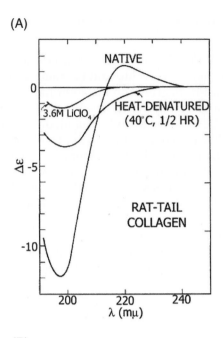

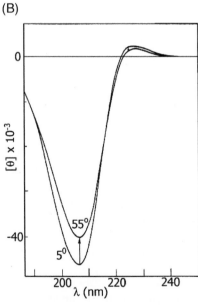

FIG. 1.   (A) CD spectra of rat-tail collagen in native state, heat-denatured state, and in 3.6 M LiClO₄. (From Tiffany and Krimm, 1969, *Biopolymers* **8,** 347–359, © 1969. Reprinted by permission of John Wiley & Sons, Inc.) (B) CD spectrum of poly-L-proline II (sigma, molecular weight 55,000) as a function of temperature. (C) CD spectrum of polyglutamic acid at pH 7 as a function of temperature. (D) CD spectrum of poly-L-lysine at pH 7 as a function of temperature. Parts (B–D) are from Tiffany and Krimm (1972). *Biopolymers* **11,** 2309–2316, © 1972. Reprinted by permission of John Wiley & Sons, Inc.

(C)

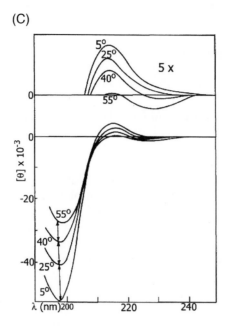

(D)

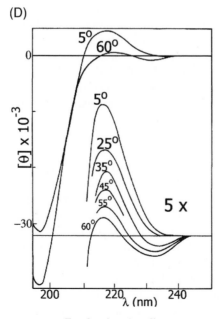

FIG. 1.   (*continued*)

such as $CaCl_2$ (Tiffany and Krimm, 1968a, 1969). The latter were then assigned to something approximating a random coil form of the polypeptide chain, a point that is less obvious. That there is less residual structure in PP peptides at high temperatures and/or high salt concentration is plausible, but it does not necessarily correspond to an ideal random coil system.

2. Tanford stressed that 6 M Gdm·HCl unfolds and thus disorders the structure of most proteins (Tanford, 1968). Based on this observation and the fact that the CD spectra of charged forms of PGA and PL are very similar to those of proteins in 6 M Gdm·HCl, the conclusion that the charged forms of PGA and PL adopt random coil structure was logical. The inference that Gdm·HCl leads to complete local disordering of peptides is gratuitous, however. One puzzling report by Blout and Fasman showed that the negative optical rotation and the intrinsic viscosity of PP in form II are increased by urea (Blout and Fasman, 1967). These observations are consistent with the idea that the solvent induces some kind of structural ordering. Tiffany and Krimm observed an increased intensity of the positive CD band at 218 nm in urea and both positive (218 nm) and negative (198 nm) bands in Gdm·HCl for PP peptides. They rationalized the result as due to the presence of locally ordered structure such as P_{II} lacking any corresponding long-range order (Tiffany and Krimm, 1973). This argument could also be used to explain results on poly[N^5-(ω-hydroxylalkyl)-L-glutamine] [with ethyl (PHEG) and propyl (PHPG) alkyl groups] (Adler *et al.*, 1968; Fasman *et al.*, 1970; Krimm and Tiffany, 1974; Lupulota *et al.*, 1965; Mattice *et al.*, 1972). Hydrodynamic evidence pointed to the fact that these polypeptides exist in a disordered structure, and they also show the characteristic "random coil" CD spectra. However the overall random coil conformation is not incompatible with local P_{II} structure, and this argument for existence of a "random coil" conformation is not convincing.

3. The assignment of the structure of the charged form of PGA to random coil is inconsistent with hydrodynamic experimental results (Doty *et al.*, 1957; Iizuka and Yang, 1965). The viscosity of PGA goes through a minimum near pH 5–6, rising at both lower and higher pHs (Doty *et al.*, 1957). The increased viscosity at low pH can be explained by formation of α-helical structure, but the increase at high pH cannot be explained by a random coil structure (Holzwarth and Doty, 1965). It is consistent with formation of another kind of structure such as extended threefold left-handed helical structure, P_{II} (Tiffany and Krimm, 1968b). In the high pH region the viscosity also increases with decreasing salt concentration (Iizuka and Yang, 1965), paralleling the structural changes that were observed from CD measurements.

4. Effects of temperature and salt concentration on the CD spectra of PGA and PL are in the same direction as for collagen and PP peptides, which can be attributed to P$_{II}$ structure (see Fig. 1) (Tiffany and Krimm, 1969, 1972).

5. Calculations by Krimm and Mark on PGA and PL showed that repulsive electrostatic interactions should favor a locally ordered structure with approximate threefold left-handed helical symmetry, designated P$_{II}$ (Krimm and Mark, 1968). Later, Krimm pointed out that electrostatic repulsion is not the only interaction that can give rise to extended P$_{II}$ structure (Tiffany and Krimm, 1972), so that there is no contradiction in finding the P$_{II}$ type of CD spectrum for systems in which electrostatic repulsion is not a factor (Tiffany and Krimm, 1973). In this paper, Tiffany and Krimm also proposed that water should play a special role in stabilizing the P$_{II}$ structure by hydrogen bonding to exposed carbonyl groups.

Corroborating results came from a study of neutral poly(Ala-Gly-Gly) (PAGG). In dilute aqueous solution PAGG exists in a polyglycine II structure (Andries et al., 1971; Lotz and Keith, 1971), a threefold left-handed helix similar to P$_{II}$, as characterized in the solid state by X-ray diffraction. CD spectra of PAGG can be obtained both in aqueous solution and from films made by casting (Rippon and Walton, 1971). These spectra are essentially the same, and similar to that of charged PGA, assigned to P$_{II}$. The effects of heating (Rippon and Walton, 1971, 1972) and salts (Rippon and Walton, 1972) on the CD spectrum of PAGG are similar to those on the spectra of charged forms of PGA and PL or PP peptides.

At this point, Mattice obtained striking results from circular dichroism studies of Ac-Ala-NHMe, Ac-Ala-OMe, Ac-Ala-Ala-OMe, and Ac-Ala-Ala-Ala-OMe that he claimed provided conclusive evidence that the characteristic CD spectra of the charged forms of PGA and PL are representative of random coil, and not P$_{II}$ or any specific structure (Mattice, 1974). His reasoning was as follows. First, the shape of the CD bands for each of these short linear peptides corresponds to the characteristic CD spectrum of PGA and PL in their charged forms, namely, a positive band in the range of 208–218 nm at 15°C in water (Fig. 2a–e). Second, heating reduces the intensity of the positive CD band of these short linear peptides (Fig. 2); this behavior is similar to that of PP peptides and was used by Tiffany and Krimm (Tiffany and Krimm, 1969, 1972) as part of the evidence to assign the structure of PGA and PL to extended P$_{II}$ conformation (see Fig. 1). Third, isothermal addition of salts produces changes in the circular dichroism of short linear peptides that parallel those seen in PGA and PL (Fig. 3). Crucial to this argument, Mattice believed that these short linear peptides can exist only as statistical random

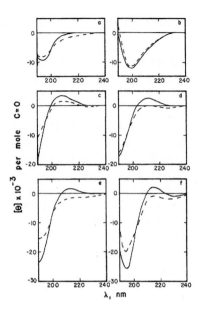

FIG. 2.   Circular dichroism in water at 15°C (solid line) and 75°C (dashed line). (a) Ac-Ala-NHMe. (b) cyclo(-Ala-Ala-). (c) Ac-Ala-OMe. (d) Ac-Ala-Ala-OMe. (e) Ac-Ala-Ala-Ala-OMe. (f) Poly-(N^5-ω-hydroxyethyl-L-glutamine). From Mattice (1974), *Biopolymers* **13**, 169–183, © 1974. Reprinted by permission of John Wiley & Sons, Inc.

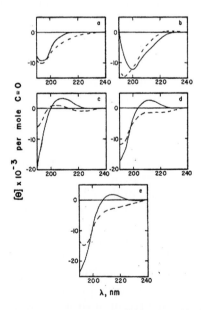

FIG. 3.   Circular dichroism at 15°C in water (solid line) and in 6 M sodium perchlorate (dashed line). (a) Ac-Ala-NHMe. (b) cyclo(-Ala-Ala-). (c) Ac-Ala-OMe. (d) Ac-Ala-Ala-OMe. (e) Ac-Ala-Ala-Ala-OMe. From Mattice (1974), *Biopolymers* **13**, 169–183, © 1974. Reprinted by permission of John Wiley & Sons, Inc.

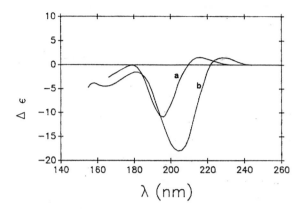

FIG. 4. Vacuum UV CD spectra of PGA (Johnson and Tinoco, 1972) and PP (Jenness *et al.*, 1976). The PGA spectrum was obtained in water at pH 8, and that of PP in trifluoroethanol. Both spectra were obtained at room temperature.

coils. Each line of evidence supporting Krimm's proposal seemed to collapse in the light of Mattice's results. However, his case rests on the unsubstantiated assumption that short uncharged linear Ala peptides do not form any kind of ordered structure. Recent new evidence from several sources shows that, in fact, they do (Han *et al.*, 1998; Poon *et al.*, 2000; Schweitzer-Stenner *et al.*, 2001; Woutersen and Hamm, 2000, 2001). This evidence is reviewed in detail below. It turns out that Mattice's spectra provide one more line of evidence to support Krimm's assignments of the CD spectra of P$_{II}$!

It is worth mentioning that two older experiments independently support Tiffany and Krimm's proposal. Johnson and co-workers extended the measurement of CD spectra of both PGA and PP into the vacuum UV region (below 170 nm) (Jenness *et al.*, 1976; Johnson and Tinoco, 1972). The resemblance between the CD spectra of these polypeptides extends to the vacuum UV region as shown in Figure 4. Both peptides show a second negative band, about one-fourth as intense as the negative band near 200 nm. In PGA, this band is at 175 nm while in PP it is at 165 nm. This difference probably reflects that between secondary and tertiary amides. Subsequently, Drake *et al.* (1988) reported the CD spectra of PL in 1,2-ethanediol/water mixtures over a temperature range from −105° to +82°C (Fig. 5). Over this extended temperature range, a tight isodichroic point at 203 nm is observed (Fig. 5B). The existence of this isodichroic point is consistent with a system that occupies two different states, one of which predominates at low temperature and the other at high temperature (Fig. 5C). Since it is unlikely that the peptide will become

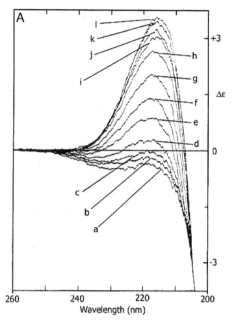

FIG. 5.   CD spectra of poly(L-Lys)•HBr in ethanediol/water (2:1), pH 7.6. (A) 260–200 nm region (note backward direction of wavelength scale): (a) 82, (b) 69, (c) 60, (d) 45, (e) 21, (f) 2, (g) −26, (h) −55, (i) −75, (j) −85, (k) −95, (l) −105°C. (B) 250–185 nm region: (a) 84, (b) 60, (c) 22, (d) −28, (e) −80, (f) −102°C. (C) Temperature variation of CD at 216 nm (solid line) ethanediol/water (2 : 1); (dotted line) water, pH 7.6; (dashed line) water, pH 7.6 with 5 M NaCl. Reprinted from Drake *et al.* (1988). *Biophys. Chem.* **31**, 143–146, © 1988, with permission from Elsevier Science.

more disordered on decreasing the temperature to −105°C, Drake *et al.* (1988) suggested that the low-temperature form must be close to that of a P_{II} conformation as assigned by Tiffany and Krimm (1968b).

B.   *Theoretical Studies of Aqueous N-Acetyl-L-alanine N'-Methylamide*

Tiffany and Krimm's proposal requires that the P_{II} region of the Ramachandran map be strongly favored over other regions (Tiffany and Krimm, 1968b). This raises the following question: If P_{II} helix is an extended conformation lacking intramolecular H-bonds (unlike the α-helix in which H-bonds between adjacent turns provide a potential source of stabilization), what interaction might stabilize the P_{II} conformation in the absence of the constraints of a pyrrolidine ring or the electrostatic repulsion imposed by the charged side chains of polyelectrolytes

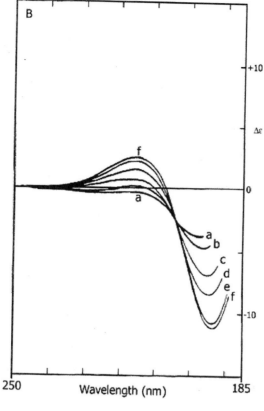

FIG. 5. (*continued*)

such as PGA and PL? Several theoretical calculations have been directed toward characterizing the simplest model of a polypeptide chain, N-acetylalanine-N'-methylamide (AAMA). Calculations in which solvent is neglected show that the minimum lies in the region of β-structure (Brant *et al.*, 1967; Jalkanen and Suhai, 1996; Rossky *et al.*, 1979; Scott and Scheraga, 1966). However, in aqueous solution P_{II} and α_R structures are stabilized relative to other structures (Table I) (Anderson and Hermans, 1988; Brooks and Case, 1993; Grant *et al.*, 1990; Hodes *et al.*, 1979a,b; Jalkanen and Suhai, 1996; Lau and Pettitt, 1987; Madison and Kopple, 1980; Mezei *et al.*, 1985; Pettitt *et al.*, 1986; Schmidt and Fine, 1994).

A recent *ab initio* quantum mechanical study (Han *et al.*, 1998) used B3LYP/6-31G* density functional theory to examine the relative stabilities of eight conformers of AAMA with four explicit water

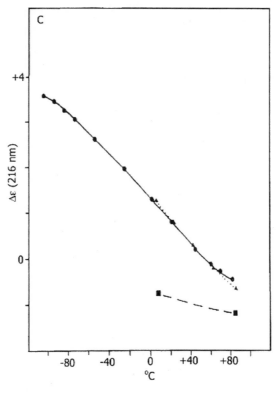

Fig. 5. (*continued*)

molecules H-bonded to its peptide NH and CO groups in an Onsager cavity (Fig. 6). The result was that the P_{II} backbone conformation is the most stable, followed by β-structure (1.9 kcal/mol less stable) and then right-handed α-helix (Table I). By applying a solvent continuum model to the AAMA + $4H_2O$ complex, they further evaluated the effect of solvent on the structure of the AAMA + $4H_2O$ complex and its influence on the vibrational modes and intensities. Figure 7 shows the experimental vibrational Raman and VROA spectra of AAMA in H_2O (Deng *et al.*, 1996) together with calculated Raman and VROA spectra of four lowest energy conformers—β_2', α_L', α_R' and P_{II}— of AAMA in an Onsager cavity model. The results clearly reveal that the spectra of the P_{II} structure account best for the experimental data. Therefore, both energetic and spectral calculations suggest that the P_{II} conformation ($\phi \sim -93°$, $\psi \sim +128°$)

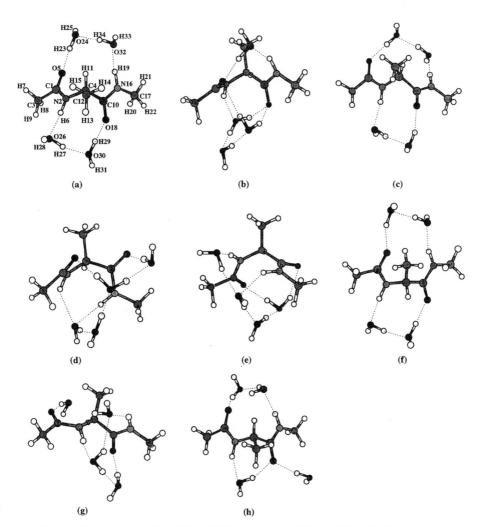

FIG. 6.   Eight optimized AAMA + 4H$_2$O conformers: (a) P_{II}, (b) C_7^{ax}, (c) β_2, (d) α_L, (e) α_R, (f) α_D, (g) α_P, (h) crystal. From Han *et al.* (1998). *J. Phys. Chem. B* **102**, 2587–2602, © 1998. Reprinted with permission from American Chemical Society.

dominates in an aqueous solution of AAMA. These results are in agreement with several earlier simulations or free-energy calculations that suggest P_{II} conformers in AAMA have relatively lower free energies in H$_2$O solution (Anderson and Hermans, 1988; Brooks and Case, 1993; Madison and Kopple, 1980; Schmidt and Fine, 1994; Smith, 1999).

TABLE I

Comparison of the ϕ and ψ (deg) Angles and the Relative Energies ΔE (kcal/mol) of the Eight B3LYP/6-31G* Optimized AAMA Structures in Different Models[a]

		C_7^{eq}	C_5^{ext}	β_2	C_7^{ax}	α_R	α_L	α_D	α_P
(1) AAMA[a]	ϕ	-81.91	-157.31	-135.86	73.77	-60.00	68.20	57.25	-169.36
	ψ	72.27	165.26	23.40	-60.02	-40.00	24.73	-133.45	-37.79
	ΔE	0.000	1.433	3.181	2.612	5.652	5.817	6.467	6.853
		P_{II}	Crystal	β_2'	$C_7^{ax'}$	α_R'	α_L'	α_D'	α_P'
(2) AAMA + $4H_2O$	ϕ	-93.55	-97.90	-150.88	58.88	-82.08	60.73	66.68	-152.94
	ψ	127.62	111.75	116.47	-121.92	-44.11	52.44	-111.27	-92.01
	ΔE	0.000	5.864	1.886	4.134	2.465	2.754	3.715	15.140
(3) AAMA + $4H_2O$ in Onsager Model	ϕ	-92.74	-98.48	-151.95	58.61	-80.27	59.08	66.61	-117.74
	ψ	128.61	107.82	118.41	-122.46	-47.61	52.45	-111.10	-110.34
	ΔE	0.018	4.791	1.870	3.592	0.000	1.160	3.761	7.784

[a] Han et al. (1998). J. Phys. Chem. B 102, 2587–2602, © 1998. Reprinted with permission from American Chemical Society.

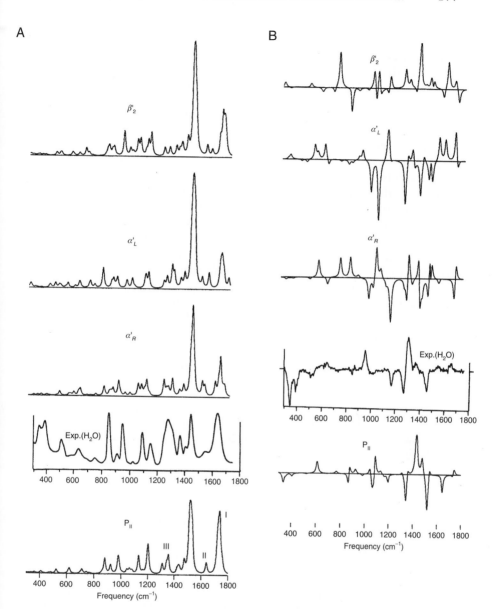

FIG. 7.   (A) Experimental vibrational Raman spectrum of AAMA in H_2O (Deng *et al.*, 1996) and calculated Raman spectra of $\beta_2 + 4H_2O$, $\alpha_L + 4H_2O$, $\alpha_R + 4H_2O$, and $P_{II} + 4H_2O$, which are optimized in the Onsager solvent model (B) Experimental ROA spectrum of AAMA in H_2O (Deng *et al.*, 1996) and calculated ROA spectra for the four AAMA structures obtained from the AAMA + $4H_2O$ + Onsager complexes. From Han *et al.* (1998). *J. Phys. Chem. B* **102**, 2587–2602, © 1998. Reprinted with permission from American Chemical Society.

C. Evidence from Nuclear Magnetic Resonance Studies

Recently, some of us investigated the backbone conformation of a seven-residue alanine peptide using NMR (Shi *et al.*, 2002). The peptide, referred to as XAO, is an 11-mer with the sequence.

$$\text{AcXXA}_1\text{A}_2\text{A}_3\text{A}_4\text{A}_5\text{A}_6\text{A}_7\text{OOamide}$$

where X is diaminobutyric acid, A is alanine, and O is ornithine. This peptide presents a soluble model system for a structureless denatured protein. It is too short to form any detectable α-helix in water via intra-peptide hydrogen bonds, whereas the $-\text{CH}_3$ side chain is too short to form nonpolar clusters and too inert to form other side chain interactions. To characterize the backbone conformations of this peptide, we measured system properties that are related directly to the ϕ and the ψ backbone angles. The 1D NMR spectrum of XAO is well-resolved in the NH region as shown in Figure 8. We could derive the ϕ angle of each alanine residue by measuring the $^3J_{\text{HN}\alpha}$ coupling constants. Using multiple samples containing single or double substitutions of ^{15}N-labeled alanine at Ala 2, 4, and 6, values of $^3J_{\text{HN}\alpha}$ for A_2, A_3, A_4, and A_5 could be determined from the well-resolved spectra of the peptide, as shown in the lower panel of Figure 8 (using the XAO26 sample ^{15}N-labeled at 2 and 6 positions with ^{15}N decoupling in the pulse sequence);

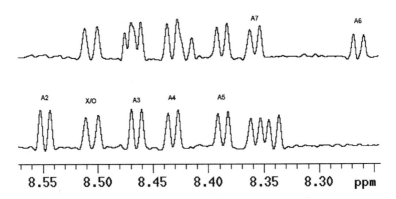

FIG. 8. Amide region of ^1H NMR spectra for XAO peptide. Each NH peak in the spectrum of XAO is cleanly resolved and allows accurate determination of the coupling constants. The values of $^3J_{\text{HN}\alpha}$ for A_2, A_3, A_4, and A_5 can be determined from unlabeled sample or ^{15}N-labeled sample (with ^{15}N decoupling in the pulse sequence), as seen from the lower panel; those for A_6 and A_7 can be obtained from the sample XAO26 (^{15}N-labeled at 2 and 6 positions) without ^{15}N decoupling in the pulse sequence, as seen from upper panel. From Shi *et al.* (2002). *Proc. Natl. Acad. Sci. USA* **99**, 9190–9195, © 2002 National Academy of Sciences, USA.

the values for A_6 and A_7 could be obtained from the sample XAO26 (without ^{15}N decoupling in the pulse sequence), as seen in the upper panel of Figure 8.

The coupling constants for different residues determined in this way change markedly with temperature (Fig. 9), a fact that alone precludes the possibility that the peptide is a random coil. Superimposing $^3J_{HN\alpha}$ values for each site in the chain reveals a common behavior in their temperature dependence: a monotonic increase in coupling constant between 2° and 56°C (Fig. 10A). The cooling curves are superimposable on the heating curves (Fig. 9), which shows that the increase in coupling constant with temperature is not the result of forming irreversible aggregates at higher temperatures.

Together with the ϕ backbone angles information derived from $^3J_{HN\alpha}$ coupling constants, NOE information provided ψ backbone angle information. A sample of XAO was prepared in which the methyl groups of Ala 1–3 and Ala 5–7 were deuterated; NOESY spectra with mixing times of 200, 300, and 400 ms were recorded on this sample. This deuteration allowed selective detection of NOEs between methyl protons of Ala 4 and proximal protons, such as its own NH proton and the NH proton of the neighboring Ala 5 residue. In each NOESY spectrum, there is a strong NOE between the methyl protons of Ala 4 and its own NH proton, and a weaker NOE from its methyl protons to the NH of the succeeding residue Ala 5 (Fig. 11D). The quantitative ratio of these NOEs was determined by taking the ratio of the respective integrated NOE volume, $NOE_{\beta i-NHi}/NOE_{\beta i-NHi+1} \sim 4$. In the amide–amide region there were no measurable NOEs observed between any pair of successive amides in the chain. This ruled out the presence of appreciable α-helix in the peptide (Fig. 11B).

The CD spectra of the XAO peptide at temperatures of 1°, 35°, 45°, and 55°C are shown in Figure 12B. For comparison, the CD spectra of Ac-(Gly-Pro-Hyp)$_n$-NH$_2$ ($n = 1, 3, 5, 6, 9$) are also shown (Fig. 12A) (Feng *et al.*, 1996). The CD spectrum of XAO at 1°C closely resembles those of Ac-(Gly-Pro-Hyp)$_n$-NH$_2$ peptides known to adopt P_{II} conformation, aside from a small wavelength shift due to the difference between secondary and tertiary amides and a smaller amplitude. The smaller amplitude is largely due to the presence of the solubilizing X and O residues in the peptide. Four of the twelve residues in XAO lie outside the Ala$_7$ sequence, and these contribute to the CD spectrum. These residues apparently have a conformation other than P_{II}, probably β. By subtracting the spectrum of XAO at 1°C from that at 55°C, the difference spectrum shown in the inset to Figure 12B is obtained. This spectrum resembles the CD spectrum of β-strand (Bhatnagar and Gough, 1996) and differs

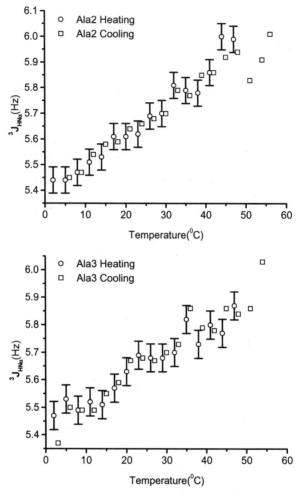

FIG. 9. The $^3J_{HN\alpha}$ coupling constants versus temperature for different alanine residues in XAO peptide. The reversibility of $^3J_{HN\alpha}$ coupling constant versus temperature was checked by measurements with temperature increasing from 2° to 56°C (labeled as heating) or decreasing from 56° to 6°C (labeled as cooling). The errors are the same in heating and cooling measurements; the error bars are shown only for heating measurements for clarity. The conditions were temperature from 2° to 56°C, concentration *ca.* 4 mM, in 30 mM sodium acetate buffer (pH = 4.6, 10% D_2O). From Shi *et al.* (2002). *Proc. Natl. Acad. Sci. USA* **99,** 9190–9195, © 2002 National Academy of Sciences, USA.

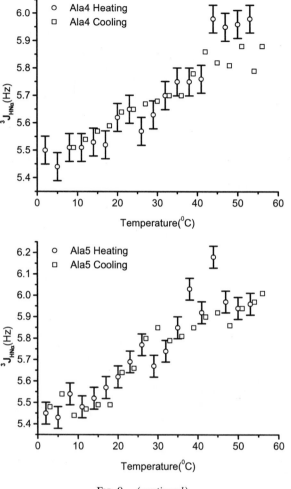

FIG. 9.   (*continued*)

markedly from the 1°C spectrum of XAO or those of Ac-(Gly-Pro-Hyp)$_n$-NH$_2$ peptides (Fig. 12).

The combined evidence from $^3J_{HN\alpha}$ coupling constants, NOE measurement, and CD spectra argues that the XAO peptide is not a random coil, and that P$_{II}$ is the dominant conformation at 2°C. In detail, the values of $^3J_{HN\alpha}$ are seen to increase monotonically with T between 2° and 56°C (Figs. 9 and 10), and the CD spectrum changes character between 1°C and 55°C. This behavior is consistent with a conversion from one type of structure to another as T increases, certainly not what one

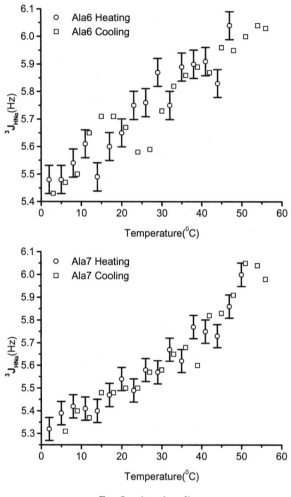

FIG. 9.   (*continued*)

expects for a classical "random coil." The ϕ angle, $-70° \pm 10°$, derived from the $^3J_{HN\alpha}$ value of 5.45 ± 0.05 at $2°C$, is consistent with either α-helix or P_{II} conformation (Fig. 11A). The CD spectrum at $1°C$ is consistent with the Ala_7 sequence consisting of $\sim$90% P_{II} conformation (Bhatnagar and Gough, 1996; Sreerama and Woody, 1994; Tiffany and Krimm, 1968b; Woody, 1992), and only small amounts of α- and β-structures. The lack of i, $i + 1$ NH–NH NOEs confirms the absence of significant α-helix or the "nascent helix" observed by Dyson and co-workers (Dyson *et al.*, 1988). Consequently, there appears to be a dominant conformation in

(A)

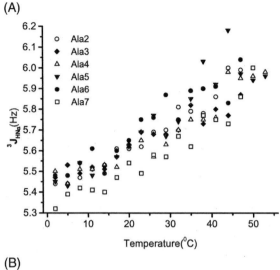

(B)

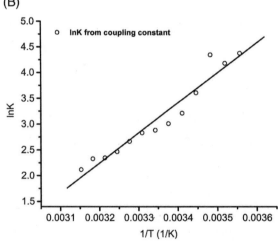

FIG. 10. (A) General trend for the change of $^3J_{HN\alpha}$ coupling constants with temperature as shown by overlaying the plots in Figure 2 for different alanine residues. Only measurements from heating are shown. The errors are the same those as in Figure 2. (B) van't Hoff plot of ln K versus $1/T$, used to determine the enthalpy of the transition from P_{II} to β-strand. Data are shown for the variation of $^3J_{HN\alpha}$ with temperature. The assumptions are as follows (see text): (1) The percentage of P_{II} conformation at $2°C$ is normalized to 100% in order to estimate ΔH for the transition; (2) ΔH is independent of temperature within the temperature range studied; (3) the $^3J_{HN\alpha}$ coupling constant for the β-conformation is 9.81 Hz and for P_{II} it is 5.45 Hz. To calculate ln K from $^3J_{HN\alpha}$ data at different temperatures, the mean coupling constants for six alanine residues were averaged. By fitting the results to a straight line, the coupling constant data give $\Delta H = 11.6 \pm 0.9$ kcal/mol. From Shi *et al.* (2002). *Proc. Natl. Acad. Sci. USA* **99**, 9190–9195, © 2002 National Academy of Sciences, USA.

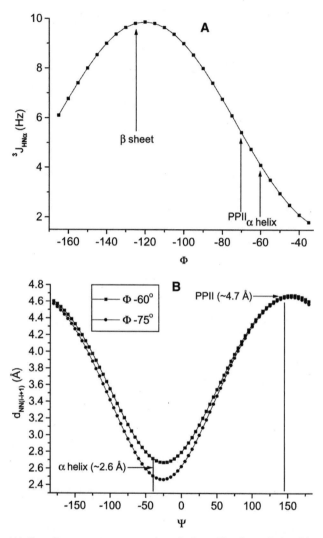

FIG. 11. (A) Coupling constant versus ϕ angle from Karplus relationship. (B) Simulation of the dependence of $d_{N,N(i,i+1)}$ on ψ, with the ϕ value restricted to the range of α-helix and P_{II}, $-60°$ and $-75°$, respectively. (C) Simulated contour plot showing the value of the ratio, $NOE_{\beta i-NHi}/NOE_{\beta i-NHi+1}$ versus ϕ and ψ angles. (D) Section of the NOESY spectrum of XAO deuterated except at alanine 4. The conditions were temperature 5°C, concentration *ca.* 4 mM, mixing time 200 ms in 30 mM sodium acetate buffer (pH = 4.6, 10% D_2O). The strip of NOESY spectrum shows a strong NOE between the methyl side chain protons of alanine 4 and its own amide proton; a weak NOE is observed between the methyl group and the amide proton of alanine 5. In the same spectrum, no NOEs are observed between succeeding amide NH protons (data not shown). From Shi *et al.* (2002). *Proc. Natl. Acad. Sci. USA* **99**, 9190–9195, © 2002 National Academy of Sciences, USA.

C

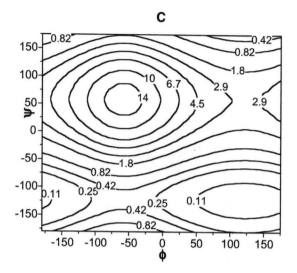

D

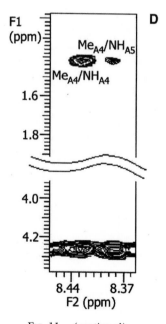

FIG. 11. (*continued*)

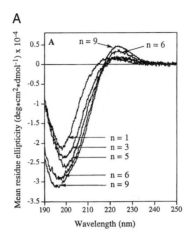

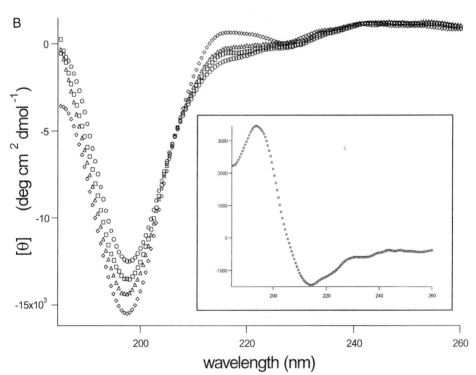

FIG. 12.   (A) CD spectra of Ac-(Gly-Pro-Hyp)$_n$-NH$_2$ ($n = 1, 3, 5, 6, 9$). (From Feng *et al.,* 1996. *J. Am. Chem. Soc.* **118,** 10351–10358, © 1996 American Chemical Society.) (B) CD spectra of XAO at (◇) 1°C, (△) 35°C, (□) 45°C, and (○) 55°C. Peptide concentrations were 50 μM, in 10 mM KH$_2$PO$_4$ buffer, pH 7.0. Inset: CD difference spectrum for XAO between 55° and 1°C. From Shi *et al.* (2002). *Proc. Natl. Acad. Sci. USA* **99,** 9190–9195, © 2002 National Academy of Sciences, USA.

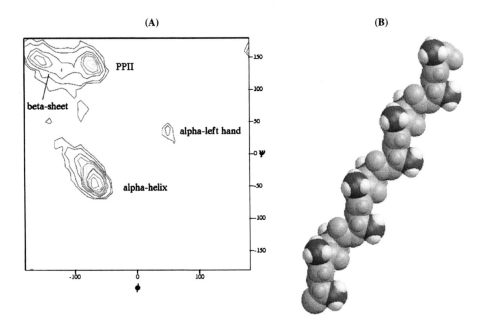

(A)

PPII

beta-sheet

alpha-left hand

alpha-helix

ψ

150

100

50

0

-50

-100

-150

-100 0 100

ϕ

(B)

FIG. 13.   (A) Ramachandran plot constructed from ϕ,ψ dihedral angle distribution of Ala in protein structures (Kleywegt and Jones, 1996; Serrano, 1995; Stapley and Creamer, 1999). (B) A 12-residue segment of polyalanine P_{II} helix, shown for reference. The backbone is shown in blue-gray; β-carbons are in red, and their hydrogens are in white. Unlike the more familiar α-helix, the P_{II} helix is left-handed ($\phi,\psi = -75, +145$). It has three residues per helical turn; that is, every third side chain is colinear, forming three parallel columns spaced uniformly around the long axis of the helix. In solution we imagine that significant fluctuations from the idealized structure shown occur. (From Shi *et al.* 2002, © 2002, National Academy of Sciences, USA)

the Ala$_7$ block at 1°C, and it is P$_{II}$. The assignment to P$_{II}$ is also consistent with the weak NOE observed between the methyl side chain protons of Ala 4 and the NH proton of Ala 5 compared with the strong NOE of the methyl protons of Ala 4 to its own NH proton (Fig. 11D). The ratio of these NOEs, NOE$_{\beta i-NHi}$/NOE$_{\beta i-NHi+1} \sim 4$, gives a ψ angle of $+145° \pm 20°$ when $\phi = -70°$ (Fig. 11C), again consistent with the P$_{II}$ structure.

A model of a P$_{II}$ helix formed by alanine side chains is illustrated for reference in Figure 13B (see color insert), while Figure 13A illustrates the common occurrence of the P$_{II}$ backbone conformation among residues outside regions of regular secondary structure (Kleywegt and Jones, 1996; Serrano, 1995; Stapley and Creamer, 1999) in protein structures from the Protein Data Bank.

The increase in $^3J_{HN\alpha}$ coupling constant with temperature, together with the CD difference spectrum between 55° and 1°C (inset to Fig. 12B), indicates partial conversion of the P$_{II}$ conformation to β-structure at 55°C. The coupling constant expected for β-structure is near 10 Hz (Smith *et al.*, 1996) and the increase between 2° and 56°C (Fig. 10) is 0.5 ± 0.1 Hz; if x is the change in fraction of β-structure between 2° and 56°C, then $5.45(1 - x) + 9.81x = 5.97$ and $x = 0.12$. The partial conversion of P$_{II}$ to β-structure with increasing temperature indicates that an enthalpy difference favors P$_{II}$ over β-structure at 2°C. The NMR data are compatible with no more than 10% non-P$_{II}$ conformer at 2°C, and this is consistent with the CD data that indicate up to 90% P$_{II}$. If one assumes 100% P$_{II}$ in the Ala$_7$ block at 2°, there would be a large apparent enthalpy difference between the P$_{II}$ and β backbone conformations, 11.6 ± 0.9 kcal/mol, in this seven-residue alanine peptide (see Fig. 10B). However, if one assumes 10% non-P$_{II}$ conformer at 2°, the van't Hoff enthalpy difference falls to 3.3 kcal/mol, a more likely estimate.

A recent report on the conformation of AcAlaNHMe by measuring dipolar couplings in the proton nuclear NMR spectrum of the peptide dissolved in a water-based liquid crystal indicates that the result can be understood in terms of one dominant conformation (Poon *et al.*, 2000). The backbone angles determined are $\phi \sim -85°$ and $\psi \sim +160°$, similar to those of P$_{II}$. Exhaustive exploration yielded two sets of dipolar coupling constants with different relative sign combinations; these fit the spectrum qualitatively better than all others. Figure 14A shows the value of the best-fit dipolar coupling constants together with the structure of AcAlaNHMe. After fitting two sets of dipolar coupling constants with different geometric models, Poon *et al.* found that for the first set of dipolar coupling constants, the best fit gave $\phi = -85 \pm 5°$ and $\psi = +160 \pm 20°$, consistent with P$_{II}$ conformation. For the second set of dipolar coupling

A

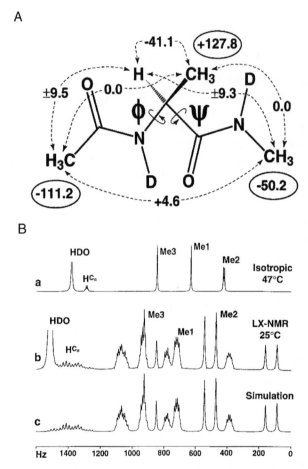

FIG. 14. (A) Ac-L-Ala-NHMe structure with best-fit dipolar coupling constants (Hz). Intramethyl couplings in ellipses. Upper and lower set of sign combinations fit spectrum equally well. (B) NMR spectra of AcAlaNHMe in 42%/wt CsPFO in D_2O. Frequency scale with 0 Hz = 0 ppm. (a) Isotropic spectrum at 47°C. (b) Oriented LX-NMR spectrum at 25°C. (c) Simulated spectrum using best-fit coupling constants shown in (A). From Poon *et al.*, (2000). *J. Am. Chem. Soc.* **122,** 5642–5643, © 2000, Reprinted with permission from the American Chemical Society.

constants with alternative sign combination, no single conformation fit reasonably. Figure 14B shows the observed isotropic and anisotropic spectra as well as the simulated anisotropic spectrum using best-fit dipolar coupling constants shown in Figure 14A.

Recently, aqueous solution conformational studies of three short peptides from 9 to 12 residues corresponding to sequences present in titin

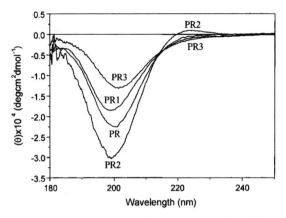

FIG. 15.   CD spectra of PR, PR1, PR2, and PR3. Peptides (0.15, 0.35, 0.4, and 0.46 mg/ml for PR, PR1, PR2, and PR3, respectively) in 10 mM KPi, pH 6.9, 2°C, were measured, and molar ellipticity (θ) per residue was plotted. From Ma *et al.* (2001). *Biochemistry* **40,** 3427–3438, © 2001, Reprinted with permission from the American Chemical Society.

reveal that these short peptides adopt mainly the P$_{II}$ conformation interspersed with unordered, presumably flexible spacer regions (Ma *et al.*, 2001). These peptides all display the characteristic P$_{II}$ CD spectra (Fig. 15). A new CD analysis of a seven-residue lysine peptide finds P$_{II}$ conformation as well (Fig. 16) (Rucker and Creamer, 2002). More recent studies by Shi further show that XVO, OFO, OQO, and OSO are also

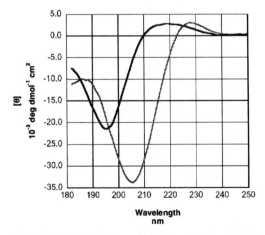

FIG. 16.   CD spectra of the K7 (black spectrum) and P7 (gray spectrum) peptides in pH 7 solutions at 20°C. From Rucker and Creamer (2002). *Protein Sci.* **11,** 980–985, © 2002, Reprinted with permission from the Protein Society.

in P_{II} conformation in aqueous solution (Shi, 2002). This is consistent with P_{II} being a dominant conformation for many short peptides with different amino acid compositions, not just alanine.

D. Evidence from IR Vibrational Spectroscopy

Hamm *et al.* (1999) have recently shown that two-dimensional vibrational spectroscopy can define the secondary structure of a peptide from analysis of the cross-peaks in 2D spectra. Since the secondary structure of a peptide determines the relative positions and orientations of its peptide ($-CO-NH-$) groups, it determines the strength of the couplings between the amide I vibrational bands, which mainly involve the stretching of the C=O bonds of the peptide groups. Coupling strengths, which are experimentally measured as intensities and anisotropies of the cross-peaks in the 2D vibrational spectrum, can thus be used to determine the secondary structure of a peptide. However, an intrinsic problem in 2D vibrational spectroscopy of peptides is that the cross-peaks usually overlap with the strong and broad diagonal peaks owing to the fact that frequency differences between the uncoupled amide I bands are of the same order as the coupling strength or the linewidth. Woutersen and Hamm (2000, 2001) recently succeeded in separating the cross-peaks in the 2D vibrational spectrum from the strong diagonal signal by exploiting differences in their polarization dependence. The pumped and probed transition dipoles giving rise to the cross-peak signal are in general not parallel, so that the anisotropy of the cross-peaks is different from that of the diagonal peaks. By subtracting the parallel from the perpendicular scan, the diagonal peaks were eliminated and the cross-peaks clearly emerge. From the intensities and anisotropies of the cross-peaks, two parameters, β and θ, that are related to ϕ and ψ could be determined (β is the coupling strength between the two amide I oscillators, and θ is the angle between the two amide I transition dipole vectors).

Woutersen and Hamm (2000, 2001) investigated the central backbone conformation of trialanine in aqueous solution using polarization-sensitive two-dimensional vibrational spectroscopy. Figure 17 shows the molecular structure of two isotopomers of trialanine at low pD. Figure 18a shows the linear absorption spectrum of deuterated trialanine in D_2O. Figure 18b shows the 2D spectrum for parallel polarizations of the pump and probe pulses, while Figure 18c shows the 2D spectrum for perpendicularly polarized pump and probe. The difference spectrum between perpendicular and parallel signals is shown in Figure 18d. From these spectra, values of $\beta = 6$ cm^{-1} and $\theta = 106°$ (Table II) could be derived. Relating β and θ to ϕ and ψ yields $\phi = -60°$ and $\psi = +140°$

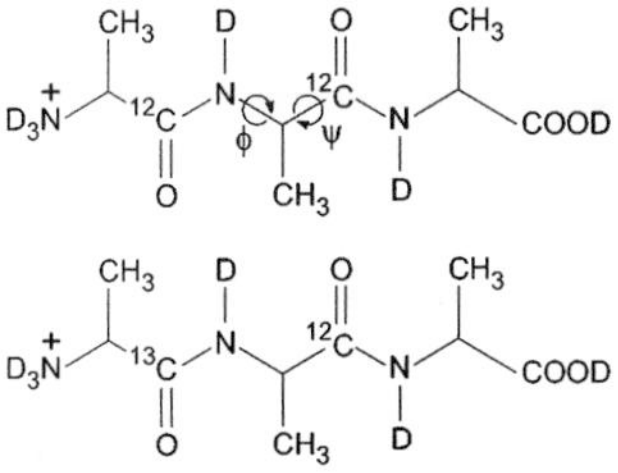

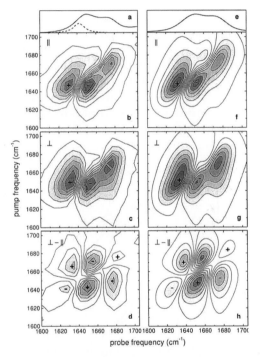

FIG. 17. Molecular structure at low pD of two labeled trialanines. In the upper structure, the dihedral angles (ϕ,ψ) are indicated by arrows. From Woutersen and Hamm (2001). *J. Chem. Phys.* **114,** 2727–2737, © 2001, Reprinted with permission from American Institute of Physics.

FIG. 18. (a) Linear absorption spectra of deuterated Ala-Ala-Ala in D_2O at pD = 1. The dotted line shows a representative pump–pulse spectrum. (b) 2D pump–probe spectrum for parallel polarizations of the pump and probe pulses, showing the absorption change as a function of pump and probe frequency. (c) 2D spectrum for perpendicularly polarized pump and probe. (d) Difference between perpendicular and parallel signals (both scaled to the maximum value occurring in the respective 2D scans). Contour intervals are 0.14 mOD (a), 0.052 mOD (b), and 0.04 mOD (c), respectively. White areas indicate zero response. The signs of the positive and negative signals are indicated. (e)–(h) Calculated signals, using parameter values $\beta = 6$ cm^{-1} and $\theta = 106°$. From Woutersen and Hamm (2001). *J. Chem. Phys.* **114,** 2727–2737, © 2001, Reprinted with permission from American Institute of Physics.

TABLE II

Values for Structural Parameters Derived from 2D
Vibrational Spectroscopy on Different ^{13}C *Isotopomers*
of Trialanine [a]

Isotopomer	β (cm^{-1})	θ (degrees)
Ala-Ala-Ala[b]	6.0	106
Ala*-Ala-Ala[b]	6.2	106
Ala-Ala*-Ala[c]		102

[a] Woutersen and Hamm (2001).
[b] Determined by least-squares fitting.
[c] Determined directly from cross-peak anisotropy.

through the contour plot shown in Figure 19. The resulting conformation is thus consistent with P_{II}. They extended this study on trialanine using site-directed ^{13}C-substituted samples and confirmed the structure assignment. Figure 20 shows the results for one of the isotopomers, Ala*-Ala-Ala. Values of $\beta = 6.2$ cm^{-1} and $\beta = 106°$ (Table II) were determined from the intensities and anisotropies of cross-peaks. Using the contour plot in Figure 19, the same set of ϕ,ψ value as those from Ala-Ala-Ala can be determined.

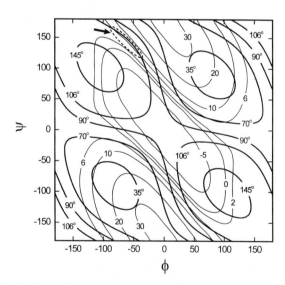

FIG. 19.   Calculated coupling strength β (in cm^{-1}, thin line) (Torri and Tasumi, 1998) and angle θ (in degrees, thick line) between the two amide I transition dipoles as a function of the dihedral angles ϕ and ψ. From Woutersen and Hamm (2001). *J. Chem. Phys.* **114**, 2727–2737, © 2001, Reprinted with permission from American Institute of Physics.

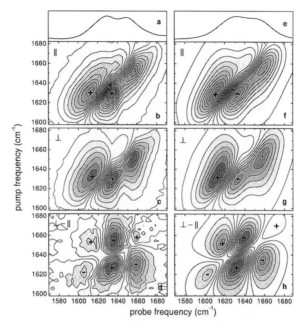

FIG. 20.   (a) Linear absorption spectra of deuterated Ala*-Ala-Ala in D$_2$O at pD = 1.
(b) 2D spectrum for parallel polarizations of the pump and probe pulses. (c) 2D spectrum
for perpendicularly polarized pump and probe. (d) Difference between perpendicular
and parallel signals (both scaled to the maximum value occurring in the respective 2D
scans). Contour intervals are 0.055 mOD (a), 0.029 mOD (b), and 0.017 mOD (c). White
areas indicate zero response. The signs of the positive and negative signals are indicated.
(e)–(h) Calculated signals, using parameter values $\beta = 6.2$ cm^{-1} and $\theta = 106°$. Contour
intervals in the 2D plots are the same as on the left. From Woutersen and Hamm (2001).
J. Chem. Phys. **114**, 2727–2737, © 2001, Reprinted with permission from American Institute
of Physics.

For comparison, the calculated linear and 2D spectra using $\beta =$
12.3 cm^{-1} and $\theta = 52°$, which correspond to an α-helical structure
(see the contour plot Fig. 19) for the isotopomer Ala*-Ala-Ala are
shown in Figure 21. The observed spectra for Ala*-Ala-Ala are strikingly different from the calculated spectra for a molecule in an
α-helical conformation. We emphasize here an important point: In
contrast to the NMR results on oligo(Ala), in which averaging of
different backbone conformations might be present because measurements are made on a time scale that is slow compared to that
of conformational motions, these vibrational spectroscopy results are
detected on a very fast time scale (Hamm *et al.*, 1999; Woutersen
and Hamm, 2000, 2001). This rules out conformational averaging.

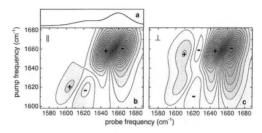

Fig. 21. Calculated linear and 2D spectra for Ala*-Ala-Ala using $\beta = 12.3$ cm^{-1} and $\theta = 52°$, which corresponds to an α-helical structure. Linear spectrum (a) and 2D scans for parallel (b) and perpendicular (c) polarizations of the pump and probe pulses. From Woutersen and Hamm (2001). *J. Chem. Phys.* **114**, 2727–2737, © 2001, Reprinted with permission from American Institute of Physics.

More recently, Schweitzer-Stenner *et al.* (2001) have determined the dihedral angles of trialanine in D$_2$O at three different pD values by combining FTIR and polarized visible Raman spectroscopy. They found that the trialanine peptide adopts a P$_{II}$ conformation in all cases. From the measured polarized visible Raman and FTIR spectra of trialanine in D$_2$O at acid, neutral, and alkaline pDs, they determined the orientational angle θ between the peptide groups and the strength of excitonic coupling Δ between the corresponding amide I modes. Figure 22 shows FTIR and isotropic and anisotropic Raman spectra of trialanine at the indicated pDs. Spectroscopic and structural parameters they obtained from analysis of the FTIR, isotropic and anisotropic Raman spectra of trialanine at three pDs are summarized in Table III. A comparison of results from different spectroscopic techniques is summarized in Table IV.

E. *Evidence from IR Vibrational Circular Dichroism*

Vibrational circular dichroism (VCD) has emerged as a powerful technique for characterizing the secondary structure of peptides. A more detailed account of the method and its application to unfolded proteins is presented in Ch. 7 of this volume. The characteristic spectra for different types of secondary structure and their application have been reviewed by Keiderling *et al.* (1999). The VCD spectrum for PP (P$_{II}$) has an intense negative amide I couplet centered near 1645 cm^{-1} (intense negative band followed by a broad positive band at higher frequency, Fig. 23A) (Dukor and Keiderling, 1991). PGA at neutral pH turns out to have a similar VCD spectrum (Fig. 23). Ionized PL in D$_2$O also has a similar VCD spectrum (Fig. 24A) (Dukor and Keiderling, 1991). All of these spectra

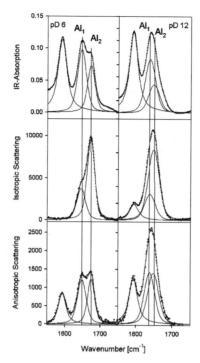

FIG. 22. FTIR (upper panel), isotropic, and anisotropic Raman spectra ($\lambda_{exc} = 457$ nm) of trialanine measured at the indicated pD. The solid lines and the band profiles arise from the fitting procedure described in the reference. From Schweitzer-Stenner *et al.*, (2001). *J. Am. Chem. Soc.* **123,** 9628–9633, © 2001, Reprinted with permission from American Chemical Society.

are opposite in sense to that assigned to α-helix as shown in Figure 24B. This led Dukor and Keiderling (1991) to conclude that these peptides all have a left-handed helical structure. This observation is clearly consistent with earlier proposals by Tiffany and Krimm (1968c) that the corresponding structures can be viewed as an extended left-handed helix,

TABLE III

Spectroscopic and Structural Parameters Obtained from the Analysis of the FTIR, Isotropic, and Anisotropic Raman Spectra of Trialanine at the PD Indicated[a]

	$\bar{\nu}_1$ (cm^{-1})	$\bar{\nu}_2$ (cm^{-1})	Γ_{G1} (cm^{-1})	Γ_{G2} (cm^{-1})	Δ (cm^{-1})	θ (deg)	ϕ (deg)	ψ (deg)
pD 1	1652	1676	21.3	18.9	4.0	119	−66	140
pD 6	1649	1675	23.4	17.2	4.2	122	−70	145
pD 12	1638	1649	29.6	30.5	1.8	128	−95	150

[a] Schweitzer-Stenner *et al.* (2001). *J. Am. Chem. Soc.* **123,** 9628–9633, © 2001, Reprinted with permission from American Chemical Society.

TABLE IV
Summary and Comparison of Results from Different Spectroscopic Techniques

Angle (deg)	Han *et al.* (2002)	Shi (2002)	Poon *et al.* (2000)	Hamm *et al.* (1999)		Schweitzer-Stenner *et al.* (2001)		
				First	Second	pD 1	pD 6	pD 12
ϕ	−93	−70	−85	−60	−60	−66	−70	−95
ψ	128	145	160	140	140	140	145	150

at least locally, rather than a true random coil. Based on assignments of VCD spectra for α-helix and P_{II} conformations, Yoder *et al.* (1997) have studied amide I′ VCD spectra for Ac-(AAKAA)$_4$-GY-NH$_2$ in D$_2$O as a function of temperature from 5° to 50°C. The spectrum changes shape on heating, from that of a typical α-helix to that of typical P_{II} from 5° to 50°C as shown in Figure 25A. A similar study of the amide I′ VCD spectra at 5°C for the peptide series Ac-(AAKAA)$_n$-GY-NH$_2$ ($n = 1$–4) in D$_2$O showed that the shape of these spectra changes from typical P_{II} to that of α-helix as n increases from 1 to 4 (Fig. 25B). These results imply that the alanine peptide Ac-(AAKAA)$_4$-GY-NH$_2$ exists mainly as α-helix at 5°C and transforms to P_{II} conformation at 50°C; as n varies in the peptide series Ac-(AAKAA)$_n$-GY-NH$_2$, when the peptide is too short ($n = 1$) to form α-helix, it exists mainly in P_{II} conformation. It forms α-helix at longer chain lengths ($n = 3$ or 4).

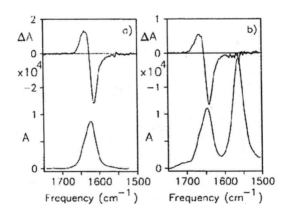

FIG. 23. IR absorption and VCD spectra for PP (a) and PGA (b) in D$_2$O solution at neutral pH. From Dukor and Keiderling (1991). *Biopolymers* **31,** 1747–1761, © 1991, Reprinted by permission of John Wiley & Sons Inc.

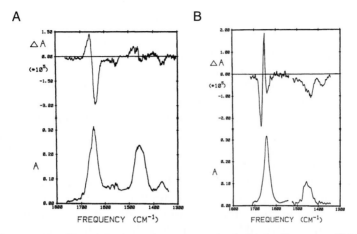

FIG. 24. (Left) VCD and IR absorption spectra of poly-Lys P$_{II}$ ("unordered" form) in D$_2$O. (Right) α-Helix in CD$_3$OD–D$_2$O. From Yasui and Keiderling (1986). *J. Am. Chem. Soc.* **108,** 5576–5581. © 1986, Reprinted with permission from American Chemical Society.

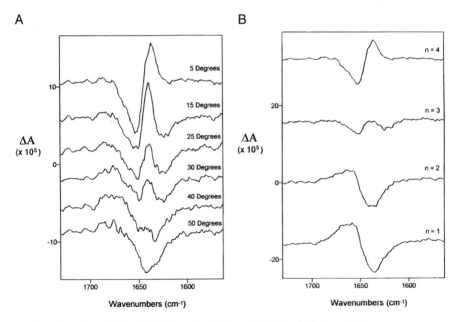

FIG. 25. (A) Amide I′ VCD spectra for Ac-(AAKAA)$_4$-GY-NH$_2$ in D$_2$O as a function of temperature from 5° to 50°C. (B) Amide I′ VCD spectra at 5°C for the peptide series Ac-(AAKAA)$_n$-GY-NH$_2$ ($n = 1$–4) in D$_2$O. From Yoder *et al.* (1997). *Biochemistry* **36,** 15123–15133. © 1997, Reprinted with permission from American Chemical Society.

F. *Evidence from Vibrational Raman Optical Activity Studies*

Vibrational Raman optical activity (VROA) measures optical activity by means of a small difference in the intensity of Raman scattering from chiral molecules in right- and left-circularly polarized incident laser light (Barron *et al.*, 2000). VROA is a relatively new technique, with exceptional potential to define the conformation of polypeptides and protein stucture in aqueous solution, especially dynamic aspects. As more and more experimental results become available, VROA bands can be assigned specifically to different types of secondary structure. VROA measurements on PL and PGA samples at different solution conditions were used to assign the characteristic VROA band for α-helix and P_{II} conformations (Smyth *et al.*, 2001). In the extended amide III region, at pH 11, PL shows a strong characteristic positive hydrated α-helix band at 1342 cm^{-1}; on lowering the pH to 3, the positive band at 1342 cm^{-1} disappears and a new characteristic positive P_{II} conformation band at 1320 cm^{-1} appears. PGA consistently gives a strong hydrated α-helix band at 1345 cm^{-1} at pH 4.8; on increasing the pH to 12.6, the positive band at 1345 cm^{-1} disappears and a new P_{II} band at 1321 cm^{-1} appears (Smyth *et al.*, 2001). By monitoring the positive bands at $\sim$1345 cm^{-1} and $\sim$1318 cm^{-1} in the VROA spectrum of AK21 peptide (Ac-AAKAAAAKAAAAKAAAAKAGY-NH_2), Blanch *et al.* (2000) observed a conversion of the structure from α-helix to P_{II} as the sample is heated from 0° to 30°C at pH 7.5 (Fig. 26). This observation parallels those from VCD spectroscopy for Ac-$(AAKAA)_4$-GY-NH_2 peptide by Yoder *et al.* (1997).

III. Circular Dichroism of Unfolded Proteins

A. *Proteins Unfolded by Different Agents*

The denaturation of proteins generally involves at least partial unfolding, with the loss of secondary and tertiary structure. In the present context, we are interested in the end point of this process — proteins that are unfolded to the maximal extent by various agents: heat, cold, acid, urea, Gdm·HCl.[1] Three major questions concerning unfolded proteins are of interest in the present chapter. Do different unfolding agents

---

[1] We shall not deal with proteins unfolded at high pH because alkaline conditions often lead to irreversible unfolding due to deamidation of Asn and Gln, chain cleavage, racemization at C_α, and β-elimination at Cys. In addition, we shall not discuss unfolding by detergents because in the most thoroughly studied example, sodium dodecyl sulfate (SDS), the protein is not extensively unfolded. Instead, SDS induces α-helix formation in proteins with little or no helix, and unfolds α-helix-rich proteins only partially, leaving a substantial fraction of helix (Mattice *et al.*, 1976).

A

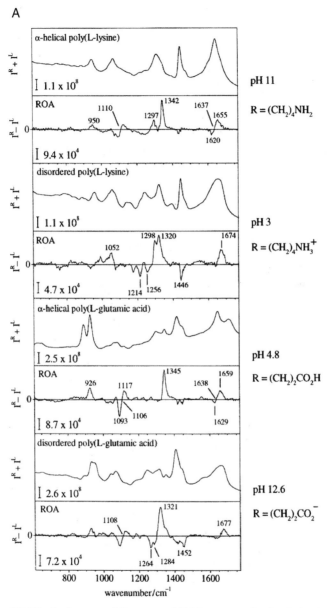

FIG. 26. (A) The backscattered Raman and ROA spectra of poly-L-lysine in α-helical (top pair) and "disordered" (P_{II}) (second pair) conformations, and of poly-L-glutamic acid in α-helical (third pair) and "disordered" (P_{II}) (bottom pair) conformations in aqueous solution. (From Smyth *et al.* (2001). *Biopolymers* **58,** 138–151, © 2001, Reprinted by permission of John Wiley & Sons, Inc.) (B) The backscattered Raman and ROA spectra of the alanine-rich α-helical peptide AK21 in 10 mM sodium chloride plus 5 mM sodium phosphate buffer (pH 7.5), measured at 0°C (top pair) and 30°C (bottom pair). From Blanch *et al.* (2000). *J. Mol. Biol.* **301,** 553–563, © 2000, with permission from Academic Press.

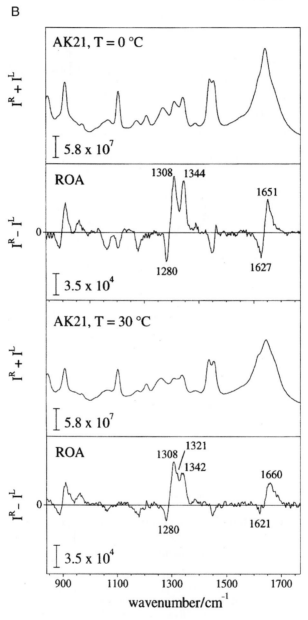

F_{IG}. 26. (*continued*)

lead to the same chain conformation? Do any unfolding agents lead to a random coil? If not, what is the predominant secondary structure in unfolded proteins?

Tanford (1968) reviewed early studies of protein denaturation and concluded that high concentrations of Gdm·HCl and, in some cases, urea are capable of unfolding proteins that lack disulfide cross-links to random coils. This conclusion was largely based on intrinsic viscosity data, but optical rotation and optical rotatory dispersion (ORD) [reviewed by Urnes and Doty (1961)] were also cited as providing supporting evidence. By these same lines of evidence, heat- and acid-unfolded proteins were held to be less completely unfolded, with some residual secondary and tertiary structure. As noted in Section II, a polypeptide chain can behave hydrodynamically as random coil and yet possess local order. Similarly, the optical rotation and ORD criteria used for a random coil by Tanford and others are not capable of excluding local order in largely unfolded polypeptides and proteins. The ability to measure the ORD, and especially the CD spectra, of unfolded polypeptides and proteins in the far UV provides much more incisive information about the conformation of proteins, folded and unfolded. The CD spectra of many unfolded proteins have been reported, but there have been few systematic studies.

Early studies provided the CD spectra of several urea-denatured proteins (Beychok, 1965, 1967; Pflumm and Beychok, 1966; Velluz and Legrand, 1965). In contrast to the spectra of poly(Glu) and poly(Lys), the then-accepted models for random coils, the CD spectra of these urea-denatured proteins lack a positive band in the 220 nm region. Instead, they show a weak negative shoulder in this region, with the CD turning sharply negative at shorter wavelengths. A negative maximum at about 202 nm was reported for urea-denatured bovine serum albumin (Velluz and Legrand, 1965).

The CD spectra of nine proteins in 6 M Gdm·HCl were studied by Cortijo *et al.* (1973). Those proteins with disulfide bridges were reduced and carboxymethylated. The spectra of individual proteins were not reported, but the range of values at wavelengths from 240 to 210 nm was given. The $[\theta]_{222}$ values ranged from -800 to -2400 deg cm²/dmol. From this substantial variation, Cortijo *et al.* (1973) concluded that the proteins studied are not true random coils in 6 M Gdm·HCl, because random coils should have CD spectra essentially independent of amino acid composition and sequence. The observed variation was attributed to differences in the conformational distribution between allowed regions of the Ramachandran map or to residual interactions between different parts of the chain that are resistant to Gdm·HCl denaturation.

Privalov *et al.* (1989) studied the unfolded forms of several globular proteins [ribonuclease A, hen egg white lysozyme, apomyoglobin (apoMb), cytochrome *c*, and staphylococcal nuclease]. Unfolding was induced by 6 M Gdm·HCl at 10°C, heating to 80°C, or by low pH at 10°C with cross-links cleaved (reduction and carboxamidomethylation or removal of heme). The unfolded forms showed CD spectra (Fig. 27)

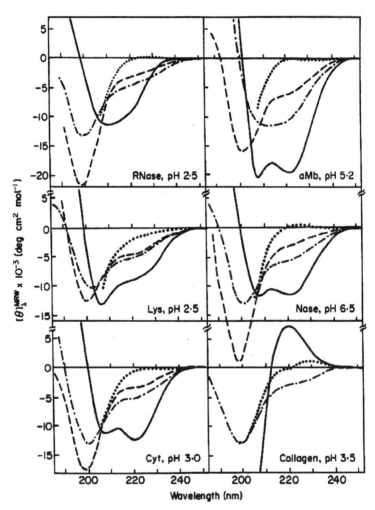

FIG. 27.   Far-UV CD spectra of six proteins in various states. Native state at 10°C at indicated pH (———); thermally denatured at 80°C at indicated pH (–·–); acid-denatured with disulfides cleaved or heme removed, at 10°C, pH 2.5 (– – –); GdmCl-denatured, 6 M GdmCl, 10°C (····). From Privalov *et al.* (1989). *J. Mol. Biol.* **205,** 737–750, with permission. ©1989, Academic Press.

consistent with unordered polypeptides, except in the case of apoMb at 80°C, which appeared to be predominantly β-sheet, probably due to aggregation.

Unfolding by Gdm·HCl yields a CD spectrum that is substantially less negative in the 220 nm region than those produced by thermal or acid unfolding. In fact, of the five globular proteins studied, unfolding by Gdm·HCl gives a CD spectrum that is nearly zero at 220 nm for four of the five proteins, lysozyme being the exception. The 222 nm CD of the five Gdm·HCl-unfolded proteins becomes more negative as the temperature increases from 10° to 80°C. The CD spectra of thermally denatured (80°C) and of acid-denatured (10°C) proteins are rather similar to each other at wavelengths above 210 nm (Privalov *et al.*, 1989). Proteins unfolded by acid have more positive CD values in this region than thermally unfolded proteins, although the difference is very small for lysozyme. The difference in the 200 nm band is substantially larger, with the acid-unfolded form having a more negative 200 nm band than the thermally unfolded form.

Privalov *et al.* (1989) also reported the temperature dependence of the ellipticity at 222 nm for the proteins studied at various pH values (Fig. 28). At the highest temperature studied (80°C), the 222 nm ellipticity value for the thermally unfolded, acid-unfolded, and Gdm·HCl-unfolded proteins appear to be converging, but show a range of ~2000 deg cm^2/dmol out of a total of ~−5000 deg cm^2/dmol. (ApoMb is an exception in that, as noted before, the thermally denatured protein is apparently an associated β-sheet. However, the acid- and Gdm·HCl-unfolded forms of apoMb have similar $[\theta]_{222}$ values at 80°C.)

Nölting *et al.* (1997) reported far- and near-UV CD spectra for the protein barstar denatured by urea, heat, and cold (Fig. 29). Heat and cold denaturation of barstar were observed in 2.2 M urea, which has a relatively small effect on the spectrum of the native protein at 25°C, but destabilizes the native protein. The CD spectra of all three denatured forms are negative in the 220 nm region. The heat-denatured form exhibits a negative shoulder near 220 nm, whereas the urea-denatured form shows a negative signal that monotonically increases in magnitude down to the low-wavelength limit of observation, ~204 nm. The CD spectrum of the cold-denatured form is intermediate, with a barely detectable shoulder near 220 nm. The magnitude of the negative 220 nm CD increases in the order urea-denatured < cold-denatured < heat-denatured, consistent with the difference between Gdm·HCl- and heat-denatured proteins (Privalov *et al.*, 1989).

The dependence of the CD of unfolded barstar on temperature and urea concentration was reported by Nölting *et al.* (1997). In concentrated urea, the CD at 222 nm shows a linear temperature dependence,

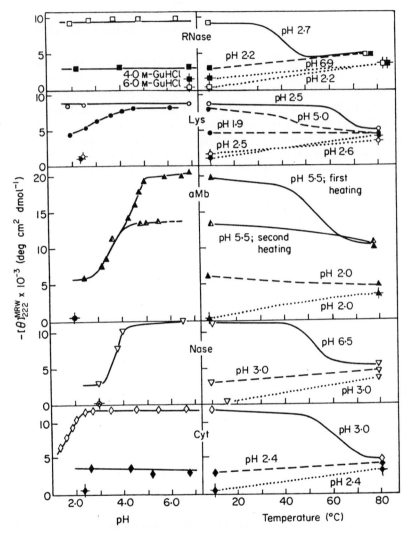

FIG. 28.   Dependence of $[\theta]_{222}$ of six proteins on pH at 10°C (left panels) and on temperature at the indicated pH (right panels). The states are indicated by solid, dotted, etc., lines as in Figure 27. The crossed symbol corresponds to data in the presence of 6 M GdmCl. From Privalov *et al.* (1989). *J. Mol. Biol.* **205,** 737–750, with permission. ©1989, Academic Press.

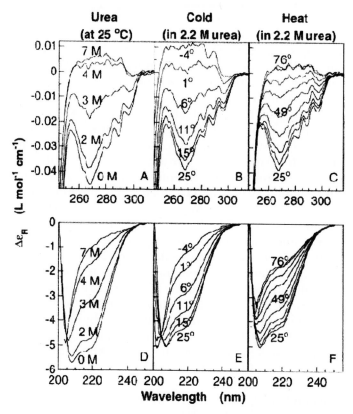

FIG. 29. Equilbrium unfolding of C40A/C82A/P27A (pseudo–wild-type) barstar monitored by $\Delta\varepsilon_R$, mean residue circular dichroism. Conditions for near-UV CD were 50 μM protein in 50 mM Tris–HCl buffer, pH 8, 0.1 M KCl, path length 1 cm. (A) Urea-induced unfolding at 25°C at urea concentrations as indicated. (B) Cold-induced unfolding in 2.2 M urea at indicated temperatures. (C) Thermal unfolding in 2.2 M urea at temperatures of 25, 35, 39, 44, 49, 54, 58, 67, and 76°C. Conditions for far-UV CD were 60 μM protein in 50 mM Tris–HCl, pH 8, 0.1 M KCl, path length 0.02 cm. (D) Urea-induced unfolding at 25°C at urea concentrations as indicated. (E) Cold-induced unfolding in 2.2 M urea at indicated temperatures. (F) Thermal unfolding in 2.2 M urea at temperatures of 25, 35, 39, 44, 49, 54, 58, 67, and 76°C. From Nölting *et al.* (1997). *Biochemistry* **36**, 9899–9905, with permission. © 1997, American Chemical Society.

becoming more negative with increasing temperature. At 4° and 25°C, $\Delta\varepsilon_{222}$ nm depends linearly on urea concentration, becoming more positive with increasing urea concentration.

Near-UV CD of denatured proteins also provides evidence for some order in the side chains, especially in urea- and cold-denatured proteins. Nölting *et al.* (1997) found a broad positive band with possible vibrational fine structure in the 260–300 nm region of barstar at 25°C in 7 M urea and at −4°C in 2.2 M urea (Fig. 29). Thermally denatured barstar (76°C in 2.2 M urea) shows at most a very weak and broad positive band in this region, which may not be above the noise level. Relatively sharp but weak bands have been reported in the near-UV CD spectrum of T7 RNA polymerase in 7 M urea at 25°C (Griko *et al.*, 2001). These weak near-UV CD bands could result from residual tertiary structure.

These observations on the CD of denatured proteins can be rationalized within the framework of the proposal (Tiffany and Krimm, 1968a) that the P_{II} conformation is a significant contributor to the conformational ensemble in unordered polypeptides and proteins. Stabilization of the P_{II} conformation by urea and Gdm·HCl (Tiffany and Krimm, 1973) accounts for the less negative $\Delta \varepsilon_{222}$ of urea and Gdm·HCl-denatured proteins (Nölting *et al.*, 1997; Privalov *et al.*, 1989). Decreased denaturant binding and concomitant melting of the P_{II} conformation explain the negative temperature coefficient of $\Delta \varepsilon_{222}$ in the spectra of urea- or Gdm·HCl-denatured proteins. At high temperatures (80°C), the denaturant-induced stabilization of the P_{II} conformer has nearly vanished and the CD at 220 nm approaches that of the heat- and acid-denatured protein. Cold denaturation, facilitated by moderate urea concentrations, leads to a CD spectrum intermediate between that of urea-denatured and heat-denatured protein, but resembling the former more closely (Nölting *et al.*, 1997). This is attributable to urea stabilization of the P_{II} form at low temperatures.

The fraction of P_{II} conformer in denatured proteins can be roughly estimated from the work of Park *et al.* (1997). They analyzed the data of Drake *et al.* (1988) on poly(Lys) over the temperature range of −100 to +80°C plus their own data on the peptide AcYEAAAKEAPAKEAAAKANMH$_2$ at temperatures from 0° to 90°C. They used a two-state model, justified by the tight isodichroic points observed in each system, and derived limiting 222 nm ellipticity values of +9500 deg cm^2 dmol^{-1} for the P_{II} conformation and −5560 deg cm^2 dmol^{-1} for the high-temperature ensemble of conformers. This leads to Eq. (1) (Bienkiewicz *et al.*, 2000):

$$[\theta]_{222} = 9500 f_{P_{II}} - 5560 f_u \tag{1}$$

where $f_{P_{II}}$ is the fraction of P_{II} helix and f_u is the fraction of the unordered high-temperature form. Although the limiting values derived by Park *et al.* (1997) are subject to uncertainty, and their application to other polypeptides increases that uncertainty, this equation should provide approximate estimates of P_{II}-helix content in various systems.

What does the P_{II}-helix fraction really mean, and what is the nature of the unordered high-temperature ensemble? The positive CD feature near 220 nm observed for poly(Pro)II only manifests itself in oligomers with three or more residues (Dukor and Keiderling, 1991). We interpret the P_{II}-helix fraction derived from Eq. (1) as the contributions of P_{II}-helical stretches with three or more residues. The unordered form that dominates at high temperatures is an ensemble of conformers in which individual residues are in P_{II}, α, or β regions of the Ramachandran map. Occasionally, two successive residues will be in the same region, and strings of three or more residues in the same region will be rare. The evidence described in Section II strongly indicates that P_{II} is at the global energy minimum for peptides in water. Therefore, individual residues in the P_{II} conformation will be more abundant in the unordered ensemble than those in the α and β regions, constituting at least one-third and perhaps as much as one-half of the residues. Thus, the total P_{II} content in an unfolded peptide will include the residues in P_{II} helices, which can be estimated from $[\theta]_{222}$ by Eq. (1), plus that part of the unordered ensemble falling in the P_{II} region of the Ramachandran map. We shall refer to $f_{P_{II}}$ as the P_{II}-helix fraction, and to the total P_{II} content as the P_{II} fraction.

Table V shows the results of this analysis for the P_{II}-helix fraction of several proteins denatured by heat, cold, acid, and Gdm·HCl/urea. There is rather good consistency among the estimated P_{II}-helix contents for proteins denatured by a given agent, except for acid-denatured proteins, which show more variability. The chemically denatured proteins have $30 \pm 5\%$ P_{II}-helix content near $0°C$. At the other extreme, heat-denatured proteins have P_{II}-helix contents near 0%, with lysozyme having the highest value (8%). Although there are only two examples of cold-denatured proteins in Table V,[2] they both have P_{II}-helix contents of about 20%. Acid-denatured proteins have P_{II}-helix contents ranging from 0 to 16%.

We can now answer the questions raised at the beginning of this section. Do different unfolding agents lead to the same chain conformation? They do not. There are distinct differences among the chain conformations produced by different unfolding agents, but consistency among different proteins unfolded by a given means. Acid denaturation appears to be an exception. It may be that apoMb, with 0% P_{II} helix, is an outlier here. ApoMb under cold denaturation conditions ($-7°C$, pH 3.72) retains a significant amount of α-helix (Privalov et al., 1986). The CD spectrum of apoMb at pH 2.5, $10°C$ (Privalov et al., 1989), is qualitatively similar, but with a smaller fraction of α-helix. If this is the case, acid-denatured proteins may, like cold-denatured proteins, generally

---

[2] The CD of myoglobin at $-7°C$, pH 3.72, indicates significant residual helix content, although by other criteria it is substantially unfolded (Privalov et al., 1986).

TABLE V
Approximate P_{II}-Helix Content of Denatured Proteins [a]

Protein	% P_{II}-helix				Ref.[g]
	Heat	Cold	Acid	Gdm·HCl/urea	
Barstar	0[b]	19[c]		30[d]	1
Cytochrome *c*	5		16	33	2
α-Lactalbumin	0			29[e]	3
β-Lactoglobulin		22[f]		32	4
Lysozyme	8		7	25	2
Apomyoglobin			0	35	2
Nuclease	0		16	33	2
Ribonuclease	4		16	33	2

[a] Determined from $[\theta]_{222}$ values (Park *et al.*, 1997; Bienkiewicz *et al.*, 2000). Fractions do not include the P_{II} component of the unordered ensemble. $[\theta]_{222}$ values estimated from published spectra. Conditions for denaturation, unless otherwise specified, are heat, 80°; cold, 10°; acid, pH 2.2–2.5; disulfides cleaved and hemes removed.

[b] 75°C, 3 M urea.
[c] 0°C, 3 M urea.
[d] 10°C, 8 M urea.
[e] Temperature not specified.
[f] 0°C, 4 M urea.
[g] Key to references: (1) Nölting *et al.* (1997); (2) Privalov *et al.* (1989); (3) Dolgikh *et al.* (1985); (4) Katou *et al.* (2001).

have conformations intermediate between heat- and Gdm·HCl(urea)-denatured proteins.

Do any unfolding agents lead to a random coil? We interpret the CD spectra of charged poly(Lys) and poly(Glu) at high temperatures as an ensemble of conformers distributed extensively over the three low-energy regions of the Ramachandran map, the P_{II}, α, and β regions. It is this ensemble that we refer to as *u* in Eq. (1). The CD spectra of thermally unfolded proteins (Fig. 27) resemble the spectra of the charged homopolymers at high temperature, and the P_{II}-helix fraction is near zero (Table V). This high-temperature conformation probably comes as close to a random coil as a polypeptide conformation can. Nevertheless, thermally denatured proteins are unlikely to be true random coils. The strong CD bands near 200 nm (Fig. 27), with mean residue ellipticities of ~−10,000 to −15,000 deg cm²/dmol, argue strongly for correlations between nearest-neighbor peptides, at the least. Interestingly, the Gdm·HCl-denatured form, regarded by Tanford (1968) as the best candidate for a random coil, appears to have the most order, with about one-third of the residues in the P$_{II}$-helix conformation at low temperatures.

What is the predominant secondary structure in unfolded proteins? In thermally unfolded proteins, there appears to be no predominant secondary structure. Individual residue conformations are distributed over the P$_{II}$, α, and β regions that constitute energy minima for single residues in aqueous solution (Section II,B). Still, since there is increasing evidence that the P$_{II}$ conformation is at the global minimum, P$_{II}$ conformers must be the most abundant in this ensemble. Short P$_{II}$- and α-helices and β-strands will be present, but will rarely exceed two or three residues in length.

Proteins unfolded by Gdm·HCl or urea will have a dominant conformation, P$_{II}$. At low temperatures we find about one-third of the residues in chemically denatured proteins in the P$_{II}$-helix conformation, with two-thirds in the form of the high-temperature ensemble. Since at least one-third of the residues in this ensemble are isolated P$_{II}$ residues or in P$_{II}$ helices of two or three residues, the total P$_{II}$ content will be ~50% or greater. The P$_{II}$ content of cold- and acid-denatured proteins will be substantial, probably >40%, but not as large as in chemically denatured proteins.

B. Proteins Unfolded in the Native State

Timasheff *et al.* (1967) reported the CD spectra of two proteins that by hydrodynamic criteria are unordered under native-like conditions: phosvitin and α_s-casein. The CD spectra of these proteins are shown in Figure 30. They show a strong negative band near 200 nm. The CD spectra of α_s-casein at neutral pH and phosvitin at pH 3.4 have a negative shoulder near 220 nm. By contrast, at pH 6.6, phosvitin has a positive band just below 220 nm and a very weak negative band at about 230 nm. Based on the behavior of model polypeptides, as discussed in Section II, the conformation of α_s-casein at neutral pH and of phosvitin at low pH is an ensemble of conformers with broad distribution of residues over the α, β, and P$_{II}$ regions of the Ramachandran map. By contrast, phosvitin at neutral pH has a substantial amount of the P$_{II}$-helix conformation, as evidenced by the positive band near 220 nm. The increased P$_{II}$-helix conformation at pH 6.6 relative to that at pH 3.4 is attributable to the increased charge density as phosphate groups ionize at higher pH. More than half of the residues in phosvitin are phosphoserine. The origin of the weak negative 230 nm band is not clear, but such a band is sometimes observed in poly(Glu) and poly(Lys) CD spectra at low ionic strength and has been attributed to the long-wavelength tail of the strong negative band near 200 nm [E. S. Pysh, quoted by Timasheff *et al.* (1967)].

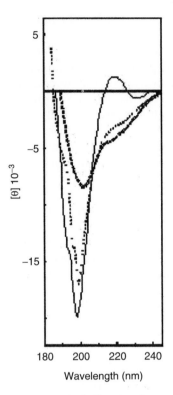

FIG. 30.   Far-UV CD spectra of two natively disordered proteins. Phosvitin, pH 6, phosphate buffer, $I = 0.1\,\mathrm{M}$ (—); phosvitin, pH 3.4, water–acetic acid (· · · ·), α-casein B, pH 7.5, 0.03 M NaF (· · · ·). Based on data of Timasheff *et al.* (1967).

Recently, there has been great interest in proteins that exhibit biological activity but lack a well-defined secondary or tertiary structure after purification (Dunker *et al.*, 1998, 2001; Schweers *et al.*, 1994; Uversky *et al.*, 2000; Wright and Dyson, 1999). Such proteins are referred to as intrinsically disordered or unstructured. An analysis in 1998 of the Swiss Protein Database revealed that about 15,000 proteins in that database are likely to contain disordered segments at least 40 residues in length (Romero *et al.*, 1998). Dyson and Wright (2002) review intrinsically disordered proteins in this volume.

CD spectroscopy, together with NMR, is one of the most useful biophysical techniques for establishing the disordered nature of intrinsically disordered proteins and domains. For example, Kriwacki *et al.* (1996) reported that p21[Waf1/Clp1/Sdi1], a cyclin-dependent kinase inhibitor, has a CD spectrum characteristic of proteins lacking well-defined structure (Fig. 31). The spectra of full-length p21-F and two C-terminally truncated

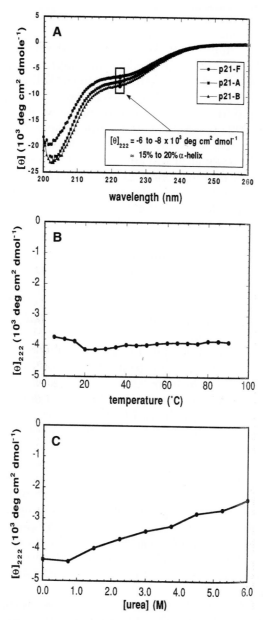

FIG. 31. CD spectra of p21 constructs. (A) Spectra for p21-F (●), p21-A (■), and p21-B (▲) in 1 mM Tris–HCl, pH 7.0/500 mM NaCl/0.5 mM EDTA, 5°C. (B) $[\theta]_{222}$ versus temperature for p21-A. (C) $[\theta]_{222}$ versus urea concentration for p21-A. From Kriwacki *et al.* (1996). *Proc. Natl. Acad. Sci. USA* **93,** 11504–11509, with permission. ©1996, National Academy of Sciences.

forms (p21-A and p21-B) all show strong negative maxima at 202–203 nm and a shoulder in the 220 nm region, with $[\theta]_{222}$ ranging from -6 to -8×10^3 deg cm^2/dmol. The red-shifted short-wavelength band (202–203) nm and the magnitude of the 222 nm ellipticity suggest that the protein contains some short stretches of α-helix. Kriwacki *et al.* (1996) estimated an α-helix content of 15–20% based on a 222 nm ellipticity of -40×10^3 deg cm^2/dmol for 100% helix. As shown in panels B and C of Figure 31, the 222 nm ellipticity shows little temperature dependence from 5° to 90°C, but shifts to more positive values with increasing urea concentration. (It is not clear why the ellipticity values shown in panels B and C are less negative than those in panel A.) The small temperature coefficient may result from cancellation of the positive slope character-istic of melting of α-helix and the negative slope observed as the P_{II} conformation melts. The increase in ellipticity with urea concentration is consistent with both a disruption of α-helix and stabilization of P_{II}.

The activation or inhibition domains of a number of proteins involved in the regulation of transcription and of the cell cycle have been shown to be unordered by CD and NMR, yet they retain biological activity (Bienkiewicz *et al.*, 2002; Campbell *et al.*, 2000; Cho *et al.*, 1996; Dahlman-Wright *et al.*, 1995; McEwan *et al.*, 1996; Schmitz *et al.*, 1994; Van Hoy *et al.*, 1993). Figure 32 shows the CD spectrum of the transactivation domain of c-Myc at three temperatures. In contrast to the spectrum of p21[Wafl/Clpl/Sdil] (Fig. 31), the spectrum of the c-Myc activation domain has its negative maximum below 200 nm and $[\theta]_{222} \sim -2 \times 10^3$ deg cm^2/dmol, indicating the absence of α-helices of any length and the presence of about 20% P_{II}-helix at 5°C. The CD spectra of the transacti-vation domains of c-Fos (Campbell *et al.*, 2000) and of p27 (Bienkiewicz *et al.*, 2002) are similar to that of the c-Myc transactivation domain.

Prothymosin α is a highly acidic protein of 109 residues that is unfolded at neutral pH, as demonstrated by CD and small-angle X-ray scattering (Gast *et al.*, 1995). Figure 33 shows the CD spectra of prothymosin α under several conditions. In phosphate-buffered saline at pH 7.4 and 20°C, it has $[\theta]_{222} \sim -1000$ deg cm^2/dmol, corresponding to $\sim$30% P_{II}-helix. At the lower pHs of 4.6 and 2.4, the 200 nm maximum is red-shifted and the ellipticity at 222 nm is more negative, indicating some conversion of P_{II}-helix to α-helix. In 50% TFE at pH 2.4, there is a strong CD signal from α-helix.

Proteins have been isolated from many cereal seeds that have exten-sive regions of repeating hexa- to nonapeptides. The consensus repeats are rich in Pro and Gln, and always contain an aromatic amino acid, generally Tyr. The CD spectra of fragments of these proteins containing the repeating sequence have been reported (DuPont *et al.*, 2000; Gilbert

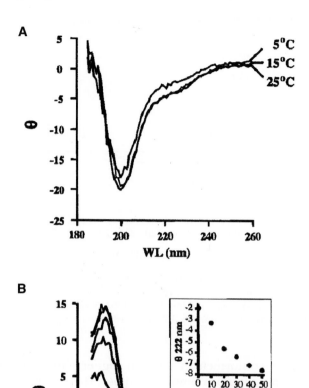

FIG. 32. CD spectra of c-myc$_{1-143}$ protein at three temperatures as indicated. The ordinate scale is the mean residue ellipticity in units of 10^3 deg cm^2/dmol. From McEwan *et al.* (1996). *Biochemistry* **35,** 9584–9593, with permission. ©1996, American Chemical Society.

et al., 2000; Tatham *et al.,* 1989, 1990a, 1990b; Van Dijk *et al.,* 1997b). The spectra in water have negative bands just below 200 nm and weak negative shoulders in the long-wavelength region, indicative of an unordered conformation. These CD spectra, supplemented by IR spectra in some cases, have been variously interpreted as supporting P_{II}-helix or β-turn conformations. In water at low temperatures, P_{II}-helix appears to be the

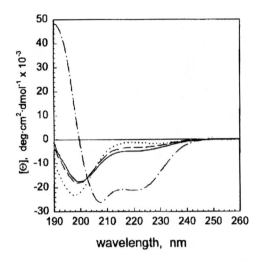

FIG. 33. CD spectra of prothymosin α, measured at 20°C in different buffers. Phosphate-buffered saline, pH 7.4 ($\cdots$); 10 mM sodium phosphate, pH 4.6 (---); 10 mM glycine/HCl, pH 2.4 (—); 5 mM glycine/HCl, pH 2.4, 50% v/v trifluoroethanol (-·-). From Gast *et al.* (1995). *Biochemistry* **34,** 13211–13218, with permission. ©1995, American Chemical Society.

most important conformer. In water at higher temperatures, β-turns have been proposed to be dominant (Gilbert *et al.*, 2000; Tatham *et al.*, 1989, 1990a; Van Dijk *et al.*, 1997a,b). DuPont *et al.* (2000) have proposed that ω-gliadin in water is a mixture of P_{II}-helix and a flexible, unordered conformation, the latter corresponding to the high-temperature ensemble in our terminology. It is clear, however, that β-turns are important in water/TFE mixtures at elevated TFE concentrations.

The C-terminal domain (CTD) of the largest subunit of RNA polymerase II contains tandem repeats of a heptapeptide with the consensus sequence YSPTSPS (Allison *et al.*, 1985; Corden *et al.*, 1985). In yeast, there are 26–27 such repeats, whereas mammals have 52 heptapeptide repeats (Corden and Ingles, 1985). This CTD is essential for cell survival, although the loss of some repeats does not compromise viability. The CD of an eight-repeat fragment has been investigated as a function of temperature and concentration of TFE (Bienkiewicz *et al.*, 2000), and some of the results are shown in Figure 34. The spectrum at low temperature in water shows a broad negative maximum at 197 nm. A tight isoelliptic point is observed at 213 nm over the temperature range 0°–60°C, indicative of a two-state equilibrium. As TFE is added, the CD spectrum changes markedly, especially above 80%. An earlier NMR and CD study (Cagas and Corden, 1995) led to the proposal of substantial β-turn conformation in 90% TFE, but a largely unordered conformation

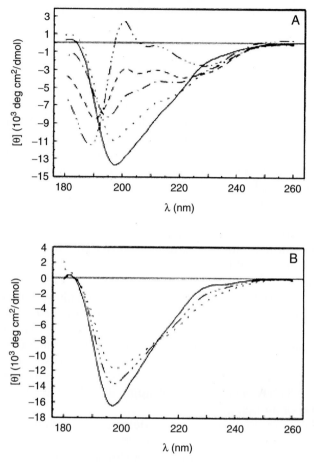

FIG. 34. (A) CD spectra of the eight-repeat fragment of CTD in water (——), 50% TFE (· · ·), 80% TFE (– · –), 90% TFE (- - - -) and 100% TFE (– · · · –) at 32°C. (B) CD spectra of the CTD eight-repeat in water at three temperatures: 2°C (——); 32°C (– · · · –); 60°C (· · · · · ·). From Bienkiewicz *et al.* (2000). *J. Mol. Biol.* **297**, 119–133, with permission. © 2000, Academic Press.

in water. Bienkiewicz *et al.* (2000) concluded that in water, both P$_{II}$ and β-turn structures are present to a modest extent: 15% P$_{II}$-helix and <10% β-turn at 32°C. At high TFE concentrations, the β-turn population increases to ~76% in 90% TFE. In addition, the Tyr side chains become ordered at high TFE, leading to the complex features centered near 190 nm and the negative band near 230 nm. Although phosphorylation of the CTD plays an important role in its function (Dahmus, 1996), phosphorylation of the Ser residues at positions 2 and 5 has little effect on the conformation of the eight-repeat, as measured by CD. The

CD spectrum of the full-length CTD was also reported (Bienkiewicz *et al.*, 2000), and it is very similar to that of the eight-repeat, supporting the relevance of studies on the shorter fragments.

C. Molten Globules

Equilibrium unfolding studies of a number of proteins have demonstrated the presence of intermediates that have secondary structure similar to that of the native protein, but lack a defined tertiary structure (Dolgikh *et al.*, 1983; Ohgushi and Wada, 1983). This type of intermediate is called a *molten globule*. A kinetic intermediate with properties like those of molten globules has been demonstrated in the folding of a number of proteins (Arai and Kuwajima, 2000; Chamberlain and Marqusee, 2000), and it has been proposed that molten globules are generally intermediates in protein folding (Kuwajima *et al.*, 1987; Ptitsyn, 1987; Ptitsyn *et al.*, 1990). Molten globules have been reviewed (Arai and Kuwajima, 2000; Christensen and Pain, 1994; Fink, 1995; Kuwajima, 1989; Ptitsyn, 1987, 1992).

The principal defining properties of the molten globule are as follows (Arai and Kuwajima, 2000): (1) substantial secondary structure; (2) no significant tertiary structure; (3) structure only slightly expanded from the native state (10–30% increase in radius of gyration); (4) a loosely packed hydrophobic core with increased solvent accessibility. The first two criteria are readily assessed by far- and near-UV CD, respectively. Therefore, CD has been extensively applied to the detection and characterization of molten globules.

Unfolding intermediates were first detected through deviations from two-state behavior in the acid denaturation of bovine carbonic anhydrase (Wong and Hamlin, 1974) and bovine α-lactalbumin (Kuwajima, 1977; Kuwajima *et al.*, 1976) or urea denaturation of growth hormone (Holladay *et al.*, 1974), *Staphylococcus aureus* penicillinase (Robson and Pain, 1976), and bovine α-lactalbumin (Kuwajima, 1977; Kuwajima *et al.*, 1976). The intermediate formed at low pH was referred to as the *A-state* (Kuwajima *et al.*, 1976). In many, but not all, cases A-states are "molten globules." We will use the term *A-state* as a general term for conformations induced by low pH, and restrict the term *molten globule* to those A-states that have been demonstrated to have molten globule properties. Figure 35 shows the far- and near-UV CD spectra of bovine α-lactalbumin in the native, molten globule, heat-, and Gdm·HCl-denatured forms (Kuwajima *et al.*, 1985). Similar spectra were reported by Dolgikh *et al.* (1985). Chyan *et al.* (1993) have reported CD spectra for the native and

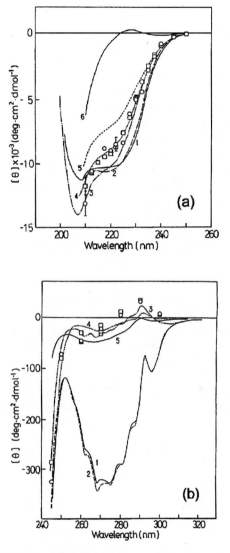

FIG. 35.   Far-UV (a) and near-UV (b) CD spectra of bovine α-lactalbumin in various states. (1 and 2) The native state of the holo and apo forms, respectively; (3) the A state; (a) thermally unfolded state at 41° (4) and 78°C (5); (6) GdmCl-unfolded state. (b) Thermally unfolded state at 62.5°C (4); GdmCl-unfolded state (5). The open circles (holo) and squares (apo) are values derived by extrapolating refolding curves to zero time. From Kuwajima et al. (1985). *Biochemistry* **24,** 874–881, with permission. ©1985, American Chemical Society.

molten globule forms of guinea pig α-lactalbumin (Fig. 36). The far-UV CD spectrum is similar to that for the bovine protein above ~215 nm, but differs in the 208 nm band. The molten globule of the guinea pig protein does not show the large increase in the 208 nm band relative to the native protein that is observed in the bovine protein. In fact, the band is somewhat weaker in the molten globule form relative to the native form.

Secondary structure analysis of the far-UV CD spectra of bovine (Dolgikh *et al.*, 1985; Kuwajima *et al.*, 1985) and guinea pig α-lactalbumin (Chyan *et al.*, 1993) indicates that there is little change (or perhaps a small increase) in the α-helix content on molten globule formation. The methods used are less reliable for β-sheet content, but this also shows little change or perhaps a small decrease. These results are consistent with only a small change in the infrared spectrum upon the native to molten globule transition (Dolgikh *et al.*, 1985). VCD analysis (Keiderling *et al.*, 1994) gave a twofold increase (18% to 39%) in α-helix for human α-lactalbumin in the transition to the molten globule and a twofold decrease (27% to 13%) in β-sheet. However, bovine and goat α-lactalbumin showed more α-helix and less β-sheet in the native form, and the changes on molten globule formation were small.

The strong far-UV CD of the molten globule in the 220 nm region supports the interpretation that the molten globule retains substantial secondary structure. How, though, are the differences in the CD spectra between the molten globule and native forms to be interpreted? The difference spectrum ($[\theta]_N - [\theta]_D$) (Fig. 37) is negative above 220 nm and positive below 220 nm, with a negative maximum near 225 nm, a positive maximum near 208 nm, and a positive shoulder near 215 nm. This negative couplet can be explained in at least two ways: (1) a blue shift of the peptide $n\pi^*$ transition and, in the case of the bovine protein, intensification of the 208 nm band; and (2) contributions of aromatic side chains, especially Trp. It may well be the case that both factors contribute to the difference spectrum.

A blue shift with little change in amplitude of the peptide $n\pi^*$ CD band is consistent with the evidence from CD and NMR studies (Baum *et al.*, 1989; Chyan *et al.*, 1993; Schulman *et al.*, 1995, 1997) that the molten globule retains most or all of the α-helical secondary structure of the native protein. The blue shift is consistent with increased solvent exposure of the peptide carbonyl groups in the molten globule. This could arise from a greater tilt of the plane of the peptide groups so that the carbonyl oxygen can form a weak hydrogen bond with the solvent (Blundell *et al.*, 1983). Hydrogen–deuterium exchange studies by NMR show that protons in the α-helices in the molten globule exchange much more rapidly with solvent than those in the native form (Baum *et al.*,

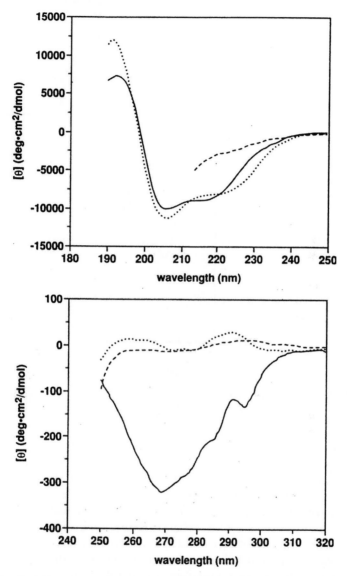

FIG. 36.   Far-UV (upper panel) and near-UV (lower panel) CD spectra of guinea pig α-lactalbumin. Native state, pH 7 (—); A state, pH 2 (···); unfolded state, 9 M urea, pH 2 (– – –). Temperature, 25°C. From Chyan *et al.* (1993). *Biochemistry* **32,** 5681–5691, with permission. ©1993, American Chemical Society.

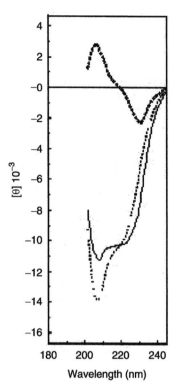

FIG. 37. Difference CD spectrum (native–molten globule) in bovine α-lactalbumin
($\bullet$ $\bullet$ $\bullet$ $\bullet$). Derived from data of Kuwajima *et al.* (1985) for native (—) and molten globule
($\cdots$) forms.

1989; Chyan *et al.*, 1993; Jeng and Englander, 1991; Schulman *et al.*,
1995, 1997).

Aromatic side-chain contributions to the native far-UV CD spectrum
were suggested by Kronman (1968) as responsible for the CD changes
on formation of the A-state. Grishina (1994) has calculated the exciton
contributions of the 220–225 nm band (B_b) of the indole chromophore
to the far-UV CD of baboon α-lactalbumin. This is predicted to give rise
to a negative couplet with the long-wavelength lobe having an amplitude
of $\sim -10^3$ deg cm^2/dmol, comparable to the difference spectrum shown
for bovine α-lactalbumin (Fig. 37). It should be noted, however, that
baboon α-lactalbumin has only three Trps, whereas bovine α-lactalbumin
has four (Acharya *et al.*, 1989). A significant contribution from aromatic
side chains is also consistent with the observation of Keiderling *et al.*
(1994) that variations in the far-UV CD among α-lactalbumins from three
species (human, bovine, goat) are larger for the native proteins than for
the molten globules.

The near-UV CD of bovine α-lactalbumin is shown in Figure 35b. The strong CD of the native protein contrasts with the weak CD of the molten globule, which is comparable to that of the heat- and Gdm·HCl-denatured protein. The weakness of the aromatic CD bands in the molten globule is attributable to the absence of a well-defined conformation and environment for the aromatic side chains, which leads to averaging of the aromatic CD contributions over many conformations and thus to extensive cancellation.

Carbonic anhydrase is another protein that forms a compact A-state at low pH (Wong and Hamlin, 1974). In this case, the far-UV CD changes on going from native protein to molten globule are quite spectacular, as illustrated in Figure 38. At neutral pH the protein has a rather weak

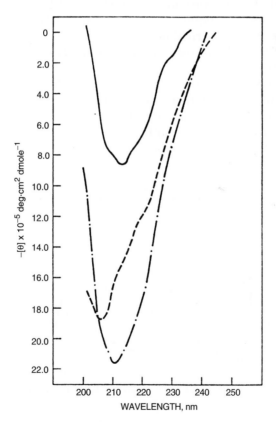

FIG. 38.   Far-UV CD spectra of bovine carbonic anhydrase B in various states. Native, pH 7 (—); molten globule, pH 3.7 (— · —); acid-denatured, pH 2 (– – –). Note that the ordinate scale is ellipticity per mole of protein. To convert to mean residue ellipticity, the values must be divided by the number of residues, 259. From Wong and Hamlin (1974). *Biochemistry* **13**, 2678–2683, with permission. ©1974, American Chemical Society.

negative band at 213 ($[\theta]_{max} = -3400$ deg cm^2 dmol^{-1}). The strength of this band increases nearly threefold on going to pH 3.7, to $[\theta]_{max} = -8600$ deg cm^2/dmol. When the pH is lowered to 2.0, the negative maximum becomes weaker and shifts to about 205 nm, consistent with acid denaturation.

The predominant secondary structure of carbonic anhydrase is antiparallel β-sheet, but this is strongly twisted (Eriksson and Liljas, 1993). Carbonic anhydrase is rich in Trp (seven in human carbonic anhydrase II), and site-directed mutagenesis has demonstrated that each Trp gives rise to a significant CD contribution in the far-UV protein (Freskgård *et al.,* 1994). Several individual Trp contributions are comparable in magnitude to the total observed CD of the wild-type. The aromatic side chain contributions are eliminated, or at least markedly reduced, on molten globule formation. This makes the amplitude of the far-UV CD more normal for a β-rich protein, although the λ_{max} is blue-shifted from the normal position, about 217 nm (Greenfield and Fasman, 1969).

The near-UV CD spectrum of carbonic anhydrase (Fig. 39) is rather strong and displays substantial fine structure at pH 7 (Wong and Hamlin, 1974). In the molten globule and the acid-denatured forms, the near-UV CD spectrum is nearly abolished, although the authors report weak residual positive CD through the aromatic region.

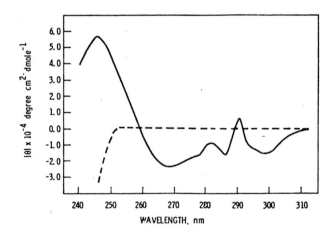

FIG. 39.   Near-UV CD spectrum of bovine carbonic anhydrase B. Native state, pH 7 (—); acid-denatured state, pH 2 (– – –). See note in legend to Fig. 38. From Wong and Hamlin (1974). *Biochemistry* **13,** 2678–2683, with permission. © 1974, American Chemical Society.

Molten globule formation has been shown to be rather common in proteins at low pH and medium to high ionic strengths (Fink *et al.*, 1994; Goto and Fink, 1989; Goto *et al.*, 1990a,b). Figure 40 shows far-UV CD for β-lactamase, apoMb, and cytochrome c under various conditions (Goto *et al.*, 1990a). The proteins in 10–20 mM HCl with no added salt are largely unfolded. If 0.45 M KCl is added, the 220 nm region becomes much more negative, approaching the native intensities for β-lactamase and cytochrome c, and more than half the native intensity for apoMb. If HCl is added instead of KCl, the 220 nm CD becomes more negative, approaching at HCl concentrations of 0.4–0.5 M the values obtained at pH 2.0 with 0.45 M KCl.

As shown in Figure 41, the near-UV CD spectra of β-lactamase and apoMb (Goto *et al.*, 1990a) in the presence of 0.45 M KCl at pH 2 are nearly identical with those for the pH 2 state in the absence of salt, and are very weak in both cases, consistent with the absence of organized tertiary structure. The near-UV CD spectrum of cytochrome c is complicated by the presence of the covalently bound heme, which gives rise to CD bands in the native, acid-denatured, and molten globule states. The authors state that sharp peaks from Tyr and/or Trp, observed at 283 and 290 nm in the native state, are absent in the acid-denatured and molten globule states, supporting the absence of well-defined tertiary structure in this case also.

Goto *et al.* (1990a) also estimated the hydrodynamic radius of the A-states for the three proteins they studied, using exclusion chromatography. Cytochrome c and β-lactamase showed a 10–20% increase in radius, whereas apoMb exhibited a 60% increase. Clearly, the A-state in β-lactamase and cytochrome c, induced by low pH and high ionic strength, meet the criteria for a molten globule. ApoMb differs in its behavior: Although tertiary structure is absent, the secondary structure in the A-state is clearly less extensive than that in the native protein, and the structure is far less compact, though more compact than the acid-denatured form. The authors propose that in apoMb there is a compact core, but that one or both of the terminal regions are unfolded.

Formation of the molten globule from acid-denatured protein as the acid (or salt) concentration increases is attributed to anion binding to the positively charged protein (Goto *et al.*, 1990a). Proteins are generally unfolded at low pH because of the large positive charge that accumulates upon protonation of buried carboxylate and imidazole side chains. Anion binding reduces the positive charge density and helps stabilize a more compact structure.

Subsequent work of Goto *et al.* (1990b) has shown that addition of various mineral acids or neutral salts to apoMb and cytochrome c

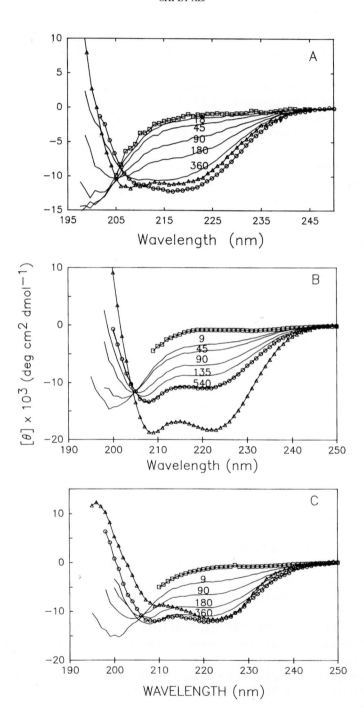

unfolded at pH 2 leads to molten globule formation. Fink *et al.* (1994) surveyed 20 proteins and found several types of behavior. Type IA proteins include cytochrome *c* and β-lactamase from two *Bacillus* species. These unfold maximally near pH 2 and then form a compact (~10% radial expansion) molten globule on further addition of HCl. Type IB includes α-amylase, apoMb and Mb (with most anions), subtilisin Carlsberg, β-lactamase from *Staphylococcus aureus,* and staphylococcal nuclease. These differ from type IA only in that the A-state is more expanded (~50% radial expansion). Type IC proteins include papain, parvalbumin, and ribonuclease A. These differ from type IA and IB proteins in that they are not fully unfolded at pH 2, according to far-UV CD. Type II proteins include α-lactalbumin and carbonic anhydrase. With these proteins, lowering the pH leads directly to the molten globule without passage through the acid-denatured state. Type III proteins, which include T4 lysozyme, ubiquitin, chicken lysozyme, chymotrypsinogen, protein A, β-lactoglobulin, and concanavalin A, are resistant to unfolding at low pH. They show no transition between pH 7 and pH 1.

A second molten globule form of apoMb has been identified (Jamin and Baldwin, 1998; Loh *et al.,* 1995). The well-characterized molten globule of apoMb (I_1) is observed at pH 4 (Eliezer *et al.,* 1998; Hughson *et al.,* 1990; Jennings and Wright, 1993). If anions such as trichloroacetate, citrate, or sulfate are added to this molten globule, the CD at 222 nm becomes more negative, indicating an increase in α-helix content in a new form, I_2, relative to I_1. NMR demonstrates that the new species lacks a well-defined tertiary structure, thus identifying it as a molten globule. Hydrogen exchange and 2D NMR show that I_2 differs from I_1 by the presence of the B helix, in addition to the A, G, and H helices that are present in I_1.

D. Implication of Intermediates in Protein Folding

Stopped-flow CD studies of protein folding have generally revealed a species that forms within the dead-time of the experiment, ~10 ms (Kuwajima *et al.,* 1985, 1987). The 222 nm ellipticity of this so-called burst-phase intermediate is generally substantially more negative than

---

FIG. 40.   Far-UV CD spectra of β-lactamase from *Bacillus cereus* (A), horse apomyoglobin (B), and horse ferricytochrome *c* (C) as a function of HCl concentration. Protein concentrations were 10 μM. The numbers refer to the HCl concentration (mM). The spectra of the native state (Δ), the A state induced by KCl, pH ≈ 2), (○) and GdmCl-unfolded state (4–5 M GdmCl, 25 mM phosphate buffer, pH 7.0) (□) are shown for comparison. From Goto *et al.* (1990a). © 1990, with permission of the authors.

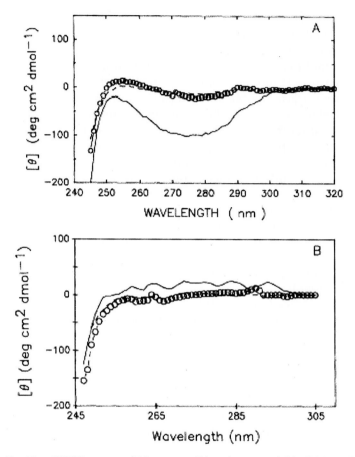

FIG. 41.   Near-UV CD spectra of β-lactamase (A) and apomyoglobin (B) in native state
(—), acid-unfolded state (– – –), and A state (○). Conditions as described in legend to
Figure 40. The GdmCl-unfolded state has a spectrum similar to that for the acid-unfolded
state. From Goto *et al.* (1990a). ©1990, with permission of the authors.

that of the unfolded species (usually in 6 M Gdm·HCl). In some cases,
the CD at 222 nm approaches or even exceeds that of the native protein,
as shown in Table VI (Roder and Elöve, 1994).

   The burst-phase intermediate has been studied intensively in the folding of bovine α-lactalbumin (Arai and Kuwajima, 1996; Ikeguchi *et al.*,
1986; Kuwajima *et al.*, 1985). The CD spectrum of the intermediate
has been determined by measuring the folding kinetics at various wavelengths and extrapolating the CD to zero time. The results are shown as
open circles (holoprotein) and squares (apoprotein) in Figure 35. The

TABLE VI

Relative Amplitude of "Burst Phase" Measured by Time-Resolved Far-UV CD Spectroscopy for Various Globular Proteins [a]

Protein	Amplitude[b] (%)	Dead time (ms)	Reference[c]
α-Lactalbumin	50	20	1
β-Lactoglobulin	200	18	2
Chymotrypsinogen A	100	18	3
Cytochrome c	44	4	4
Dihydrofolate reductase	40	18	5
Hen lysozyme	90	4	6, 7
Parvalbumin	60	18	8
Rat intestinal fatty acid binding protein	32	~100	9
Ribonuclease T$_1$	100	15	3
Staphylococcal nuclease	30	15	10
Tryptophan synthase β2	57	13	11
Tryptophan synthase β2 (F2–V8 fragment)	100	4	12

[a] © Oxford University Press, 1994. Reprinted from Mechanisms of Protein Folding edited by Roger H. Pain (1994), by permission of Oxford University Press.

[b] The burst phase amplitude represents the percentage change in the far-UV CD signal (216–225 nm) occurring in the dead time of the stopped-flow experiment relative to the total change on folding.

[c] Key to references: (1) Kuwajima *et al.* (1985); (2) Kuwajima *et al.* (1987); (3) Kiefhaber *et al.* (1992); (4) Elöve *et al.* (1992); (5) Kuwajima *et al.* (1991); (6) Radford *et al.* (1992); (7) Chafotte *et al.* (1992b); (8) Kuwajima *et al.* (1988); (9) Ropson *et al.* (1990); (10) Sugawara *et al.* (1991); (11) Goldberg *et al.* (1992); (12) Chaffotte *et al.* (1992a).

spectrum of this folding intermediate is similar to that of the equilibrium molten globule. Also shown in Figure 35 are the corresponding data for the near-UV CD. The intermediate has no well-defined tertiary structure. Arai and Kuwajima (1996) have also measured the transition curve for denaturation of the burst-phase intermediate by Gdm·HCl and found it to be in full agreement with that of the equilibrium molten globule.

The identification of the burst-phase intermediate in the folding of α-lactalbumin as a molten globule is supported by NMR. Balbach *et al.* (1995) showed that on folding from 6 M Gdm·HCl, a species with a spectrum closely resembling that of the equilibrium molten globule is formed first, which then forms the native species. Two-dimensional NMR experiments (Balbach *et al.*, 1996) were performed in which the pH of an equilibrium molten globule (pH 2.0, 3°C) solution was increased to pH 7.0 to initiate folding. The rate constants for the transition in the various secondary structural elements were essentially the same, indicating cooperative folding.

Identification of the burst-phase intermediate with a molten globule is also well-established for apoMb (Eliezer *et al.*, 1998; Hughson *et al.*, 1990; Jennings and Wright, 1993) and ribonuclease H (Dabora *et al.*, 1996; Raschke and Marqusee, 1997). The role of molten globules in protein folding and the evidence that the burst-phase intermediate is a molten globule have been discussed in several recent reviews (Arai and Kuwajima, 2000; Chamberlain and Marqusee, 2000; Ptitsyn, 1992, 1995).

The widely held view that the burst phase in stopped-flow CD is attributable to secondary structure formation in a relatively compact intermediate has been challenged. Sosnick *et al.* (1997) compared the burst phase in cytochrome *c* with those for two cytochrome fragments truncated from the C-terminus at Met 65 and Met 80. They found in both stopped-flow fluorescence and CD that the fragments gave bursts comparable to those obtained with the full-length cytochrome *c* upon dilution of 4.3 M Gdm·HCl (Fig. 42). However, the fragments are incapable of folding to a compact structure, as evidenced by their CD spectra, which are nearly independent of temperature and resemble that of thermally denatured cytochrome *c* (Fig. 43). Sosnick *et al.* (1997) attributed the burst phase in these systems to a conformational change resulting from the dilution of the denaturant. As noted in Section III,A, proteins unfolded in concentrated Gdm·HCl or urea differ in conformation from heat- or acid-denatured proteins.

Qi *et al.* (1998) have demonstrated that ribonuclease A exhibits behavior like that of cytochrome *c*. The burst phase observed on dilution of Gdm·HCl-denatured RNase A is mimicked exactly by reduced RNase A. The latter, when carboxamidomethylated to prevent oxidation, has a CD at 222 nm that is nearly independent of temperature and indicative of extensive unfolding at zero denaturant.

The studies of RNase A and cytochrome *c* (Qi *et al.*, 1998; Sosnick *et al.*, 1997) show that caution is required in interpreting burst phenomena in protein folding. However, they do not require a reinterpretation of the cases in which the molten globule character of the burst-intermediate has been established (Arai and Kuwajima, 2000; Chamberlain and Marqusee, 2000).

IV. SUMMARY AND BROADER IMPLICATIONS

Several lines of evidence indicate that oligomers of Ala in solution assume a predominantly P_{II} local conformation and that proteins unfolded by Gdm·HCl or urea also have a dominant conformation, P_{II}. Preliminary results on ubiqitin fragments and short sequences containing QQQ, SSS, FFF, and VVV in a series of 11-mers suggest that this is

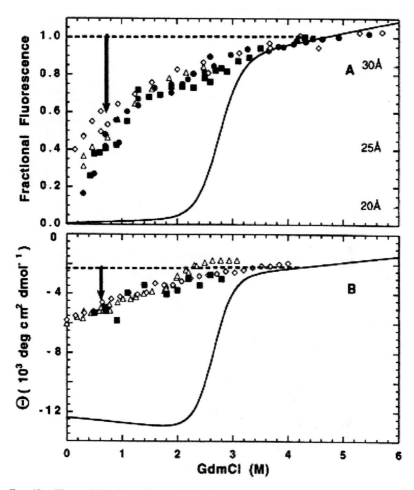

FIG. 42.   The unfolded baseline and the Cyt c burst phase. The solid curves show the equilibrium behavior of Cyt c. The equilibrium fluorescence and CD of the (unfolded) fragments ($\triangle$ and $\diamond$) match the unfolded holo Cyt c baseline at high GdmCl and define the continuation of the unfolded baseline to lower GdmCl concentrations. The horizontal dashed line shows the initial fluorescence and CD in the stopped-flow experiments (4.3 M GdmCl). The solid symbols indicate the fluorescence (A) and the ellipticity at 222 nm (B) reached by holo Cyt c in the burst phase on dilution into lower (or higher) GdmCl, as suggested by the arrows (starting from either pH 2 ($\bullet$) or pH 4.9 ($\blacksquare$)). These comparisons are made on an absolute, per-molecule basis. Förster-averaged distance (Trp-59 to heme) is at the right of A. (From Sosnick *et al.*, 1997, with permission. © 1997, National Academy of Sciences, USA.)

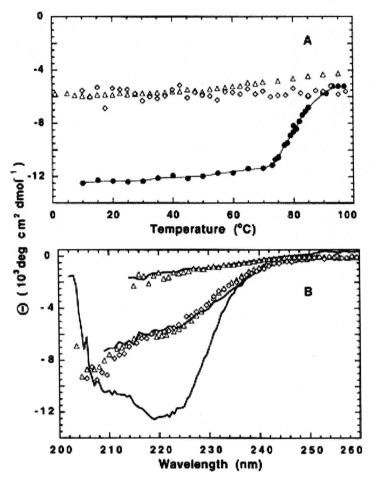

FIG. 43.   Ellipticity of the fragments (△, F1-80; ◇, F1-65), and holo Cyt *c* (● and/ or —). (A) Ellipticity (222 nm) as a function of temperature. (B) CD spectra: native Cyt *c* (bottom); the fragments at 22°C, and Cyt *c* thermally denatured at 97°C (middle); Cyt *c* and F1-80 in 4.4 M GdmCl (top). (From Sosnick *et al.,* 1997, with permission. © 1997, National Academy of Sciences, U.S.)

general for other nonrepeating sequences and amino acids other than Ala: All show the CD spectra assigned to P_{II}-helix (Shi, 2002). Many other studies of peptide fragments from native proteins show similar CD spectra (Blanco and Serrano, 1995; Dyson *et al.,* 1992; Jimenez *et al.,* 1993; Luisi *et al.,* 1999; Muñoz *et al.,* 1995; Najbar *et al.,* 2000; Viguera *et al.,* 1996). The new view we offer is that unfolded peptides and proteins have a strong tendency to be P_{II} locally, while conforming statistically to the overall dimensions of a statistical coil. Actually, statistical surveys of structures in the PDB show that P_{II} is a commonly occurring conformation in globular proteins (Adzhubei and Sternberg, 1993, 1994; Stapley and Creamer, 1999). It is estimated that up to 10% of residues that are not assigned to regular secondary structure are P_{II}. An analysis by Serrano of the ϕ and ψ angles in regions of native proteins that are outside of the regular α or β secondary structured regions shows a strong preference for the ϕ and ψ angles corresponding to P_{II} (Fig. 44), although this is not emphasized in his discussion (Serrano, 1995). Similarly, the coupling constants in unfolded proteins with or without denaturant are

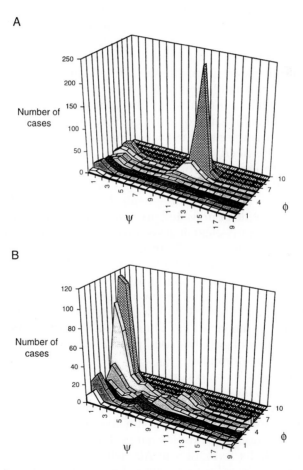

FIG. 44.   Distribution of Ala in the Ramachandran plot when using (A) all secondary structure conformations in the protein database or (B) only those Ala residues in a coil conformation. (From Serrano, 1995. © 1995, with permission from Academic Press.)

now interpreted in terms of a conformational blend in which α and β structures play a dominant role (Bai *et al.*, 2001; Dyson and Wright, 2001; Fiebig *et al.*, 1996; Penkett *et al.*, 1998; Smith *et al.*, 1996; Yao *et al.*, 2001). While this picture is not false, it misses a fundamental point: The presence of native-like or non-native secondary structure in unfolded proteins is superimposed on a background of P_{II} conformation, which is predominant in most cases for which data are available. In light of our new view on the structure in the unfolded state, we believe the following points need to be reconsidered:

1. It is common today to consider a peptide or an unfolded protein to be "random coil"—i.e., a blend of conformations—if their CD spectra resemble the traditionally accepted "random coil" CD spectrum (Blanco and Serrano, 1995; Dyson *et al.*, 1992; Jimenez *et al.*, 1993; Luisi *et al.*, 1999; Muñoz *et al.*, 1995; Najbar *et al.*, 2000; Viguera *et al.*, 1996). As we argue here, this CD spectrum does not correspond to random coil at all. Theoretical calculated spectra representing blends of different conformations have been published by Krimm's group (Ronish and Krimm, 1972, 1974). While the results are uncertain because of the unknown compositional coefficients for each conformation in the blend, these spectra are clearly unrelated to that of P_{II} and many unfolded peptides and proteins. We need to reconsider what kind of CD spectrum is an appropriate representative for a true random coil.

2. One of the methods used to assign the residual structure in unfolded proteins is based on NMR chemical shifts. If the chemical shift of a residue in a peptide or unfolded protein matches that of "random coil" instead of α-helix or β-sheet, it is assumed to sample conformations corresponding to a "random coil," although the chemical shift data actually may potentially represent a P_{II} conformation. All the NMR chemical shift index data are derived from oligo(Gly)-based series of peptides (Braun *et al.*, 1994; Plaxco *et al.*, 1997; Schwarzinger *et al.*, 2000, 2001; Wishart *et al.*, 1995). However, poly(Gly) is known to form a polyglycine II conformation that is similar to P_{II}. Thus, it is necessary to reexamine the structures of these oligo(Gly)-based peptides to check if they are really random coil or predominantly P_{II}.

3. Dobson's and Serrano's groups (Fiebig *et al.*, 1996; Serrano, 1995; Smith *et al.*, 1996) have attempted to derive a model of unfolded proteins based on the statistical distribution information of each amino acid in the PDB. Our work raises the possibility that this approach is not valid.

4. If P_{II} is the dominant conformation for oligo(Ala), the traditional description of helix formation in Ala-rich model peptides as a "helix-to-coil" transition may be a misnomer. Questions that remain to be addressed include the following: Where does the energetic barrier come

from if the loss of backbone entropy is not the dominant barrier to the transition? Experimental results actually indicate that there is an entropy difference between α-helix and P_{II} in Ala-rich model peptides, but where does this entropy come from? Does it reflect enthalpy–entropy compensation, or is it the result of reorganization of water molecules? Is the P_{II} conformation more flexible?

5. What interactions are responsible for the predominance of P_{II} in small oligo(Ala) peptides? Does it originate from the interaction of water molecules with the peptide backbone as suggested from X-ray structures of collagen peptides (Berisio *et al.*, 2000; Kramer *et al.*, 1998)? Or dipole–dipole interactions within the backbone itself? Or from steric effects as proposed by Pappu *et al.* (2000)?

6. Substantial experimental evidence has accumulated for residual structure in denatured proteins (Dill and Shortle, 1991; Dobson, 1992; Gillespie and Shortle, 1997; Shortle, 1993, 1996; Shortle and Ackerman, 2001). On the theoretical side, several recent studies have pointed to a substantially more limited range of accessible conformations than was previously believed (Baldwin and Zimm, 2000; Dinner and Karplus, 2001; Pappu *et al.*, 2000; van Gunsteren *et al.*, 2001a,b). As shown in Section III,A, unfolded proteins have a more limited range of conformations than was formerly appreciated (Tanford, 1968). The two types of denaturation that probably provide the best models for a nascent protein chain are acid and cold denaturation. Roughly half of the residues in acid- and cold-denatured proteins are in the P_{II} conformation. The concentration of the residues in the P_{II} region of conformational space lowers the entropy of the unfolded protein chain and therefore facilitates folding under appropriate conditions.

ACKNOWLEDGMENTS

RWW thanks Ms. Janice Chapman and Dr. Narasimha Sreerama for their expert assistance. ZS and NRK thank Drs. Robert L. Baldwin, Trevor Creamer, Tobin Sosnick, Michael Levitt and George Rose for helpful discussions and guidance. Preparation of this manuscript and the research described herein from the Woody and Kallenbach laboratories, published and unpublished, have been supported by NIH Grant GM 22994 (RWW) and ONR grant N00014-02-1-0125 (NRK).

REFERENCES

Acharya, K. R., Stuart, D. I., Walker, N. P., Lewis, M., and Phillips, D. C. (1989). *J. Mol. Biol.* **208**, 99–127.

Adler, A. J., Hoving, R., Potter, J., Wells, M., and Fasman, G. D. (1968). *J. Am. Chem. Soc.* **90**, 4736–4738.

Adzhubei, A. A., and Sternberg, M. J. (1993). *J. Mol. Biol.* **229**, 472–493.

Adzhubei, A. A., and Sternberg, M. J. (1994). *Protein Sci.* **3,** 2395–2410.

Allison, L. A., Moyle, M., Shales, M., and Ingles, C. J. (1985). *Cell* **42,** 599–610.

Anderson, A. G., and Hermans, J. (1988). *Proteins* **3,** 262–265.

Andries, J. C., Anderson, J. M., and Walton, A. G. (1971). *Biopolymers* **10,** 1049–1057.

Arai, M., and Kuwajima, K. (1996). *Folding Design* **1,** 275–287.

Arai, M., and Kuwajima, K. (2000). *Adv. Protein Chem.* **53,** 209–282.

Arcus, V. L., Vuilleumier, S., Freund, S. M., Bycroft, M., and Fersht, A. R. (1994). *Proc. Natl. Acad. Sci. USA* **91,** 9412–9416.

Arcus, V. L., Vuilleumier, S., Freund, S. M., Bycroft, M., and Fersht, A. R. (1995). *J. Mol. Biol.* **254,** 305–321.

Bai, Y., Chung, J., Dyson, H. J., and Wright, P. E. (2001). *Protein Sci.* **10,** 1056–1066.

Balbach, J., Forge, V., Lau, W. S., van Nuland, N. A., Brew, K., and Dobson, C. M. (1996). *Science* **274,** 1161–1163.

Balbach, J., Forge, V., van Nuland, N. A., Winder, S. L., Hore, P. J., and Dobson, C. M. (1995). *Nat. Struct. Biol.* **2,** 865–870.

Baldwin, R. L., and Zimm, B. H. (2000). *Proc. Natl. Acad. Sci. USA* **97,** 12391–12392.

Barron, L. D., Hecht, L., Blanch, E. W., and Bell, A. F. (2000). *Prog. Biophys. Mol. Biol.* **73,** 1–49.

Baum, J., Dobson, C. M., Evans, P. A., and Hanley, C. (1989). *Biochemistry* **28,** 7–13.

Berisio, R., Vitagliano, L., Mazzarella, L., and Zagari, A. (2000). *Biopolymers* **56,** 8–13.

Beychok, S. (1965). *Proc. Natl. Acad. Sci. USA* **53,** 999–1006.

Beychok, S. (1967). In *Poly-α-amino Acids* (G. D. Fasman, ed.), pp. 293–337. Marcel Dekker, New York.

Bhatnagar, R. S., and Gough, C. A. (1996). In *Circular Dichroism and the Conformational Analysis of Biomolecules* (G. D. Fasman, ed.), pp. 183–199. Plenum Press, New York.

Bienkiewicz, E. A., Adkins, J. N., and Lumb, K. J. (2002). *Biochemistry* **41,** 752–759.

Bienkiewicz, E. A., Woody, A.-Y. M., and Woody, R. W. (2000). *J. Mol. Biol.* **297,** 119–133.

Blanch, E. W., Morozova-Roche, L. A., Cochran, D. A., Doig, A. J., Hecht, L., and Barron, L. D. (2000). *J. Mol. Biol.* **301,** 553–563.

Blanco, F. J., and Serrano, L. (1995). *Eur. J. Biochem.* **230,** 634–649.

Blout, E. R., and Fasman, G. D. (1967). *Rec. Adv. Gelatin Glue Res.* **1,** 122.

Blundell, T., Barlow, D., Borkakoti, N., and Thornton, J. (1983). *Nature* **306,** 281–283.

Bond, C. J., Wong, K. B., Clarke, J., Fersht, A. R., and Daggett, V. (1997). *Proc. Natl. Acad. Sci. USA* **94,** 13409–13413.

Brant, D. A., Miller, W. G., and Flory, P. J. (1967). *J. Mol. Biol.* **23,** 47–65.

Braun, D., Wider, G., and Wüthrich, K. (1994). *J. Am. Chem. Soc.* **116,** 8466–8469.

Brooks, C. L., and Case, D. A. (1993). *Chem. Rev.* **93,** 2487–2502.

Cagas, P. M., and Corden, J. L. (1995). *Proteins* **21,** 149–160.

Campbell, K. M., Terrell, A. R., Laybourn, P. J., and Lumb, K. J. (2000). *Biochemistry* **39,** 2708–2713.

Carver, J. P., Shechter, E., and Blout, E. R. (1966). *J. Am. Chem. Soc.* **88,** 2550–2561.

Chaffotte, A. F., Cadieux, C., Guillou, Y., and Goldberg, M. E. (1992a). *Biochemistry* **31,** 4303–4308.

Chaffotte, A. F., Guillou, Y., and Goldberg, M. E. (1992b). *Biochemistry* **31,** 9694–9702.

Chamberlain, A. K., and Marqusee, S. (2000). *Adv. Protein Chem.* **53,** 283–328.

Cho, H. S., Liu, C. W., Damberger, F. F., Pelton, J. G., Nelson, H. C., and Wemmer, D. E. (1996). *Protein Sci.* **5,** 262–269.

Christensen, H., and Pain, R. H. (1994). In *Mechanisms of Protein Folding* (R. H. Pain, ed.), pp. 55–79. Oxford University Press, Oxford.

Chyan, C. L., Wormald, C., Dobson, C. M., Evans, P. A., and Baum, J. (1993). *Biochemistry* **32,** 5681–5691.

Corden, J. L., Cadena, D. L., Ahearn, J. M., Jr., and Dahmus, M. E. (1985). *Proc. Natl. Acad. Sci. USA* **82,** 7934–7938.

Corden, J. L., and Ingles, C. J. (1985). In *Transcriptional Regulation* (S. L. McKnight and K. R. Yamamoto, eds.), pp. 81–107. Cold Spring Harbor Laboratory, Plainview, NY.

Cortijo, M., Panijpan, B., and Gratzer, W. B. (1973). *Int. J. Pept. Protein Res.* **5,** 179–186.

Cowan, P. M., and McGavin, S. (1955). *Nature* **176,** 501–503.

Dabora, J. M., Pelton, J. G., and Marqusee, S. (1996). *Biochemistry* **35,** 11951–11958.

Dahlman-Wright, K., Baumann, H., McEwan, I. J., Almlof, T., Wright, A. P., Gustafsson, J. A., and Härd, T. (1995). *Proc. Natl. Acad. Sci. USA* **92,** 1699–1703.

Dahmus, M. E. (1996). *J. Biol. Chem.* **271,** 19009–19012.

Deng, Z., Polavarapu, P. L., Ford, S. J., Hecht, L., Barron, L. D., Ewig, C. S., and Jalkanen, K. (1996). *J. Phys. Chem.* **100,** 2025–2034.

Dill, K. A., and Shortle, D. (1991). *Annu. Rev. Biochem.* **60,** 795–825.

Dinner, A. R., and Karplus, M. (2001). *Angew. Chem. Int. Ed.* **40,** 4615–4616.

Dobson, C. M. (1992). *Curr. Opin. Struct. Biol.* **2,** 6–12.

Dolgikh, D. A., Abaturov, L. V., Bolotina, I. A., Brazhnikov, E. V., Bychkova, V. E., Gilmanshin, R. I., Lebedev, Yu. O., Semisotnov, G. V., Tiktopulo, E. I., and Ptitsyn, O. B. (1985). *Eur. Biophys. J.* **13,** 109–121.

Dolgikh, D. A., Abaturov, L. V., Brazhnikov, E. V., Lebedev, I. O., Chirgadze, I. N., and Ptitsyn, O. B. (1983). *Dokl. Akad. Nauk SSSR* **272,** 1481–1484.

Doty, P., Wada, A., Yang, J. T., and Blout, E. R. (1957). *J. Polym. Sci.* **23,** 851–861.

Drake, A. F., Siligardi, G., and Gibbons, W. A. (1988). *Biophys. Chem.* **31,** 143–146.

Dukor, R. K., and Keiderling, T. A. (1991). *Biopolymers* **31,** 1747–1761.

Dunker, A. K., Garner, E., Guilliot, S., Romero, P., Albrecht, K., Hart, J., Obradovic, Z., Kissinger, C., and Villafranca, J. E. (1998). *Pac. Symp. Biocomput.* **3,** 473–484.

Dunker, A. K., Lawson, J. D., Brown, C. J., Williams, R. M., Romero, P., Oh, J. S., Oldfield, C. J., Campen, A. M., Ratliff, C. M., Hipps, K. W., Ausio, J., Nissen, M. S., Reeves, R., Kang, C., Kissinger, C. R., Bailey, R. W., Griswold, M. D., Chiu, W., Garner, E. C., and Obradovic, Z. (2001). *J. Mol. Graph. Model* **19,** 26–59.

DuPont, F. M., Vensel, W. H., Chan, R., and Kasarda, D. D. (2000). *Cereal Chem.* **77,** 607–614.

Dyson, H. J., Merutka, G., Waltho, J. P., Lerner, R. A., and Wright, P. E. (1992). *J. Mol. Biol.* **226,** 795–817.

Dyson, H. J., Rance, M., Houghten, R. A., Wright, P. E., and Lerner, R. A. (1988). *J. Mol. Biol.* **201,** 201–217.

Dyson, H. J., and Wright, P. E. (2001). *Methods Enzymol.* **339,** 258–270.

Dyson, H. J., and Wright, P. E. (2002). *Adv. Protein Chem.* **62,** 311–340.

Eliezer, D., Yao, J., Dyson, H. J., and Wright, P. E. (1998). *Nat. Struct. Biol.* **5,** 148–155.

Elöve, G. A., Chaffotte, A., Roder, H., and Goldberg, M. E. (1992). *Biochemistry* **31,** 6876–6883.

Eriksson, A. E., and Liljas, A. (1993). *Proteins* **16,** 29–42.

Fasman, G. D., Hoving, H., and Timasheff, S. N. (1970). *Biochemistry* **9,** 3316–3324.

Feng, Y. B., Melacini, G., Taulane, J. P., and Goodman, M. (1996). *J. Am. Chem. Soc.* **118,** 10351–10358.

Fiebig, K. M., Schwalbe, H., Buck, M., Smith, L. J., and Dobson, C. M. (1996). *J. Phys. Chem.* **100,** 2661–2666.

Fink, A. L. (1995). *Annu. Rev. Biophys. Biomol. Struct.* **24,** 495–522.

Fink, A. L., Calciano, L. J., Goto, Y., Kurotsu, T., and Palleros, D. R. (1994). *Biochemistry* **33,** 12504–12511.

Flory, P. J. (1969). *Statistical Mechanics of Chain Molecules.* Interscience, New York.

Freskgård, P. O., Mårtensson, L. G., Jonasson, P., Jonsson, B. H., and Carlsson, U. (1994). *Biochemistry* **33,** 14281–14288.

Gast, K., Damaschun, H., Eckert, K., Schulze-Forster, K., Maurer, H. R., Muller-Frohne, M., Zirwer, D., Czarnecki, J., and Damaschun, G. (1995). *Biochemistry* **34,** 13211–13218.

Gilbert, S. M., Wellner, N., Belton, P. S., Greenfield, J. A., Siligardi, G., Shewry, P. R., and Tatham, A. S. (2000). *Biochim. Biophys. Acta* **1479,** 135–146.

Gillespie, J. R., and Shortle, D. (1997). *J. Mol. Biol.* **268,** 170–184.

Goldberg, M. E., Semisotnov, G. V., Friguet, B., Kuwajima, K., Ptitsyn, O. B., and Suga, S. (1992). *FEBS. Lett.* **263,** 51–56.

Goto, Y., Calciano, L. J., and Fink, A. L. (1990a). *Proc. Natl. Acad. Sci. USA* **87,** 573–577.

Goto, Y., and Fink, A. L. (1989). *Biochemistry* **28,** 945–952.

Goto, Y., Takahashi, N., and Fink, A. L. (1990b). *Biochemistry* **29,** 3480–3488.

Grant, J. A., Williams, R. L., and Scheraga, H. A. (1990). *Biopolymers* **30,** 929–949.

Greenfield, N., and Fasman, G. D. (1969). *Biochemistry* **8,** 4108–4116.

Griko, Y., Sreerama, N., Osumi-Davis, P., Woody, R. W., and Woody, A.-Y. M. (2001). *Protein Sci.* **10,** 845–853.

Grishina, I. B. (1994). Ph.D. Thesis. Colorado State University.

Hamm, P., Lim, M., DeGrado, W. F., and Hochstrasser, R. M. (1999). *Proc. Natl. Acad. Sci. USA* **96,** 2036–2041.

Han, W. G., Jalkanen, K. J., Elstner, M., and Suhai, S. (1998). *J. Phys. Chem. B* **102,** 2587–2602.

Hodes, Z. I., Nemethy, G., and Scheraga, H. A. (1979a). *Biopolymers* **18,** 1611–1634.

Hodes, Z. I., Nemethy, G., and Scheraga, H. A. (1979b). *Biopolymers* **18,** 1565–1610.

Holladay, L. A., Hammonds, R. G., Jr., and Puett, D. (1974). *Biochemistry* **13,** 1653–1661.

Holzwarth, G., and Doty, P. (1965). *J. Am. Chem. Soc.* **87,** 218–228.

Hughson, F. M., Wright, P. E., and Baldwin, R. L. (1990). *Science* **249,** 1544–1548.

Iizuka, E., and Yang, J. T. (1965). *Biochemistry* **4,** 1249–1257.

Ikeguchi, M., Kuwajima, K., Mitani, M., and Sugai, S. (1986). *Biochemistry* **25,** 6965–6972.

Jalkanen, K. J., and Suhai, S. (1996). *Chem. Phys.* **208,** 81–116.

Jamin, M., and Baldwin, R. L. (1998). *J. Mol. Biol.* **276,** 491–504.

Jeng, M.-F., and Englander, S. W. (1991). *J. Mol. Biol.* **221,** 1045–1061.

Jenness, D. D., Sprecher, C., and Johnson, W. C., Jr. (1976). *Biopolymers* **15,** 513–521.

Jennings, P. A., and Wright, P. E. (1993). *Science* **262,** 892–896.

Jimenez, M. A., Bruix, M., Gonzalez, C., Blanco, F. J., Nieto, J. L., Herranz, J., and Rico, M. (1993). *Eur. J. Biochem.* **211,** 569–581.

Johnson, W. C., and Tinoco, I. (1972). *J. Am. Chem. Soc.* **94,** 4389–4390.

Katou, H., Hoshino, M., Kamikubo, H., Batt, C. A., and Goto, Y. (2001). *J. Mol. Biol.* **310,** 471–484.

Keiderling, T. A., Silva, R. A., Yoder, G., and Dukor, R. K. (1999). *Bioorg. Med. Chem.* **7,** 133–141.

Keiderling, T. A., Wang, B. L., Urbanova, M., Pancoska, P., and Dukor, R. K. (1994). *Faraday Discuss.* **99,** 263–285.

Kiefhaber, T., Schmid, F. X., Willert, K., Engelborghs, Y., and Chaffotte, A. (1992). *Protein Sci.* **1,** 1162–1172.

Kleywegt, G. J., and Jones, T. A. (1996). *Structure* **4,** 1395–1400.

Kramer, R. Z., Vitagliano, L., Bella, J., Berisio, R., Mazzarella, L., Brodsky, B., Zagari, A., and Berman, H. M. (1998). *J. Mol. Biol.* **280,** 623–638.

Krimm, S., and Mark, J. E. (1968). *Proc. Natl. Acad. Sci. USA* **60**, 1122–1129.
Krimm, S., and Tiffany, M. L. (1974). *Isr. J. Chem.* **12**, 189–200.
Kriwacki, R. W., Hengst, L., Tennant, L., Reed, S. I., and Wright, P. E. (1996). *Proc. Natl. Acad. Sci. USA* **93**, 11504–11509.
Kronman, M. J. (1968). *Biochem. Biophys. Res. Commun.* **33**, 535–541.
Kuwajima, K. (1977). *J. Mol. Biol.* **114**, 241–258.
Kuwajima, K. (1989). *Proteins* **6**, 87–103.
Kuwajima, K., Garvey, E. P., Finn, B. E., Matthews, C. R., and Sugai, S. (1991). *Biochemistry* **30**, 7693–7703.
Kuwajima, K., Hiraoka, Y., Ikeguchi, M., and Sugai, S. (1985). *Biochemistry* **24**, 874–881.
Kuwajima, K., Nitta, K., Yoneyama, M., and Sugai, S. (1976). *J. Mol. Biol.* **106**, 359–373.
Kuwajima, K., Sakuraoka, A., Fueki, S., Yoneyama, M., and Sugai, S. (1988). *Biochemistry* **27**, 7419–7428.
Kuwajima, K., Yamaya, H., Miwa, S., Sugai, S., and Nagamura, T. (1987). *FEBS Lett.* **221**, 115–118.
Lapanje, S., and Tanford, C. (1967). *J. Am. Chem. Soc.* **89**, 5030–5033.
Lau, W. F., and Pettitt, B. M. (1987). *Biopolymers* **26**, 1817–1831.
Legrand, M., and Viennet, R. (1964). *C. R. Hebd. Seances Acad. Sci.* **259**, 4277–4280.
Levinthal, C. (1968). *J. Chim. Phys.-Chim. Biol.* **65**, 44–45.
Loh, S. N., Kay, M. S., and Baldwin, R. L. (1995). *Proc. Natl. Acad. Sci. USA* **92**, 5446–5450.
Lotz, B., and Keith, H. D. (1971). *J. Mol. Biol.* **61**, 201–215.
Luisi, D. L., Wu, W. J., and Raleigh, D. P. (1999). *J. Mol. Biol.* **287**, 395–407.
Lupu-Lotan, N., Yaron, A., Berger, A., and Sela, M. (1965). *Biopolymers* **3**, 625–655.
Ma, K., Kan, L., and Wang, K. (2001). *Biochemistry* **40**, 3427–3438.
Madison, V., and Kopple, K. D. (1980). *J. Am. Chem. Soc.* **102**, 4855–4863.
Matouschek, A., Kellis, J. T., Jr., Serrano, L., and Fersht, A. R. (1989). *Nature* **340**, 122–126.
Mattice, W. L. (1974). *Biopolymers* **13**, 169–183.
Mattice, W. L., Mandelkern, L., and Lo, J. T. (1972). *Macromolecules* **5**, 729–734.
Mattice, W. L., Riser, J. M., and Clark, D. S. (1976). *Biochemistry* **15**, 4264–4272.
McEwan, I. J., Dahlman-Wright, K., Ford, J., and Wright, A. P. (1996). *Biochemistry* **35**, 9584–9593.
Mezei, M., Mehrotra, P. K., and Beveridge, D. L. (1985). *J. Am. Chem. Soc.* **107**, 2239–2245.
Muñoz, V., Serrano, L., Jimenez, M. A., and Rico, M. (1995). *J. Mol. Biol.* **247**, 648–669.
Najbar, L. V., Craik, D. J., Wade, J. D., and McLeish, M. J. (2000). *Biochemistry* **39**, 5911–5920.
Neri, D., Billeter, M., Wider, G., and Wüthrich, K. (1992a). *Science* **257**, 1559–1563.
Neri, D., Wider, G., and Wuthrich, K. (1992b). *Proc. Natl. Acad. Sci. USA* **89**, 4397–4401.
Nölting, B., Golbik, R., Soler-Gonzalez, A. S., and Fersht, A. R. (1997). *Biochemistry* **36**, 9899–9905.
Nozaki, Y., and Tanford, C. (1967). *J. Am. Chem. Soc.* **89**, 742–749.
Ohgushi, M., and Wada, A. (1983). *FEBS Lett.* **164**, 21–24.
Pappu, R. V., Srinivasan, R., and Rose, G. D. (2000). *Proc. Natl. Acad. Sci. USA* **97**, 12565–12570.
Park, S. H., Shalongo, W., and Stellwagen, E. (1997). *Protein Sci.* **6**, 1694–1700.
Penkett, C. J., Redfield, C., Jones, J. A., Dodd, I., Hubbard, J., Smith, R. A., Smith, L. J., and Dobson, C. M. (1998). *Biochemistry* **37**, 17054–17067.
Pettitt, B. M., Karplus, M., and Rossky, P. J. (1986). *J. Phys. Chem.* **90**, 6335–6345.
Pflumm, M. N., and Beychok, S. (1966). Abstr. 152nd Meeting, American Chemical Society, New York, Sept. 1966, Abstr. 233.

Plaxco, K. W., Morton, C. J., Grimshaw, S. B., Jones, J. A., Pitkeathly, M., Campbell, I. D., and Dobson, C. M. (1997). *J. Biomol. NMR* **10,** 221–230.

Poon, C. D., Samulski, E. T., Weise, C. F., and Weisshaar, J. C. (2000). *J. Am. Chem. Soc.* **122,** 5642–5643.

Privalov, P. L., Griko, Yu. V., Venyaminov, S., and Kutyshenko, V. P. (1986). *J. Mol. Biol.* **190,** 487–498.

Privalov, P. L., Tiktopulo, E. I., Venyaminov, S., Griko, Yu. V., Makhatadze, G. I., and Khechinashvili, N. N. (1989). *J. Mol. Biol.* **205,** 737–750.

Ptitsyn, O. B. (1987). *J. Protein Chem.* **6,** 273–293.

Ptitsyn, O. B. (1992). In *Protein Folding* (T. E. Creighton, ed.), pp. 243–300. W. H. Freeman, New York.

Ptitsyn, O. B. (1995). *Adv. Protein Chem.* **47,** 83–229.

Ptitsyn, O. B., Pain, R. H., Semisotnov, G. V., Zerovnik, E., and Razgulyaev, O. I. (1990). *FEBS Lett.* **262,** 20–24.

Qi, P. X., Sosnick, T. R., and Englander, S. W. (1998). *Nat. Struct. Biol.* **5,** 882–884.

Radford, S. E., Dobson, C. M., and Evans, P. A. (1992). *Nature* **358,** 302–307.

Raschke, T. M., and Marqusee, S. (1997). *Nat. Struct. Biol.* **4,** 298–304.

Rippon, W. B., and Walton, A. G. (1971). *Biopolymers* **10,** 1207–1212.

Rippon, W. B., and Walton, A. G. (1972). *J. Am. Chem. Soc.* **94,** 4319–4324.

Robson, B., and Pain, R. H. (1976). *Biochem. J.* **155,** 331–344.

Roder, H., and Elöve, G. A. (1994). In *Mechanisms of Protein Folding* (R. H. Pain, ed.), pp. 26–54. Oxford University Press, Oxford.

Romero, P., Obradovic, Z., Kissinger, C. R., Villafranca, J. E., Garner, E., Guilliot, S., and Dunker, A. K. (1998). *Pac. Symp. Biocomput.* **3,** 437–448.

Ronish, E. W., and Krimm, S. (1972). *Biopolymers* **11,** 1919–1928.

Ronish, E. W., and Krimm, S. (1974). *Biopolymers* **13,** 1635–1651.

Ropson, I. J., Gordon, J. I., and Frieden, C. (1990). *Biochemistry* **29,** 9591–9599.

Rossky, P. J., Karplus, M., and Rahman, A. (1979). *Biopolymers* **18,** 825–854.

Rucker, A. L., and Creamer, T. P. (2002). *Protein Sci.* **11,** 980–985.

Schellman, J. A., and Schellman, C. G. (1964). In *The Proteins* 2nd ed. (H. Neurath, ed.), vol. 2, pp 1–37. Academic Press, New York.

Schmidt, A. B., and Fine, R. M. (1994). *Mol. Simul.* **13,** 347–365.

Schmitz, M. L., dos Santos Silva, M. A., Altmann, H., Czisch, M., Holak, T. A., and Baeuerle, P. A. (1994). *J. Biol. Chem.* **269,** 25613–25620.

Schulman, B. A., Kim, P. S., Dobson, C. M., and Redfield, C. (1997). *Nat. Struct. Biol.* **4,** 630–634.

Schulman, B. A., Redfield, C., Peng, Z. Y., Dobson, C. M., and Kim, P. S. (1995). *J. Mol. Biol.* **253,** 651–657.

Schwalbe, H., Fiebig, K. M., Buck, M., Jones, J. A., Grimshaw, S. B., Spencer, A., Glaser, S. J., Smith, L. J., and Dobson, C. M. (1997). *Biochemistry* **36,** 8977–8991.

Schwarzinger, S., Kroon, G. J. A., Foss, T. R., Chung, J., Wright, P. E., and Dyson, H. J. (2001). *J. Am. Chem. Soc.* **123,** 2970–2978.

Schwarzinger, S., Kroon, G. J. A., Foss, T. R., Wright, P. E., and Dyson, H. J. (2000). *J. Biomol. NMR* **18,** 43–48.

Schweers, O., Schonbrunn-Hanebeck, E., Marx, A., and Mandelkow, E. (1994). *J. Biol. Chem.* **269,** 24290–24297.

Schweitzer-Stenner, R., Eker, F., Huang, Q., and Griebenow, K. (2001). *J. Am. Chem. Soc.* **123,** 9628–9633.

Scott, R. A., and Scheraga, H. A. (1966). *J. Chem. Phys.* **45,** 2091–2101.

Serrano, L. (1995). *J. Mol. Biol.* **254,** 322–333.

Shi, Z., Olson, C. A., Rose, G. D., Baldwin, R. L., and Kallenbach, N. R. (2002). *Proc. Natl. Acad. Sci. USA* **99,** 9190–9195.

Shi, Z. (2002). Ph.D. Thesis. New York University, New York.

Shortle, D. (1993). *Curr. Opin. Struct. Biol.* **3,** 66–74.

Shortle, D. (1996). *FASEB J.* **10,** 27–34.

Shortle, D., and Ackerman, M. S. (2001). *Science* **293,** 487–489.

Smith, L. J., Bolin, K. A., Schwalbe, H., MacArthur, M. W., Thornton, J. M., and Dobson, C. M. (1996). *J. Mol. Biol.* **255,** 494–506.

Smith, P. E. (1999). *J. Chem. Phys.* **111,** 5568–5579.

Smyth, E., Syme, C. D., Blanch, E. W., Hecht, L., Vasak, M., and Barron, L. D. (2001). *Biopolymers* **58,** 138–151.

Sosnick, T. R., Shtilerman, M. D., Mayne, L., and Englander, S. W. (1997). *Proc. Natl. Acad. Sci. USA* **94,** 8545–8550.

Sreerama, N., and Woody, R. W. (1994). *Biochemistry* **33,** 10022–10025.

Stapley, B. J., and Creamer, T. P. (1999). *Protein Sci.* **8,** 587–595.

Sugawara, T., Kuwajima, K., and Sugai, S. (1991). *Biochemistry* **30,** 2698–2706.

Tanford, C. (1968). *Adv. Protein Chem.* **23,** 121–282.

Tanford, C., Kawahara, K., and Lapanje, S. (1967a). *J. Am. Chem. Soc.* **89,** 729–736.

Tanford, C., Kawahara, K., Lapanje, S., Hooker, T. M., Zarlengo, M. H., Salahudd, A., Aune, K. C., and Takagi, T. (1967b). *J. Am. Chem. Soc.* **89,** 5023–5029.

Tatham, A. S., Drake, A. F., and Shewry, P. R. (1989). *Biochem. J.* **259,** 471–476.

Tatham, A. S., Drake, A. F., and Shewry, P. R. (1990a). *J. Cereal Sci.* **11,** 189–200.

Tatham, A. S., Marsh, M. N., Wieser, H., and Shewry, P. R. (1990b). *Biochem. J.* **270,** 313–318.

Tiffany, M. L., and Krimm, S. (1968a). *Biopolymers* **6,** 1767–1770.

Tiffany, M. L., and Krimm, S. (1968b). *Biopolymers* **6,** 1379–1382.

Tiffany, M. L., and Krimm, S. (1969). *Biopolymers* **8,** 347–359.

Tiffany, M. L., and Krimm, S. (1972). *Biopolymers* **11,** 2309–2316.

Tiffany, M. L., and Krimm, S. (1973). *Biopolymers* **12,** 575–587.

Timasheff, S. N., Susi, H., Townend, R., Stevens, L., Gorbunoff, M. J., and Kumosinski, T. F. (1967). In *Conformation of Biopolymers* (G. N. Ramachandran, ed.), vol. 1, pp. 173–196. Academic Press, London.

Torri, H., and Tasumi, M. (1998). *J. Raman Spectrosc.* **29,** 81–86.

Urnes, P., and Doty, P. (1961). *Adv. Protein Chem.* **16,** 401–535.

Uversky, V. N., Gillespie, J. R., and Fink, A. L. (2000). *Proteins* **41,** 415–427.

Van Dijk, A. A., De Boef, E., Bekkers, A., Van Wijk, L. L., Van Swieten, E., Hamer, R. J., and Robillard, G. T. (1997a). *Protein Sci.* **6,** 649–656.

Van Dijk, A. A., Van Wijk, L. L., Van Vliet, A., Haris, P., Van Swieten, E., Tesser, G. I., and Robillard, G. T. (1997b). *Protein Sci.* **6,** 637–648.

van Gunsteren, W. F., Burgi, P., Peter, C., and Daura, X. (2001a). *Angew. Chem. Int. Ed.* **40,** 4616–4618.

van Gunsteren, W. F., Burgi, P., Peter, C., and Daura, X. (2001b). *Angew. Chem. Int. Ed.* **40,** 351–355.

Van Hoy, M., Leuther, K. K., Kodadek, T., and Johnston, S. A. (1993). *Cell* **72,** 587–594.

Velluz, L., and Legrand, M. (1965). *Angew. Chem. Int. Ed.* **4,** 838–845.

Viguera, A. R., Jimenez, M. A., Rico, M., and Serrano, L. (1996). *J. Mol. Biol.* **255,** 507–521.

Wishart, D. S., Bigam, C. G., Holm, A., Hodges, R. S., and Sykes, B. D. (1995). *J. Biomol. NMR.* **5,** 67–81.

Wong, K. B., Freund, S. M., and Fersht, A. R. (1996). *J. Mol. Biol.* **259,** 805–818.

Wong, K. P., and Hamlin, L. M. (1974). *Biochemistry* **13,** 2678–2683.

Woody, R. W. (1992). *Adv. Biophys. Chem.* **2**, 37–79.

Woutersen, S., and Hamm, P. (2000). *J. Phys. Chem. B* **104**, 11316–11320.

Woutersen, S., and Hamm, P. (2001). *J. Chem. Phys.* **114**, 2727–2737.

Wright, P. E., and Dyson, H. J. (1999). *J. Mol. Biol.* **293**, 321–331.

Yasui, S. C., and Keiderling, T. A. (1986). *J. Am. Chem. Soc.* **108**, 5576–5581.

Yao, J., Chung, J., Eliezer, D., Wright, P. E., and Dyson, H. J. (2001). *Biochemistry* **40**, 3561–3571.

Yoder, G., Pancoska, P., and Keiderling, T. A. (1997). *Biochemistry* **36**, 15123–15133.

Zhang, O., and Forman-Kay, J. D. (1997). *Biochemistry* **36**, 3959–3970.

Zhang, O., Forman-Kay, J. D., Shortle, D., and Kay, L. E. (1997a). *J. Biomol. NMR* **9**, 181–200.

Zhang, O., Kay, L. E., Shortle, D., and Forman-Kay, J. D. (1997b). *J. Mol. Biol.* **272**, 9–20.

TOWARD A TAXONOMY OF THE DENATURED STATE: SMALL ANGLE SCATTERING STUDIES OF UNFOLDED PROTEINS

By IAN S. MILLETT,* SEBASTIAN DONIACH,* and KEVIN W. PLAXCO[†]

*Department of Applied Physics, Stanford University, Stanford, California 92343; [†]Department of Chemistry and Biochemistry, and Interdepartmental Program in Biomolecular Science and Engineering, University of California, Santa Barbara, Santa Barbara, California 93106

I. INTRODUCTION

Despite the critical role the unfolded state plays in defining protein folding kinetics and thermodynamics (Berg *et al.*, 2002; Dunker, 2002; Shortle, 2002; Wright and Dyson, 2002), our understanding of its detailed structure remains rather rudimentary; the heterogeneity of the unfolded ensemble renders difficult or impossible its study by traditional, atomic-level structural methods. Consequently, recent years have seen a significant expansion of small-angle X-ray and neutron scattering (SAXS and SANS, respectively) techniques that provide direct, albeit rotationally and time-averaged, measures of the geometric properties of the unfolded ensemble. These studies have reached a critical mass, allowing us for the first time to define general observations regarding the nature of the geometry—and possibly the chemistry and physics—of unfolded proteins.

"That which we call a rose, by any other name would smell as sweet." The essence of the denatured state, in contrast, is rather more elusive. The difficulty stems, at least in part, from the wide variety of methods of inducing a protein to unfold. Some proteins appear to be "natively" unfolded; that is, they remain unfolded in the cell under conditions in which they retain their biological activity (Plaxco and Gross, 1997; Wright and Dyson 1999; Dunker, 2002). Other proteins unfold only under the influence of changes in pH, high or low temperatures, or

*ADVANCES IN
PROTEIN CHEMISTRY, Vol. 62*

high concentrations of chemical denaturants such as urea, guanidine hydrochloride (GuHCl), or organic cosolvents. A fundamental limitation of these approaches is that, presumably, none of them accurately recreates the ensemble populated by nascent proteins or by an unfolded polypeptide in equilibrium with the folded state under physiological conditions i.e., the unfolded state, which defines the thermodynamics and kinetics of folding where it counts—in the cell). To achieve this goal, at least partially, several groups have studied mutant proteins that are largely or entirely unfolded under physiologically relevant solvent conditions. In all, a bewildering (and potentially confounding) array of denaturation approaches have been employed in SAXS and SANS studies of the unfolded state. Fortunately, in recent years a sufficient number of small-angle scattering studies of unfolded proteins have been reported, producing a coherent, relatively general picture of the variety of denatured states studied in the laboratory.

We will limit ourselves to reviewing recent SAXS and SANS studies of putatively "fully" unfolded states formed at equilibrium. We direct readers interested in partially folded states (kinetic and equilibrium molten globules and their brethren) to a number of excellent recent articles and reviews (Kataoka and Goto, 1996; Kataoka *et al.*, 1997; Uversky *et al.*, 1998; Pollack *et al.*, 1999; Doniach, 2001). Similarly, we will not discuss in detail the technical aspects of scattering studies or the precise interpretation of scattering profiles, but instead direct the reader to the appropriate resources (Glatter and Kratky, 1982; Doniach *et al.*, 1995; Kataoka and Goto, 1996; Doniach, 2001).

II. A TAXONOMY OF UNFOLDED STATES

A. *The Chemically Denatured State*

The majority of SAXS and SANS studies of the unfolded state focus on the ensembles of states induced by chemical denaturants such as urea, GuHCl, extremes of pH, or organic cosolvents. As urea and GuHCl dominate spectroscopic studies of protein folding thermodynamics and kinetics, these denaturants have similarly been employed in the vast majority of small-angle scattering studies as well. The extraordinary solubility (and aggregation resistance) of unfolded proteins at high levels of urea or GuHCl provides an added technical benefit.

Limited SAXS studies suggest that urea and GuHCl produce indistinguishable denatured states. For example, the unfolded states of apomyoglobin, creatine kinase, and the pI3K SH2 domain induced by urea and GuHCl are experimentally indistinguishable in terms of radius of gyration (R_g) and shape (as defined by the Kratky scattering profile; see below) (Kataoka *et al.*, 1995; Zhou *et al.*, 1997; I. S. Millett and K. W. Plaxco, unpublished). Comparision of the unfolded states induced by GuHCl at around neutral pH with those induced by urea at acidic pH suggests that the unfolded ensembles produced by these denaturant conditions are indistinguishable in ubiquitin (Kamatari *et al.*, 1999; Millett *et al.*, 2002), cytochrome *c* (Kamatari *et al.*, 1996; Segel *et al.*, 1998) and lysozyme (Chen *et al.*, 1996; Millett *et al.*, 2002).

The unfolded states produced by urea and GuHCl are highly expanded, coil-like configurations. For proteins lacking disulfide bonds, the R_g of chemically unfolded states is typically 2- to 3-fold greater than that of the wild-type protein (Table I). Analysis of Kratky scattering profiles, of value because they allow one to distinguish between compact, globular configurations and coil-like extended conformations (Lattman, 1994; Doniach *et al.*, 1995; Kataoka and Goto, 1996), indicates that these chemically unfolded states adopt an extended, coil-like conformation (Fig. 1) (e.g., Semisotnov *et al.*, 1996; Segel *et al.*, 1998; Russo *et al.*, 2000). Detailed simulations support this suggestion by demonstrating that random-flight polymers accurately reproduce the observed scattering profiles of chemically denatured CheY (Garcia *et al.*, 2001) and yeast phosphoglycerate kinase (yPGK) (Calmettes *et al.*, 1994). Not surprisingly, the GuHCl- or urea-unfolded states of proteins containing intact disulfide bonds are rather more compact (Table II); but they, too, appear to populate constrained, random-walk conformations (Damaschun *et al.*, 1997; Russo *et al.*, 2000).

Well-founded theory suggests that, as temperature or denaturant concentration is increased, the denatured ensemble will expand as the solvent's ability to solvate unfolded polypeptides improves (Alonso and Dill, 1991; Dill and Shortle, 1991). While limited studies suggest that increases in denaturant concentration may produce small changes in the conformation of the unfolded ensemble (Segel *et al.*, 1998), the postulated solvent quality–linked expansion of the denatured state is apparently too subtle for current scattering techniques to detect. For example, we find no statistically significant evidence of measurable expansion of the unfolded state as denaturant concentration is increased for five proteins undergoing urea or GuHCl denaturation (Millett *et al.*, 2002). The unfolding of one, cytochrome *c*, is illustrated in Figure 2 left; over the range 3.2–4.8 M GuHCl, the R_g is scattered more or less randomly about 30.11 ± 0.05 Å. At quite low denaturant concentrations such denatured state contraction is, in contrast, sometimes observed. The denatured ensemble of the drkSH3 domain, which exhibits an unfolded

TABLE I
Proteins Lacking Disulfide Bonds

Protein	Denaturing conditions	R_g (unfolded) (Å)	R_g (native) (Å)	Ratio	N	Reference
yPGK	2 M GuHCl	71.0 ± 1.0[a]	24.8 ± 0.3	2.9	416	Receveur et al., 2000
	Cold denatured	25.3 ± 0.2		1.0		Receveur et al., 2000
Creatine kinase	4°C, 0.45 M GuHCl		22.5[b]		380	
	>4.5 M urea	46 ± 2[a]		2.0		Zhou et al., 1997
	3 M GuHCl	46 ± 2		2.0		Zhou et al., 1997
α-Subunit trp synthase	>4 M urea	34 ± 4[a]	19.1 ± 1.4	1.8	268	Gualfetti et al., 1999
Bovine carbonic anhydrase B	>2 M GuHCl	59 ± 2[a]	19 ± 2	3.1	260	Semisotnov et al., 1996
OspA	60°C	56.1 ± 1.6[a]	25.0 ± 0.3	2.2	257	Koide et al., 1999
Apomyoglobin	4 M GuHCl, pH 6.8[c]	35.8 ± 1.0[a]	19.7 ± 1.3	1.8	154	Kataoka et al., 1995
	8 M urea, pH 6.8[c]	34.2 ± 1.5		1.7		Kataoka et al., 1995
	pH 2	29.3 ± 1.0		1.5		Nishii et al., 1994
	pH 2, 5 M urea	34.2 ± 2[c]		1.7		Kamatari et al., 1999
	60% methanol, pH 2	30.6 ± 2‡		1.6		Kamatari et al., 1999
Snase wild-type	8 M urea	33.0 ± 1.0[a]		1.7		Flanagan et al., 1993
	50–70°C	42 ± 2		2.1		Panick et al., 1998
	>3000 atm	~35		1.8		Panick et al., 1998
	pH 2.5	37.2 ± 0.6		1.9		Uversky et al., 1998
Truncated Snase	Intrinsically unfolded	20.2 ± 0.4[d]	15.6 ± 0.2	1.3	136	Flanagan et al., 1992
Che Y	>5 M urea	38.0 ± 1.0[a]	14.8 ± 0.2	2.6	129	Garcia et al., 2001
pI3K SH2	6 M urea	30.4 ± 1.6[a]	N.D.[b]	N.D.	112	I. S. Millett and K. W. Plaxco, unpublished
	3 M GuHCl	29.6 ± 3.3		N.D.		I. S. Millett and K. W. Plaxco, unpublished

Prothymosin α	Natively unfolded	37.8 ± 0.9^a	N.A.	N.A.	109	Uversky et al., 1999
Cytochrome c	>3 M GuHCl, pH 7	30.2 ± 0.1^a	13.6 ± 0.1	2.2	104	Segel et al., 1998
	70°C, pH 2.8	23.1 ± 1.1		1.6		Hagihara et al., 1998
	60% methanol, pH 3	31.7 ± 2^c		2.2		Kamatari et al., 1996
	pH 2	30.1 ± 2^c		2.1		Kamatari et al., 1996
	pH 2, 4.5 M urea	32.1 ± 2^c		2.2		Kamatari et al., 1996
	pH 12	14.9 ± 0.1		1.1		Cinelli et al., 2001
pI3K SH3	2.67 M GuHCl	30.9 ± 0.3^a	N.D.^{b}	N.D.	103	I. S. Millett and K. W. Plaxco unpublished
Acyl phosphatase	>6.5 M urea	30.4 ± 1.3^a	13.9 ± 1.5	2.2	98	Millett et al., 2002
Protein L	>4 M GuHCl	26.0 ± 0.6^a	16.2 ± 0.5	1.6	79	Plaxco et al., 1999
	1.4 M GuHCl	25.9 ± 1.1		1.6		Plaxco et al., 1999
Ubiquitin	>5.5 M GuHCl	26.0 ± 1.2^a	13.2 ± 1.0	2.0	76	Millett et al., 2002
	5 M urea, pH 2	26.3 ± 2^c		2.0		Kamatari et al., 1999
	60% Methanol, pH 2	28.4 ± 2^c		2.2		Kamatari et al., 1999
drk SH3	2 M GuHCl	21.9 ± 0.5^a	11.9 ± 0.5	1.8	59	Choy et al., 2002
	Intrinsically unfolded	16.7 ± 1.4		1.4		Choy et al., 2002
Protein G	>1 M GuHCl	23 ± 1^a	10.8 ± 0.2	2.1	56	Smith et al., 1996

[a] Value employed in Figure 4.

[b] Native state is a dimer under conditions employed. Calculated monomer value presented for creatine kinase. The presence of unstructured tails in the pI3K SH2 and pI3K SH3 domains prevents accurate calculation of expected native state R_g.

[c] M. Kataoka, personal communication; Y. Kamatari, personal communication.

[d] Wild-type sequence; various point mutations alter this value from 17 to 33 Å (Flanagan et al., 1993).

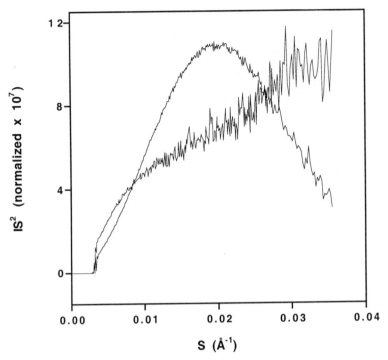

FIG. 1. *IS²* analysis, now termed "Kratky plots," was pioneered in the late 1950s by Otto Kratky, one of the early proponents of small-angle X-ray scattering (Glatter and Kratky, 1982; Lattman, 1994; Doniach *et al.,* 1995; Kataoka and Goto, 1996; Doniach, 2001). Kratky plots provide an easy, rapid mechanism for distinguishing between polymers with specific, well-defined shapes by emphasizing small changes in signal at higher *S* (angle). For example, a Kratky plot of the scattering profile of a globular particle—shown here is native cytochrome *c*—yields an inverted parabola. In contrast, a Kratky plot of an expanded, unstructured coil—here cytochrome *c* in 4.3 M GuHCl—linearly increases with increasing *S* (I. S. Millett., unpublished data).

state that can be studied in the absence of chemical denaturants, is significantly more compact in water than in 2 *M* GuHCl (Choy *et al.,* 2002). The low-denaturant unfolded states of other proteins, however, do not expand on the addition of further denaturant. For example, the R_g of reduced lysozyme in water is quite close to that observed for the reduced protein in 4 M GuHCl (Hoshino *et al.,* 1997). Similarly, the ensemble transiently populated when unfolded protein L is rapidly transferred to conditions that strongly favor folding (1.4 M GuHCl) is indistinguishable from that observed at high denaturant concentrations (Plaxco *et al.,* 1999). Apparently, the denatured state R_g of many proteins remains relatively fixed over a broad range of solvent quality.

TABLE II
Disulfide-Bonded Proteins

Protein	Denaturing conditions	R_g (unfolded) (Å)	R_g (native) (Å)	Ratio	N	Reference
α-Lactalbumin	8 M urea, pH 7.5[c]	24±2	15±2	1.6	142	Semisotnov et al., 1996
	8 M urea, pH 2[d]	30.0±0.7		2.0		Kataoka et al., 1997
	60% methanol, pH 2	24.3±2[c]		1.6		Kamatari et al., 1999
	>40% DMSO, pH 7	~24		~1.6		Iwase et al., 1999
Lysozyme	>5.5 M GuHCl, pH 7	22.5±0.4	15.1±0.7	1.5	129	Millett et al., 2002
	>8 M urea, pH 2.9	21.8±0.3		1.4		Chen et al., 1996
	80°C	>19.8[e]		>1.3		Arai and Hirai, 1999
	pH 2, 4 M GuHCl	22.9±1.0		1.5		Hoshino et al., 1997
	pH 2, TFE	21.0±0.6		1.4		Hoshino et al., 1997
	60% methanol, pH 2	24.9±2[c]		1.6		Kamatari et al., 1999
	60% DMSO, pH 7	>25[e]		>1.7		Iwase et al., 1999
	Reduced in water	37.9±1.0		2.5		Hoshino et al., 1997
	Reduced; 4 M GuHCl	35.8±0.5[a]		2.4		Hoshino et al., 1997
Ribonuclease A	4 M GuHCl, pH 2.8	25.9±0.2	14.5±0.1	1.6	124	Hagihara et al., 1998
	5 M urea, pH 2	27.3±2[c]		1.9		Kamatari et al., 1999
	70°C, pH 2.4	20.1±0.3		1.3		Hagihara et al., 1998
	>67°C, pH 5.7	~19		1.3		Sosnick and Terwhella, 1992
	60% methanol, pH 2	25.3±2[c]		1.7		Kamatari et al., 1999
	Reduced, 6 M GuHCl	24.1±1.0[a]		1.7		Sosnik and Terwhella, 1992
	Reduced, >56°C	28.0±1.0		1.9		Sosnik and Terwhella, 1992
	Reduced, <2.5 M urea	~21		~1.4		Zhou et al., 1998
	Reduced, >5 M urea	24.0±0.1		1.7		Zhou et al., 1998

continues

TABLE II (*continued*)

Protein	Denaturing conditions	R_g (unfolded) (Å)	R_g (native) (Å)	Ratio	N	Reference
β-Lactoglobulin	8 M urea, 25°C	37.2 ± 0.5	19.7 ± 0.5	1.9	123	Katou et al., 2001
	8 M urea, 0°C	36.1 ± 1.7		1.8		Katou et al., 2001
	Cold denatured 0°C, 4 M urea	24.9 ± 0.5		1.2		Katou et al., 2001
	60% methanol, pH 2	39.5 ± 2[c]		2.0		Kamatari et al., 1999
Subtilisin inhibitor	8 M urea, 35°C, pH 1.8	29.8	14.8[b]	2.0	113	Konno et al., 1997
	35°C, pH 1.8	25.8 ± 1.5		1.7		Konno et al., 1995
	5 M urea 20°C, pH 3	32 ± 2[c]		2.2		Kamatari et al., 1999
	8 M urea 3°C, pH 1.8	29.3		2.0		Konno et al., 1997
	3°C, pH 1.8	20.7 ± 1.3		1.4		Konno et al., 1995
	60% methanol, pH 2	28.3 ± 2[c]		1.9		Kamatari et al., 1999
Neocarzinostatin	>5 M GuHCl	33 ± 1	12.5 ± 0.2	2.5	113	Russo et al., 2001
	Heat (78°C)	26.3		2.0		Pérez et al., 2001
Ribonuclease TI	60°C	22.1 ± 1.0	14.3 ± 0.3	1.5	104	Damaschun et al., 1997

[a] Value employed in Figure 4.

[b] Native state is a dimer under conditions employed. Calculated monomer value presented (I. S. Millett, unpublished).

[c] G. Semisotnov, personal communication; Y. Kamatari, personal communication.

[d] Loss of a calcium cross-link at low pH may account for the expansion relative to the GuHCl- and DMSO-induced states.

[e] It is not clear that complete denaturation has been achieved under these conditions.

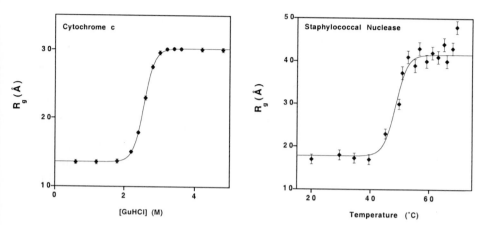

FIG. 2. It appears a general rule that the R_g of the chemically or thermally unfolded states are largely independent of denaturant concentration or temperature. (Left) The GuHCl-induced unfolding of cytochrome c is well-fitted as a two-state process in which protein state is converted into an unfolded ensemble whose dimensions remain fixed (to within a precision of ~0.3%) over a wide range of denaturant concentrations (Millett *et al.*, 2002). (Right) Similarly, the thermal unfolding of Snase produces a fixed-R_g denatured ensemble (to within a precision of ~2%) at temperatures below 68°C. Above this temperature (e.g., rightmost data point) irreversible aggregation precludes further analysis (Panick *et al.*, 1998). Data kindly provided by D. J. Segel and C. A. Royer. (Segel *et al.*, 1998; Panick *et al.*, 1998). Left-hand figure reproduced with permission from Millett *et al.* (2002) *Biochemistry* **41,** 321–325 © 1998 American Chemical Society. Right-hand figure adapted from Panick *et al.* (1998). *J. Mol. Biol.* **275,** 389–402, with permission of Academic Press.

Scattering techniques have been applied to the acid-induced unfolded states of a number of proteins and the base-induced unfolding of cytochrome c. Acid unfolding has been employed in the study of the cytochrome c, staphylococcal nuclease (Snase), and apomyoglobin unfolded states, all of which are significantly expanded (Kamatari *et al.*, 1996, 1999; Uversky *et al.*, 1998). In the case of cytochrome c and Snase, the acid-unfolded state is as expanded as the urea- or GuHCl-induced states, whereas that of apomyoglobin is rather more compact (Nishii *et al.*, 1994; Kataoka *et al.*, 1995). For several proteins (cytochrome c, subtilisin inhibitor, α-lactalbumin), the state produced by low pH in the presence of a high concentration of urea or GuHCl is somewhat expanded compared to the acid- and/or denaturant-only unfolded state. Obviously this expansion may be due to charge repulsion in the highly charged states produced at low pH; however, the calculated charge state changes of the various denatured states are difficult to reconcile with this suggestion (K. W. Plaxco, unpublished calculations). For example, despite a significant increase in charge (from +8 to +18), lysozyme exhibits nearly identical R_g at neutral pH in GuHCl and at acidic pH in urea or GuHCl (Chen *et al.*, 1996; Hoshino *et al.*, 1997; Millett *et al.*, 2002). In contrast to the expanded, acid-induced unfolded state observed for a half dozen proteins, the single example of base-induced unfolding reported to date is the extremely compact denatured state observed at pH 12 for cytochrome *c* (Cinelli *et al.*, 2001). The R_g of this unfolded ensemble is only 10% expanded relative to the native protein.

Unfolded states induced by a small number of organic cosolvents have been characterized by SAXS and SANS, including trifluorethanol (Hoshino *et al.*, 1997), methanol (Kamatari *et al.*, 1999), and dimethyl sulfoxide (DMSO) (Iwase *et al.*, 1999). Although spectroscopic probes clearly demonstrate that the two alcohols induce unfolded states characterized by significant α-helix content (and thus denature proteins via mechanisms quite different from those of GuHCl or urea), both denaturants generally produce unfolded states with R_g indistinguishable

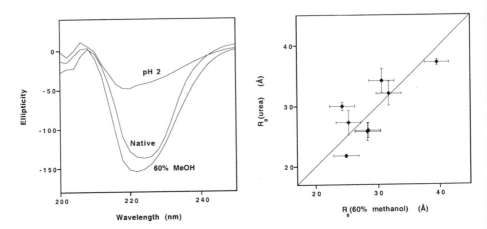

FIG. 3. While methanol induces an exceedingly helical unfolded state, the R_g of the methanol-induced unfolded state is effectively indistinguishable from that of the urea-induced unfolded state. (Left) Far-UV circular dichroism spectra of native, acid-denatured (pH 2), and 60% methanol-denatured (pH 2) cytochrome *c* (unpublished data courtesy E. McCarney). The strong dip at 220 nm for the methanol-induced state demonstrates that the ϕ and ψ torsion angles of this ensemble populate the α-helical region of the Ramachandran plot slightly more often than they are populated in the highly helical native state of the protein. Other spectroscopic probes (near-UV CD, fluorescence, NMR) confirm that the tertiary structure of the protein is completely abolished under these conditions (Kamatari *et al.*, 1996). In contrast to the methanol-induced state, the acid-denatured ensemble produces a spectrum characteristic of a random coil. (Right) The R_g of the urea- and methanol-induced unfolded states of the eight proteins for which both values have been reported (Kamatari *et al.*, 1996, 1999). The solid line represents a best-fit proportionality with a slope within error of unity, suggesting that the two sets of R_g are effectively indistinguishable.

from those produced by urea (Fig. 3). The general characteristics of the DMSO-induced denatured ensemble are less easily defined: The DMSO-induced denatured state of lysozyme is expanded significantly relative to the GuHCl-denatured state, whereas for α-lactalbumin the two are effectively equivalently expanded (Iwase *et al.*, 1999).

B. The Thermally Denatured State

Studies of thermally denatured proteins remain technically challenging owing to the propensity of thermally unfolded proteins to aggregate. Despite this potential difficulty, small-angle scattering techniques have been employed in the characterization of a number of thermally unfolded states.

In comparison to the chemically denatured state, the thermally denatured state is somewhat more difficult to classify. In general, thermally unfolded states are more compact than the equivalent chemically unfolded state. Examples reported to date include cytochrome c (Hagihara *et al.*, 1998), neocarzinostatin (Pérez *et al.*, 2001), and subtilisin inhibitor (Konno *et al.*, 1997), which are ~10–25% more compact than the equivalent chemically unfolded states. The R_g of the thermally denatured state of lysozyme, however, is indistinguishable from the equivalent chemically unfolded state, and those of Snase and ribonuclease A are significantly *larger* (Panick *et al.*, 1998; Arai and Hirai, 1999; Millett *et al.*, 2002; Sosnick and Trewhella, 1992). There are also hints that, as is the case for chemical denaturation, once unfolding has been achieved, the thermally denatured states of most proteins do not expand measurably as solvent quality decreases (Sosnick and Trewhella, 1992; Hagihara *et al.*, 1998; Panick *et al.*, 1998; Koide *et al.*, 1999; Pérez *et al.*, 2001). The R_g of the thermally unfolded Snase, for example, is effectively independent of temperature from 53° to 68°C, beyond which irreversible aggregation precludes further analysis (see Fig. 2 right). Still more limited data suggest that chemically unfolded proteins do not expand significantly upon heating; both ribonuclease A and subtilisin inhibitor exhibit indistinguishable R_g over at least moderate ranges of temperatures (Sosnick and Terhwella, 1992; Konno *et al.*, 1997).

C. The Pressure-Denatured State

Because the molar volume of an unfolded protein is less than that of the native state, increasing pressure leads to denaturation (Gross and Jaenicke, 1994). Royer and co-workers have employed high-pressure SAXS to monitor the pressure-induced unfolding of Snase. They find that the unfolded ensemble achieves a pressure-independent R_g of

~35 Å (Panick *et al.*, 1998), which is somewhat more compact than the thermally denatured ensemble (Panick *et al.*, 1998) but quite close to the R_g of the acid- and urea-denatured states (Flanagan *et al.*, 1993; Uversky *et al.*, 1998).

D. Cold-Unfolded States

Because of the temperature dependence of the hydrophobic effect, all proteins unfold at sufficiently low temperatures; most, however, do not unfold in water until temperatures well below the freezing point are reached. In the presence of chemical denaturants (urea, GuHCl, or low pH), the cold-unfolding point can sometimes be raised to experimentally accessible temperatures. Using this approach, the cold-denatured states of yPGK (4°C; 0.45 M GuHCl), β-lactoglobin (0°C; 4 M urea), and subtilisin inhibitor (3°C; pH 1.8) have been characterized (Konno *et al.*, 1997; Receveur *et al.*, 2000; Katou *et al.*, 2001). While spectroscopic studies suggest that these cold-denatured ensembles are highly unstructured, small-angle scattering studies demonstrate that all three are quite compact. The R_g of cold-denatured yPGK, for example, is *indistinguishable* from that of the native protein (in contrast to the R_g of the chemically denatured state, which is almost three times larger). Similarly, the R_g of the cold-unfolded state of subtilisin inhibitor is within 40% of that calculated for the native monomer (the native protein is a homodimer) (Konno *et al.*, 1997). Consistent with its near-native compactness, the Kratky plot of this cold-unfolded state exhibits a clear peak, indicating the ensemble is globular. On addition of urea, the R_g of this cold-unfolded state expands to twice that of the native state, with a concomitant change in the Kratky profile to that of an extended-coil conformation. The cold-denatured state of β-lactoglobulin is also compact, with a R_g only 20% greater than that of the native protein (Katou *et al.*, 2001). The Kratky scattering profile of this protein exhibits a minor peak, suggesting that the average geometry of the ensemble lies somewhere between that of a compact globule and an extended-coil configuration. It thus appears a general rule that cold-unfolded ensembles are significantly more compact and more globular than those induced thermally or by chemical denaturants.

E. Intrinsically Unfolded Proteins: The Physiological Unfolded State

Chemical denaturants, excessive heat, or low temperatures are typically not the reasons proteins unfold in the cell. Several groups have studied denatured ensembles populated in the absence of these harsh, nonphysiological denaturants by studying proteins that are either unfolded physiologically or unfolded under physiologically relevant conditions via mutation or the reduction of disulfide bonds.

In recent years, it has been established that a large number of proteins are natively unfolded; that is, they are unfolded in their active form inside of the cell (Plaxco and Gross, 1997; Wright and Dyson, 1999). Not surprisingly, hydrophilic residues typically dominate the sequence composition of natively unfolded proteins, thus, there are few of the hydrophobic interactions thought to induce compact denatured states (Uversky *et al.*, 2000; Dunker *et al.*, 2000; Dunker, 2002). Consistent with this observation, the single natively unfolded protein studied to date by SAXS, prothymosin α, is highly expanded and exhibits a Kratky scattering profile consistent with an extended-coil configuration (Uversky *et al.*, 1999).

Prothymosin α and many other natively unfolded proteins exhibit distinct sequence composition patterns that suggest they cannot fold to a well-packed native structure under any conditions. In contrast, some other intrinsically unfolded proteins closely resemble sequences that are stable and well folded in the absence of denaturant. Two of these, the drk SH3 domain and a truncation mutation of Snase, have been examined via SAXS. The drk SH3 domain is a 59-residue, predominantly β-sheet protein that has a folding equilibrium constant of approximately unity in water and thus populates a (well-studied) unfolded state in the absence of denaturant (Zhang and Forman-Kay, 1997). Kay and co-workers employed singular value decomposition to define the scattering profile and R_g of the unfolded state of this molecule in equilibrium with the native conformation (Choy *et al.*, 2002). This unfolded state is relatively compact, with a R_g only 1.4-fold greater than that of the native state. The addition of 2 M GuHCl expands this denatured state significantly (R_g increases to 1.8 times the native state) and produces a Kratky scattering profile indicative of an extended-coil conformation (I. S. Millett and S. Doniach, unpublished observations). Similarly, truncated Snase, spectroscopically well established as an unfolded state (Alexandrescu *et al.*, 1994), is also relatively compact (R_g only 1.3-fold greater than the native state) (Flanagan *et al.*, 1992). Heat (35°C) or cold (4°C) induces the expansion of this unfolded state to a R_g indistinguishable from the chemically unfolded, full-length protein (Flanagan *et al.*, 1993). Curiously, point mutations have been reported that significantly alter the dimensions of this generally compact unfolded state, rendering it somewhat more compact (only 10% expanded relative to the native R_g) or expanding it to chemically denatured dimensions (Flanagan *et al.*, 1993).

Like mutation, the reduction of disulfide bonds often causes normally
native proteins to unfold. This approach has been used to study the
unfolded states of lysozyme and ribonuclease A in water. The reduced
(and carboxymethylated) state of lysozyme is highly expanded, with a
R_g indistinguishable from that of the reduced protein in 4 M GuHCl.
Apparently, not all "natively folded" proteins populate a compact unfolded state in water. In contrast, the reduced, unfolded form of ribonuclease A is moderately compact in water, with Zhou *et al.* reporting
$R_g \sim 21$ Å versus 14.5 Å and 24 Å for the native and chemically denatured
forms respectively (Zhou *et al.*, 1998). Sosnick and Trewhella report the
apparently cooperative thermal unfolding of this chemically unfolded
state to $R_g \sim 28$ Å at higher temperatures. There is, however, some con-
cern regarding these values: Goto and co-workers (Hagihara *et al.*, 1998)
have reported a R_g for GuHCl-unfolded, *disulfide-intact* RNase that is
greater than the two values reported for *reduced* RNase in urea or GuHCl,
and (using hydrodynamic techniques) Tanford derived a R_g of 41 Å for
the reduced, denatured protein (Tanford *et al.*, 1966; see also Nöppert
et al., 1996)!

III. A RANDOM-COIL DENATURED STATE?

The cold-denatured state clearly has residual structure in that it is collapsed and exhibits a fairly globular Kratky scattering profile. Truncated,
equilibrium unfolded states appear similarly compact (and have been
shown by other spectroscopic means to have structure). But what of the
highly expanded, urea, GuHCl, or thermally denatured states?

An excluded-volume random-coil conformation will be achieved when
the solvent quality exceeds the theta point, the temperature or denaturant concentration at which the solvent–monomer interactions exactly
balance the monomer–monomer interactions that cause the polymer
to collapse into a globule under more benign solvent conditions. A
number of lines of small-angle scattering–based evidence are consistent
with the suggestion that typical chemical or thermal denaturation conditions are "good" solvents (i.e., are beyond the theta point) and thus
that chemically or thermally unfolded proteins adopt a near random-coil
conformation.

The geometric properties of highly denatured states appear to be consistent with those expected for a random-coil polymer. For example,
proteins unfolded at high temperatures or in high concentrations of
denaturant invariably produce Kratky scattering profiles exhibiting the
monotonic increase indicative of an expanded, coil-like conformation
(Fig. 1) (Hagihara *et al.*, 1998; see also Doniach *et al.*, 1995). Consistent

with this, the scattering profiles of chemically denatured yeast phosphoglycerate kinase (yPGK) and neocarzinostatin closely match those calculated for excluded-volume random coils (Calmettes *et al.*, 1994; Petrescu *et al.*, 1998; Russo *et al.*, 2000). Perhaps notably, the thermally denatured state of neocarzinostatin at 78°C (which is somewhat less expanded than the chemically unfolded state) is accurately described by a Kratky–Porod chain model lacking excluded-volume effects (Pérez *et al.*, 2001; see also Russo *et al.*, 2000). Thus the observation that thermally unfolded proteins are generally more compact than chemically unfolded proteins may arise because thermal denaturation more closely approximates theta conditions (where excluded-volume effects go to zero and the chain contracts slightly) than does chemical denaturation.

Additional evidence that chemically or thermally denaturing conditions are typically good solvents for the unfolded state stems from the observation that R_g is generally fixed over a broad range of temperatures or denaturant concentrations. While Kratky scattering profiles sometimes (but not always; see Garcia *et al.*, 2001; Pérez *et al.*, 2001) undergo small changes that reflect subtle conformational shifts (Segel *et al.*, 1998), no measurable expansion is observed after the unfolded state is reached (i.e., the conformational changes are sufficiently subtle that they do not measurably alter R_g) (Hagihara *et al.*, 1998; Panick *et al.*, 1998; Plaxco *et al.*, 1999; Millett *et al.*, 2002). As described above, this "fixed" R_g can (transiently) hold even under conditions at which equilibrium folding is effectively complete (Plaxco *et al.*, 1999; Woenckhaus *et al.*, 2001).

A simple scaling law has been postulated to define the relationship between polymer length and R_g under various solvent conditions (Flory, 1953):

$$R_g = R_0 N^\nu \tag{1}$$

where N is the number of monomers in the polymer chain, R_0 is a constant that is a function of the persistence length of the polymer, and ν is an exponential scaling factor that is a function of solvent quality. Idealized values of ν range from $\frac{1}{3}$ for a folded, spherical molecule (i.e., a collapsed polymer in a "bad" solvent) through $\frac{1}{2}$ for a chain at the theta point to $\frac{3}{5}$ for an excluded-volume, random-coil polymer in a good solvent [more precisely, 0.588 (LeGuillou and Zinn-Justin, 1977)]. And do chemically and thermally unfolded proteins obey the expected random-coil relationship? Previous studies have suggested that they do (Wilkins *et al.*, 1999), a conclusion largely supported by the data presented here:

Of the 20 reduced or disulfide bond–free proteins for which the R_g of the chemically or thermally unfolded state has ben reported, 17 fall on a single curve (Fig. 4). Fitting the R_g of these 17 proteins produces a strong, statistically significant correlation ($r^2 = 0.96$) and an exponent, $\nu = 0.61 \pm 0.03$, startlingly close to the expected value for an excluded-volume random coil chain.

Why, then, the outliers? With the possible exception of reduced ribonuclease A (as described above, there appears to be some discrepancy

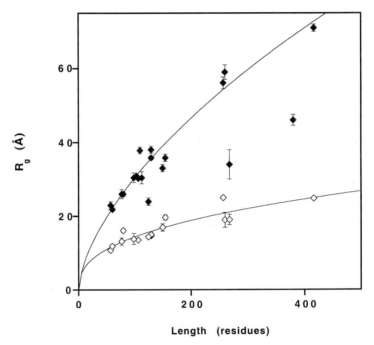

FIG. 4. The R_g of 20 disulfide-free proteins under native (open symbols) and highly denaturing (filled symbols) conditions (proteins and conditions denoted as footnote a in Table I). Note that the majority of the unfolded proteins appear to fall on a single curve. Omitting the three obvious denatured-state outliers, both the denatured and native data sets are well-fitted as power laws ($R_g = R_0 \, N^\nu$). The best-fit exponent for the denatured states, $\nu = 0.61 \pm 0.03$, is indistinguishable from the 3/5 expected for the length dependence of an excluded-volume random coil. Given the nonspherical shape of many proteins, the best-fit exponent for the native proteins, $\nu = 0.38 \pm 0.05$, is, in turn, surprisingly close to the 1/3 expected for a compact, spherical globules. The denatured outliers (from the left: reduced ribonuclease A, tryptophan synthase α-subunit, and creatine kinase) could represent systematic experimental error (e.g., see text discussion of ribonuclease A) or significant denatured state structure (Zitzewitz *et al.*, 1999). Consistent with the latter suggestion, all three outliers are significantly more compact than other denatured proteins of similar length.

in the literature regarding its R_g, there is little justification for excluding the three outliers from the fit. Still, it is notable that all three are systematically more compact than would be expected for a random-coil unfolded state; no outliers are observed that are s more expanded than "expected" based on the R_g of other proteins of similar length. Consistent with this, residual structure persisting at 5 M urea (the conditions under which SAXS was performed) but cooperatively lost as denaturant concentration is raised still higher has been reported for one outlier, the α-subunit of tryptophan synthase (Zitzewitz et al., 1999). Thus, it is quite feasible that, while the majority of proteins unfolded at high temperatures or high concentrations of denaturant closely resemble a random coil, at least a few contain sufficient residual structure to deviate significantly from this simple trend.

IV. RECONCILING THE RANDOM COIL WITH A STRUCTURED DENATURED STATE

The observation that highly denatured proteins behave like random-coil polymers recapitulates studies performed starting decades ago using intrinsic viscosity methods and continuing to the present using hydrodynamic radii determinations (but SAXS uses a *synchrotron*, so it *must* be better!) (Tanford et al., 1966; Tanford, 1968, 1970; Wilkins et al., 1999). We believe they bear repeating, however, both because the large number of recent small-angle scattering studies provide a more complete and direct measure of the denatured state and because recent advances in spectroscopic and theoretical techniques have shifted the field's perspective largely onto the nature and role of residual structure persistent in even the most highly denaturing conditions (see, for example, the following excellent papers: Kazmirski et al., 2001; Shortle and Ackerman, 2001; Garcia et al., 2001). In the justifiable excitement over transiently populated helices, hydrophobically collapsed elements, and long-range order in even the most highly unfolded proteins, it is perhaps understandable that relatively little emphasis has been placed on the more general observation that chemically or thermally denatured states are well approximated as random-coil conformations. We believe, however, that the reconciliation of these apparently mutually exclusive but equally compelling sets of observations remains an important missing element in our understanding of the unfolded state.

Why is this reconciliation important? After all, a 6 M GuHCl solution hardly recapitulates the physiologically milieu, and proteins unfolded under more physiologically relevant conditions are often (though not always; see Hoshino et al., 1997) much more compact than expected for an excluded-volume random coil. Still, the chemically or thermally denatured state is the "gold standard" unfolded state almost universally employed in defining the thermodynamics and kinetics of protein folding in the laboratory. Moreover, it has been shown that a number of proteins can refold from the chemically unfolded state without first passing through a well-populated, collapsed-but-unfolded state (Nöppert et al., 1998; Plaxco et al., 1999; Woenckhaus et al., 2001), and that the aqueous unfolded state is thermodynamically equivalent to that obtained via chemical denaturation (Bai et al., 1995). Thus, while residual denatured state structure and hydrophobic collapse have been postulated to play a role in the folding of many proteins (Agashe et al., 1995; Pollack et al., 1999; Yao et al., 2001), it is possible for "biologically relevant" folding to arise even from the highly denatured, random-coil ensemble.

How, then, do we reconcile the random-coil results of scattering experiments with the observation of residual denatured state structure? Reconciliation of short-range (i.e., sequence-local) order with near random-coil behavior may lie in the observation that the R_g of a denatured ensemble can be surprisingly independent of the detailed structure of the unfolded chain at the residue-to-residue level. For example, trifluoroethanol- and methanol-induced denatured states are highly helical (as determined by circular dichroism), but produce R_g effectively within error (of $\sim$3–10%) of those observed in GuHCl and/or urea (Fig. 3). Similarly, while the denatured-state R_g is generally fixed as a function of denaturant concentration or temperature (Fig. 2), the Kratky scattering profiles at larger angles sometimes change over the same range of solvent conditions. Because the proportionality constant for the relationship between R_g and length [R_0 in Eq. (1)] is a function of the chain's persistence length and of the potential of mean force between the residues, cosolvent-induced helicity (which should change persistence length) and changes in the Kratky at larger angles (indicating changes in short-range order) should correspond to changes in R_g. A possible explanation is that R_g scales with persistence length to the $\frac{2}{5}$ power (or weaker; T. C. B. McLeish and K. W. Plaxco, unpublished) and potential of mean force to the $\frac{1}{5}$ power (Flory, 1953) and thus, given these weak dependencies, that residual structure does not change either of these parameters enough to measurably affect molecular dimensions. Simulations of the R_g of plausible unfolded ensembles as ϕ–ψ preferences and/or monomer–monomer attraction are altered should provide a ready test of this suggestion that apparent random-coil behavior can occur even in the presence of short-range order.

In contrast to the reconciliation of short-range structure with net random coil–like behavior, the presence of long-range order in even the most highly denatured states is perhaps more problematic (Shortle and

Ackerman, 2001; see also Plaxco and Gross, 2001; Shortle, 2002). A partial reconciliation may stem from the ability of expanded but native-like topologies to produce scattering profiles reasonably consistent with those observed experimentally. For example, Smith and co-workers have used simulations to model the scattering of denatured yPGK and found a best-fit conformation that is perhaps (very) crudely consistent with the global topology of the native state (Callmettes *et al.*, 1993). Other simulations of unfolded yPGK, however, have been used by the same group to argue against the presence of long-range order (Petrescu *et al.*, 1998). The observation of fixed R_g over a broad range of denaturing conditions also raises the question of what forces might account for such order. If this ordering is due to residual hydrophobic or electrostatic interactions, then, for example, increasing GuHCl concentrations might be expected to reduce their efficacy and lead to expansion of the denatured state (Alonso and Dill, 1991; Dill and Shortle, 1991). As noted above, such expansion is not observed. An alternative suggestion is that long-range order is "encoded" by excluded volume effects and interactions between adjacent residues (see discussions in Pappu *et al.*, 2000; Plaxco and Gross, 2001; Shortle, 2002). How these effects might influence the "random"-coil properties of an unfolded protein has not been determined; the "expected" random-coil properties described here are derived from the physics of *achiral* homopolymers or random heteropolymers and thus might differ significantly from those of a sequence-specific, homochiral polypeptide. More detailed analysis of the magnitudes of these effects and their relationship to putative long-range order in the denatured state should prove a fruitful direction for future studies of the unfolded ensemble.

ACKNOWLEDGMENTS

The authors are indebted to E. McCarney, C. A. Royer, D. Segel, M. Kataoka, Y. Kamatari, and G. V. Semisotnov for providing unpublished or previously published data and T. C. B. McLeish, G. Fredrickson, D. Shortle, R. Pappu, and G. Rose for many stimulating conversations regarding the reconciliation "problem."

REFERENCES

Agashe, V. R., Shastry, M. C. R., and Udgaonkar, J. B. (1995). *Nature* **377**, 754–757.
Alexandrescu, A. T., Abeygunawardana, C., and Shortle, D. (1994). *Biochemistry* **33**, 1063–1072.
Alonso, D. O., and Dill, K. A. (1991). *Biochemistry* **30**, 5974–5985.
Arai, S., and Hirai, M. (1999). *Biophys. J.* **76**, 2192–2197.
Bai, Y. W., Sosnick, T. R., Mayne, L., and Englander, S. W. (1995). *Science* **269**, 192–197.
Berg, J., Amzel, M., and Lattman, E. (2002). *Adv. Prot. Chem.* In press.

Calmettes, P., Durand, D., Desmadril, M., Minard, P., Receveur, V., and Smith, J. C. (1994). *Biophys. Chem.* **53**, 105–114.

Calmettes, P., Roux, B., Durand, D., Desmadril, M., and Smith, J. C. (1993). *J. Mol. Biol.* **231**, 840–848.

Chen, L. L., Hodgson, K. O., and Doniach, S. (1996). *J. Mol. Biol.* **261**, 658–671.

Choy, W.-Y., Mulder, F. A. A., Crowhurst, K. A., Muhandiram, D. R., Millet, I. S., Doniach, S., Forman-Kay, J. D., and Kay, L. E. (2002). *J. Mol. Biol.* **316**, 101–112.

Cinelli, S., Spinozzi, F., Itri, R., Finet, S., Carsughi, F., Onori, G., and Mariani, P. (2001). *Biophys, J.* **81**, 3522–3533.

Damaschun, H., Gast, K., Hahn, U., Krober, R., MullerFrohne, M., Zirwer, D., and Damaschun, G. (1997). *BBA-Prot. Struct. Mol. Enz.* **1340**, 235–244.

Dill, K. A., and Shortle, D. (1991). *Annu. Rev. Biochem.* **60**, 795–825.

Doniach, S. (2001). *Chem. Rev.* **101**, 1763–1778.

Doniach, S., Bascle, J., Garel, T., and Orland, H. (1995). *J. Mol. Biol.* **254**, 960–967.

Dunker, A. K. (2002). *Adv. Protein Chem.* **62**, 25–49.

Dunker, A. K., Lawson, J. D., Brown, C. J., Williams, R. M., Romero, P., Oh, J. S., Oldfield, C. J., Campen, A. M., Ratliff, C. R., Hipps, K. W., Ausio, J., Nissen, M. S., Reeves, R., Kang, C. H., Kissinger, C. R., Bailey, R. W., Griswold, M. D., Chiu, M., Garner, E. C., and Obradovic, Z. (2001). *J. Mol. Graph. Mod.* **19**, 26–59.

Flanagan, J. M., Kataoka, M., Fujisawa, T., and Engelman, D. M. (1993). *Biochemistry* **32**, 10359–10370.

Flanagan, J. M., Kataoka, M., Shortle, D., and Engelman, D. M. (1992). *Proc. Natl. Acad. Sci. USA* **89**, 748–752.

Flory, P. J. (1953). *Principles of Polymer Chemistry.* Cornell University Press, Ithaca, NY.

Garcia, P., Serrano, L., Durand, D., Rico, M., and Bruix, M. (2001). *Protein Sci.* **10**, 1100–1112.

Glatter, O., and Kratky, O. (1982). *Small Angle X-Ray Scattering.* Academic Press, London.

Gross, M., and Jaenicke, R. (1994). *Eur. J. Biochem.* **221**, 617–630.

Gualfetti, P. J., Iwakura, M., Lee, J. C., Kihara, H., Bilsel, O., Zitzewitz, J. A., and Matthews, C. R. (1999). *Biochemistry* **38**, 13367–13378.

Hagihara, Y., Hoshino, M., Hamada, D., Kataoka, M., and Goto, Y. (1998). *Fold. Des.* **3**, 195–201.

Hoshino, M., Hagihara, Y., Hamada, D., Kataoka, M., and Goto, Y. (1997). *FEBS Lett.* **416**, 72–76.

Iwase, H., Hirai, M., Arai, S., Mitsuya, S., Shimizu, S., Otomo, T., and Furusaka, M. (1999). *J. Phys. Chem. Solids* **60**, 1379–1381.

Kamatari, Y. O., Konno, T., Kataoka, M., and Akasaka, K. (1996). *J. Mol. Biol.* **259**, 512–523.

Kamatari, Y., Ohji, S., Konno, T., Seki, Y., Soda, K., Kataoka, M., and Akasaka, K. (1999). *Protein Sci.* **8**, 873–882.

Kataoka, M., and Goto, Y. (1996). *Fold. Des.* **1**, R107–R114.

Kataoka, M., Kuwajima, K., Tokunaga, F., and Goto, Y. (1997). *Prot. Sci.* **6**, 422–430.

Kataoka, M., Nishii, I., Fujisawa, T., Ueki, T., Tokunaga, F., and Goto, Y. (1995). *J. Mol. Biol.* **249**, 215–228.

Katou, H., Hoshino, M., Kamikubo, H., Batt, C. A., and Goto, Y. (2001). *J. Mol. Biol.* **310**, 471–484.

Kazmirski, S. L., Wong, K. B., Freund, S. M. V., Tan, Y. J., Fersht, A. R., and Daggett, V. (2001). *Proc. Natl. Acad. Sci. USA* **98**, 4349–4354.

Koide, S., Bu, Z. M., Risal, D., Pham, T. N., Nakagawa, T., Tamura, A., and Engelman, D. M. (1999). *Biochemistry* **38**, 4757–4767.

Konno, T., Kataoka, M., Kamatari, Y., Kanaori, K., Nosaka, A., and Akasaka, K. (1995). *J. Mol. Biol.* **251**, 95–103.

Konno, T., Kamatari, Y. O., Kataoka, M., and Akasaka, K. (1997). *Protein Sci.* **6**, 2242–2249.

Lattman, E. E. (1994). *Curr. Opin. Struct. Biol.* **4**, 87–92.

LeGuillou, J. C., and Zinn-Justin, J. (1977). *Phys. Rev. Lett.* **39**, 95–98.

Millet, I. S., Townsley, L., Chiti, F., Doniach, S., and Plaxco, K. W. (2002). *Biochemistry* **41**, 321–325.

Nishii, I., Kataoka, M., Tokunaga, F., and Goto, Y. (1994). *Biochemistry* **33**, 4903–4909.

Nöppert, A., Gast, K., Muller-Frohne, M., Zirwer, D., and Damaschun, G. (1996). *FEBS Lett.* **380**, 179–182.

Nöppert, A., Gast, K., Zirwer, D., and Damaschun, G. (1998). *Fold. Des.* **3**, 213–221.

Panick, G., Malessa, R., Winter, R., Rapp, G., Frye, K. J., and Royer, C. A. (1998). *J. Mol. Biol.* **275**, 389–402.

Pappu, R. V., Srinivasan, R., and Rose, G. D. (2000). *Proc. Natl. Acad. Sci. USA* **97**, 12565–12570.

Pérez, J., Vachette, P., Russo, D., Desmadril, M., and Durand, D. (2001). *J. Mol. Biol.* **308**, 721–743.

Petrescu, A. J., Receveur, V., Calmettes, P., Durand, D., and Smith, J. C. (1998). *Protein Sci.* **7**, 1396–1403.

Plaxco, K. W., and Gross, M. (1997). *Nature* **386**, 657–659.

Plaxco, K. W., and Gross, M. (2001). *Nat. Struct. Biol.* **8**, 659–660.

Plaxco, K. W., Millett, I. S., Segel, D. J., Doniach, S., and Baker, D. (1999). *Nat. Struct. Biol.* **6**, 554–556.

Pollack, L., Tate, M. W., Darnton, N. C., Knight, J. B., Gruner, S. M., Eaton, W. A., and Austin, R. H. (1999). *Proc. Natl. Acad. Sci. USA* **96**, 10115–10117.

Receveur, V., Garcia, P., Durand, D., Vachette, P., and Desmadril, M. (2000). *Proteins* **38**, 226–238.

Russo, D., Durand, D., Calmettes, P., and Desmadril, M. (2001). *Biochemistry* **40**, 3958–3966.

Russo, D., Durand, D., Desmadril, M., and Calmettes, P. (2000). *Physica B,* **276**, 520–521.

Segel, D. J., Fink, A. L., Hodgson, K. O., and Doniach, S. (1998). *Biochemistry* **37**, 12443–12451.

Semisotnov, G. V., Kihara, H., Kotova, N. V., Kimura, K., Amemiya, Y., Wakabayashi, K., Serdyuk, I. N., Timchenko, A. A., Chiba, K., Nikaido, K., Ikura, T., and Kuwajima, K. (1996). *J. Mol. Biol.* **262**, 559–574.

Shortle, D. (2002). *Adv. Protein Chem.* **62**, 1–23.

Shortle, D., and Ackerman, M. S. (2001). *Science* **293**, 487–489.

Smith, C. K., Bu, Z., Anderson, K. S., Sturtevant, J. M., Engelman, D. M., and Regan, L. (1996). *Prot. Sci.,* **5**, 2009–2019.

Sosnick, T. R., and Trewhella, J. (1992). *Biochemistry* **31**, 8329–8335.

Tanford, C. (1968). *Adv. Protein Chem.* **23**, 121–282.

Tanford, C. (1970). *Adv. Protein Chem.* **24**, 1–95.

Tanford, C., Kawahara, K., and Lapanjes, S. (1966). *J. Biol. Chem.* **241**, 1921–1923.

Uversky, V. N., Gillespie, J. R., and Fink, A. L. (2000). *Prot. Struct. Funct. Gen.* **41**, 415–427.

Uversky, V. N., Gillespie, J. R., Millett, I. S., Khodyakova, A. V., Vasiliev, A. M., Chernovskaya, T. V., Vasilenko, R. N., Kozovskaya, G. D., Dolgikh, D. A., Fink, A. L., Doniach, S., and Abramov, V. M. (1999). *Biochemistry* **38**, 15009–15016.

Uversky, V. N., Karnoup, A. S., Segel, D. J., Seshadri, S., Doniach, S., and Fink, A. L. (1998). *J. Mol. Biol.* **278**, 897–894.

Wilkins, D. K., Grimshaw, S. B., Receveur, V., Dobson, C. M., Jones, J. A., and Smith, L. J. (1999). *Biochemistry* **38,** 16424–16431.

Woenckhaus, J., Kohling, R., Thiyagarajan, P., Littrell, K. C., Seifert, S., Royer, C. A., and Winter, R. (2001). *Biophys. J.* **80,** 1518–1523.

Wright, P. E., and Dyson, H. J. (1999). *J. Mol. Biol.* **293,** 321–331.

Wright, P. E., and Dyson, H. J. (2002). *Adv. Protein Chem.* **62,** 311–340.

Yao, J., Chung, J., Eliezer, D., Wright, P. E., and Dyson, H. J. (2001). *Biochemistry* **40,** 3561–3571.

Zhang, O. W., and Forman-Kay, J. D. (1997). *Biochemistry* **36,** 3959–3970.

Zhou, J. M., Fan, Y. X., Kihara, H., Kimura, K., and Amemiya, Y. (1997). *FEBS Lett.* **415,** 183–185.

Zhou, J. M., Fan, Y. X., Kihara, H., Kimura, K., and Amemiya, Y. (1998). *FEBS Lett.* **430,** 257–277.

Zitzewitz, J. A., Gualfetti, P. J., Perkons, I. A., Wasta, S. A., and Matthews, C. R. (1999). *Protein Sci.* **8,** 1200–1209.

DETERMINANTS OF THE POLYPROLINE II HELIX FROM MODELING STUDIES

By TREVOR P. CREAMER and MARGARET N. CAMPBELL

Center for Structural Biology, Department of Molecular and Cellular Biochemistry,
University of Kentucky, Lexington, Kentucky 40536

I. Introduction

More than three decades ago, Tiffany and Krimm suggested that many protein unfolded states possess significant amounts of extended helical conformations very similar to the left-handed polyproline II (PPII) helix (Tiffany and Krimm, 1968a,b; Krimm and Tiffany, 1974). This hypothesis was based in large part on similarities between the circular dichroism (CD) spectra of denatured proteins and homopolymers of proline, which are known to adopt the PPII conformation. This subject was reviewed by Woody (1992), who concluded that, while some of Tiffany and Krimm's arguments were perhaps invalid, the evidence for significant quantities of PPII helix-like conformations in denatured proteins was strong. Over the years, a number of groups have examined this hypothesis using a variety of peptide systems and several biophysical methods (Drake *et al.*, 1988; Dukor and Keiderling, 1991; Wilson *et al.*, 1996; Park *et al.*, 1997). Each of these groups found evidence in support of Tiffany and Krimm's hypothesis.

Kallenbach and co-workers have recently demonstrated via CD spectropolarimetry and NMR spectrometry that a seven-residue alanine peptide adopts predominantly the PPII helical conformation in aqueous solution (Shi *et al.*, 2002). Since alanine is nothing but backbone, such a finding indicates that the polypeptide backbone possesses an intrinsic

ADVANCES IN
PROTEIN CHEMISTRY, Vol. 62

propensity to adopt this conformation. Following up on this result, Rucker and Creamer (2002) have demonstrated that a seven-residue lysine peptide adopts this conformation at a pH of 12. It has previously been suggested that peptides and homopolymers of lysine will adopt a PPII-like conformation as a result of electrostatic repulsion between side chains (Tiffany and Krimm, 1968a,b; Makarov *et al.*, 1994; Arunkumar *et al.*, 1997). At a pH of 12 the lysine side chains are essentially uncharged, indicating that electrostatic repulsion is not the only driving force for PPII helix formation. Once again, this is evidence for the polypeptide backbone possessing an intrinsic propensity to be in the PPII helical conformation.

Kelly *et al.* (2001) have demonstrated that each residue possesses its own propensity to adopt the PPII helical conformation. This was determined via host–guest experiments employing a polyproline host peptide known to adopt the PPII structure. Undoubtedly the measured propensities are affected by steric interactions with the proline immediately following the guest site in this host peptide. However, it remains clear from this work that each side chain has some effect on the propensity for the backbone to adopt this structure. Notably, alanine was found to have a high propensity (Kelly *et al.*, 2001), in keeping with the work of Kallenbach and co-workers on a polyalanine peptide (Shi *et al.*, 2002). Once again, the backbone is seen to be an important determinant of this structure, with side chains modulating the extent to which it is adopted.

This model, an intrinsic backbone propensity modulated by the side chains, is in essence identical to that espoused by Aurora *et al.* (1997) for the α-helix. These authors argued that, since alanine has the strongest propensity to adopt the α-helical conformation and alanine is solely backbone in nature, the polypeptide backbone must favor the α-helix. Side chains modulate this propensity, leading to each residue having its own intrinsic propensity to adopt the α-helical conformation. How does one reconcile this seeming contradiction of α-helices and PPII helices both arising from backbone propensities? In fact, there is no real contradiction here. Aurora *et al.* (1997) went on to postulate that residues in unfolded protein chains reside almost exclusively in two conformational regions: the α- and β-regions. Residues adopt the α-conformation as a result of an intrinsic propensity that arises from the favorable intrachain hydrogen bonding and van der Waals interactions that occur in an α-helix. When not in this conformation, the residues are predominantly in the β-region, where such favorable interactions between residues are far less likely. For this region, which includes the PPII helical conformation, to be populated to any significant extent, other favorable interactions must exist. It is these other interactions, which are discussed below, that make up the intrinsic propensity of the backbone to adopt the PPII helical conformation. The polypeptide backbone possesses both types of propensity. The conformations adopted by a residue in an unfolded protein will be determined by the nature of the side chain and the local environment.

Using the above arguments, it is postulated that the conformational distributions of residues in unfolded proteins consist of just the α- and β-conformations, with the β-region being dominated by the PPII conformation. This model is a combination of those proposed by Tiffany and Krimm, who suggested that unfolded states consist of short stretches of PPII helix interspersed with bends (Tiffany and Krimm, 1968a,b; Krimm and Tiffany, 1974), and that of Aurora *et al.* (1997). In the current model, bend-like regions would be dominated by α-helical conformations, which could consist of more than a single turn of α-helix. This model leads to a very simple picture of protein unfolded states, one with far fewer conformations than commonly believed. Such a model does not exclude conformations other than the two types of helix, and is entirely consistent with recent work indicating not only that unfolded proteins occupy a relatively small portion of conformational space (Lattman *et al.*, 1994), but also that even highly denatured proteins can possess residual structure (Lumb and Kim, 1994; Wilson *et al.*, 1996; Shortle and Ackerman, 2001). Truly disordered protein states are clearly difficult to obtain experimentally, although Krimm and Tiffany (1974) did propose a model CD spectrum for such states.

The validity of this model for unfolded proteins rests on elucidation of the physical determinants of the two types of helix and the determination of the energetic favorability of these conformations relative to all other possible conformations in unfolded proteins. The determinants of α-helix formation have received significant attention over the past 15 years, and are thought to be mostly understood (reviewed in part by Aurora *et al.*, 1997). The determinants of PPII helix formation have received far less attention and are only now beginning to be understood.

Here, the determinants of PPII helix formation are considered in light of computational modeling. Surveys of known protein structures for the occurrence of PPII helices and factors leading to their formation are discussed. More complex modeling is then considered, ranging from simple hard-sphere computer simulations to calculations involving more detailed energetics. Finally, the implications of these data for modeling protein unfolded states are summarized.

II. The Left-Handed Polyproline II Conformation

The PPII conformation, as suggested by its name, was originally observed in homopolymers of proline (Isemura *et al.*, 1968; Tiffany and

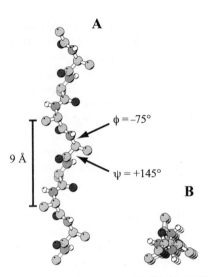

FIG. 1.   Cartoon of seven-residue alanine peptide in the PPII helical conformation.
(A) Viewed along the long axis; (B) viewed down the long axis. Figure generated using
MOLSCRIPT (Kraulis, 1991).

Krimm, 1968a,b; Mattice and Mandelkern, 1971; Krimm and Tiffany,
1974). This conformation is similar to that of a single strand from
collagen, with average backbone dihedrals of $(\phi, \psi) = (-75°, +145°)$.
These dihedrals lead to an extended left-handed helical conformation
with precisely three residues per turn and 9 Å between residues i and
$i + 3$ (measured $C\beta$ to $C\beta$). A cartoon of a seven-residue alanine peptide
in this conformation is shown in Figure 1. Notably, backbone carbonyl
and amide groups point perpendicularly out from the helical axis into
the solvent and are well-exposed.

A polyproline I (PPI) helical conformation also exists. It is right-
handed with average backbone dihedrals around $(\phi, \psi) = (-75°,$
$+160°)$, leading to 3.3 residues per turn. The major difference between
this conformation and the PPII helix is that all peptide bonds must be
cis, which results in the right-handed conformation and a much more
compact structure. The PPI conformation is energetically unfavorable
in aqueous solution, being induced in large quantities only by use of
organic solvents (Mutter *et al.*, 1999).

III.   PHYSICAL DETERMINANTS OF THE POLYPROLINE II CONFORMATION

What are the physical determinants of the PPII helical conformation?
Pappu *et al.* (2000) have shown that the PPII region of (ϕ, ψ)-space is

highly populated in a model employing only steric interactions and hydrogen bonds to describe intramolecular interactions. Their data indicate that this region is favored by a lack of unfavorable steric interactions, which follows from the extended nature of the conformation. In addition, based on the results of host–guest studies employing a polyproline host peptide, Kelly *et al.* (2001) suggested that backbone solvation is a major determinant. Such a hypothesis is not new, having been suggested by Adzhubei and Sternberg (1993) and anticipated much earlier by Krimm and Tiffany (1974). This hypothesis is supported by the work of Kallenbach and co-workers on a polyalanine peptide (Shi *et al.*, 2002) and by molecular dynamics computer simulations (Sreerama and Woody, 1999). Moreover, Rucker and Creamer (2002) have determined that a polylysine peptide at pH 12 adopts predominantly the PPII conformation and that this is due to the propensity for the backbone to adopt this structure, not to some property of the lysine side chains.

In the PPII conformation, the backbone amide and carbonyl groups are well-exposed in alanine (Fig. 1), which has a high propensity to adopt the structure (Kelly *et al.*, 2001; Shi *et al.*, 2002). Bulky side chains such as valine and isoleucine can occlude the backbone, leading to partial desolvation. Desolvation of the polypeptide backbone is very unfavorable. Wimley and White (1996) estimated that desolvation of a peptide unit costs 1.25 kcal mol^{-1} Partial desolvation due to bulky side chains will lead to low PPII helix-forming propensities for such residues. Longer, more flexible side chains, such as leucine and methionine, will occlude backbone from solvent when occupying some rotamers and leave it exposed in others, leading to intermediate propensities. Polar side chains could interact with the backbone, perhaps in part substituting for solvent, leading to further variations in PPII helix-forming propensity (Stapley and Creamer, 1999). Evidence obtained from modeling studies that is relevant to these suggested determinants is discussed in the following sections.

IV.   SURVEYS OF KNOWN PROTEIN STRUCTURES

A.   *Polyproline II Helices in Known Protein Structures*

A common approach in the study of secondary structure or small pieces of structure is to survey known protein structures for all instances of such structures and then to analyze those found. Such an approach has been taken for PPII helices (Adzhubei and Sternberg, 1993; Stapley and Creamer, 1999) and for residues that adopt the PPII conformation but are not necessarily part of PPII helices (Sreerama and Woody,

1994). Adzhubei and Sternberg (1993) surveyed a relatively small data set of protein structures (80 protein chains) using a somewhat complex definition of PPII structure. Stapley and Creamer (1999) used a larger data set (274 protein chains) with a simpler definition of PPII structure that involves all stretches of four or more residues in a PPII-like conformation with no β-sheet-like backbone–backbone hydrogen bonding. Although the differences in data set size and PPII helix definition led to some differences in results from these two surveys, the main findings were similar.

1. Occurrence of Polyproline II Helices

The PPII conformation is abundant in known protein structures, although PPII helices are not particularly common. Sreerama and Woody (1994) found that around 10% of all protein residues are in the PPII helical conformation. However, the majority of those are not part of a PPII helix. Stapley and Creamer (1999) and Adzhubei and Sternberg (1993) found that only 2% of the residues in the proteins examined were part of PPII helices four residues or longer in length. Moreover, on average, each protein possesses just one such PPII helix. The PPII helices found tend to be very short. Stapley and Creamer (1999) found that 95% of the PPII helices in their protein data set were only four, five, or six residues long.

In light of the hypothesis that backbone solvation is an important driving force for PPII helix formation, the finding that PPII helices in known protein structures are short and relatively rare is not at all surprising. The proteins surveyed were compact and well-ordered since these are the characteristics necessary for determining protein structures to high resolution. Compact proteins are well-packed, leading to very good van der Waals interactions between residues (Richards and Lim, 1994). Furthermore, >90% of the residues in such high-resolution structures have their backbone hydrogen-bonding capacity satisfied (Stickle *et al.*, 1992; McDonald and Thornton, 1994). These are precisely the conditions under which PPII helix formation would be disfavored since hydrogen-bonding and van der Waals interactions will compensate for desolvation of the polypeptide backbone. Notably, Adzhubei and Sternberg (1993) and Stapley and Creamer (1999) found that residues in PPII helices in known protein structures were significantly more accessible to solvent than the average for all residues. Polar atoms were found to be of the order of 60%, and nonpolar atoms 50%, more exposed.

2. Residues in Polyproline II Helices

As with other secondary structure types, all twenty residues are found in PPII helices in known protein structures (Adzhubei and Sternberg,

1993; Stapley and Creamer, 1999). It is with the Chou–Fasman frequencies of occurrence of each residue type that the surveys of Stapley and Creamer (1999) and Adzhubei and Sternberg (1993) disagree. The correlation coefficient between the two sets of frequencies is just 0.55. This disagreement is due both to the different definitions used to identify PPII helices and to differing sizes of the protein data sets surveyed. Not surprisingly, both surveys found that proline has by far the largest frequency, although not all PPII helices found contained proline. They also agree in that glutamine and arginine have the second and third largest frequencies. There is little agreement among the frequencies for the other residues.

Notably, there is very low correlation between the frequencies determined for residues in PPII helices by Stapley and Creamer (1999) and the PPII helix-forming propensities determined by Kelly *et al.* (2001) and A. L. Rucker, M. N. Campbell and T. P. Creamer (unpublished results). The propensities are expressed as percentage of PPII helix content of the guest-containing peptide in these host–guest experiments. The correlation between the two scales is only 0.54 [Fig. 2; proline is left out of plot and fit owing to its extremely high frequency of 5.06 (Stapley and Creamer, 1999)]. This lack of agreement highlights one of the potential pitfalls of surveys of protein structures. It is always unclear how

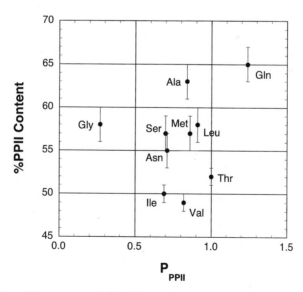

FIG. 2.   Plot of %PPII content (Kelly *et al.,* 2001; A. L. Rucker, M. N. Campbell, and T. P. Creamer, unpublished results) against Chou–Fasman frequency to be in a PPII helix, P_{PPII} (Stapley and Creamer, 1999).

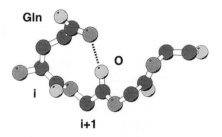

FIG. 3.   Cartoon of a glutamine side chain hydrogen-bonded to the carbonyl oxygen of the next residue in sequence. Residues are in the PPII conformation and side chains other than the glutamine are omitted for clarity. Figure generated using MOLSCRIPT (Kraulis, 1991).

much tertiary interactions influence the formation of the structure being surveyed. This problem would be particularly acute for short structures without regular hydrogen bonding such as the PPII helices surveyed by Stapley and Creamer (1999). Despite the lack of correlation, it is notable that glutamine has both the second highest frequency and second highest propensity after proline. In fact, Kelly *et al.* (2001) specifically measured the propensity of glutamine because of its high frequency. Stapley and Creamer (1999) found that a number of the glutamines in PPII helices from protein structures made a hydrogen bond from the glutamine side chain to the backbone carbonyl oxygen of the next residue in sequence. An example of such a hydrogen bond is depicted in Figure 3. These authors hypothesized that such a hydrogen bond could stabilize a PPII helix. The high propensity measured by Kelly *et al.* (2001) supports this hypothesis.

B.   *Hydrogen Bonding in the Polyproline II Conformation*

Further hints as to the importance of side chain-to-backbone hydrogen bonds in the PPII conformation can be gleaned from known protein structures. The same data set of 274 high-resolution protein chains used by Stapley and Creamer (1999) was surveyed for all occurrences of hydrogen bonds between a polar side chain at residue i and the carbonyl oxygen of the next residue, $i + 1$. The computer program DANSSR (kindly provided by Rajgopal Srinivasan, Johns Hopkins University) was used to conduct this survey. Overall, very few such hydrogen bonds were found, which should be kept in mind when considering the data (Table I). That so few were found is not entirely surprising since the vast majority of side chains that make hydrogen bonds in proteins are involved in interactions with residues more than one away in sequence (Stickle

TABLE I
Residues Making a Side Chain-to-Backbone Carbonyl $i \rightarrow i+1$ Hydrogen Bond[a]

Residue i	Total number of residues	Number making hydrogen bonds	Residues i		Residues $i+1$	
			β	PPII	β	PPII[b]
Gln	2284	34	33	16	20	19 (9)
Asn	2995	25	25	7	23	8 (3)
Ser	3868	49	44	7	46	29 (4)
Thr	3870	43	36	5	35	15 (2)
Lys	3700	42	40	11	40	24 (6)
Arg	2806	31	31	8	31	18 (3)

[a] The number of each residue occupying the β and PPII regions of (ϕ,ψ)-space is shown.

[b] Numbers in parentheses are those residues in the PPII region where residue i is also in the PPII region.

et al., 1992). Those hydrogen bonds that were identified were analyzed for the backbone conformations of residues i (donor side chain) and $i+1$ (acceptor carbonyl oxygen).

For the residues examined—glutamine, asparagine, serine, threonine, lysine, and arginine—the vast majority of those making a side chain-to-backbone hydrogen bond to the next residue are found in the β-region of (ϕ,ψ)-space (Table I). Fewer of the i residues occupy the PPII region, defined here as $(-75 \pm 25°, +145 \pm 25°)$. Glutamine has proportionally the most, with around 50% in this region. Asparagine, lysine, and arginine all have about 27% of all residues i that make the hydrogen bond falling in the PPII region. Although there are more occurrences of serine and threonine making such hydrogen bonds, relatively few do so while in the PPII region ($\sim$13%), and those that do tend to be at the edges of the region defined as PPII.

Notably, most of the residues at $i+1$ are also in the β-region, and many are found in the PPII region (more than half for all residues i, bar asparagine; Table I). However, the number of residues at $i+1$ in the PPII region with the residue at i also in the region is smaller than the number of residues at i in PPII. In other words, while it appears that forming such a hydrogen bond tends to favor having both residues i and $i+1$ in the β-region, and residue $i+1$ is often PPII, in many cases residue i is not concurrently PPII. One significant exception is glutamine, which has nine occurrences of both residues in the PPII region out of a total of sixteen in PPII at position i (Fig. 4). This is in keeping with the hypothesis that glutamine can make a side chain-to-backbone hydrogen bond that stabilizes the PPII helical conformation (Kelly et al., 2001). It is also

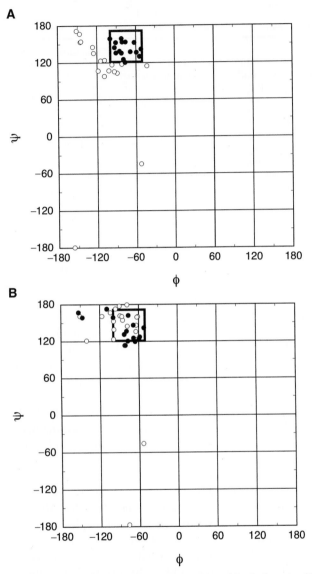

FIG. 4.   Ramachandran plots of glutamines (A) making side chain-to-backbone hydrogen bonds with the next residue in sequence (B). Filled circles denote residues where the glutamine is in the PPII conformation. Open circles denote all residues where the glutamine is not in the PPII conformation.

notable that asparagine has relatively few of the residues at $i+1$ in the PPII region when i is in the PPII conformation. Kelly *et al.* (2001) found that asparagine has a PPII helix-forming propensity comparable to that of leucine, suggesting that it does not make a stabilizing hydrogen bond while part of a PPII helix.

It is interesting that, with the exception of asparagine, all of the polar residues examined appear to restrict the following residue to the PPII region in more than half of the hydrogen bond occurrences identified. Stapley and Creamer (1999) found that glutamine, arginine, lysine, and threonine are all more common than expected as the first residue in a PPII helix. In their definition of a PPII helix, the first residue needs only its ψ dihedral to be within the range specified for a PPII helix, while the ϕ dihedral was allowed to fall outside the PPII range. Asparagine is relatively uncommon within a PPII helix, but is common in the position immediately preceding a PPII helix (Stapley and Creamer, 1999). These data suggest that these residues could fulfill a PPII helix capping role analogous to that observed for α-helices (Presta and Rose, 1988; Aurora and Rose, 1998), helping to stabilize and perhaps even initiate a PPII helix while not necessarily being a part of the helix itself.

V. MODELING STUDIES OF POLYPROLINE II HELIX DETERMINANTS

A. *Role of Sterics*

The PPII helical conformation is adopted by peptides and homopolymers of proline (Isemura *et al.,* 1968; Tiffany and Krimm, 1968a,b; Mattice and Mandelkern, 1971; Helbecque and Loucheux-Lefebvre, 1982). This is a direct result of steric interactions (Creamer, 1998). It has been known for some time that proline restricts the residue preceding it in sequence to the β-region of (ϕ,ψ)-space (MacArthur and Thornton, 1991). Proline itself is conformationally restricted since it is an imino acid with its side chain covalently bonded to its own backbone nitrogen, forming a five-membered ring. If a proline residue is immediately followed by another proline in sequence, its $C\beta$ will undergo steric interactions with the $C\delta$ of the next proline, restricting rotation about the ψ dihedral of the first. This will lead to the first proline being restricted to the small portion of the β-region that has become known as the PPII region. It follows that all prolines in a homopolymer will be restricted to this region except the last (Creamer, 1998). The last proline does not have a following proline to restrict it, and consequently will be free to adopt all ψ angles usually available to proline.

Creamer (1998) has demonstrated, using simple Monte Carlo computer simulations, that sterics are sufficient to drive formation of PPII helices by short proline peptides. In these simulations each atom was treated as a hard sphere, with no other forces being applied. Sterics are not sufficient to induce other residues to adopt the PPII conformation. In a series of host–guest simulations employing the hard-sphere potential, Creamer (1998) inserted various apolar residues into the Xaa position in an Ace-Ala-$(Pro)_3$-Xaa-$(Pro)_3$-Ala-NMe peptide (terminal alanines were included for technical reasons and have no effect on the results). All guest residues except glycine were found to be restricted to the β-region by steric interactions, but the PPII region was not particularly favored.

Of course, the vast majority of proteins do not contain large amounts of proline, so unfolded states would not contain significant quantities of PPII helix induced by such steric interactions. Sterics do, however, play a role in the formation of this structure. Pappu *et al.* (2000), via Monte Carlo computer simulations using hard spheres plus a hydrogen-bonding term, have shown that the PPII region is well-populated by short oligomers of alanine. This indicates that the PPII conformation, which has no internal backbone–backbone hydrogen bonds, is favored in part by a lack of steric interactions. This is not surprising given the extended nature of the structure (Fig. 1).

B. *Backbone Solvation*

It has been hypothesized that backbone solvation is a major, if not the dominant, driving force in the formation of PPII helices by non-proline residues (Krimm and Tiffany, 1974; Adzhubei and Sternberg, 1993; Sreerama and Woody, 1999; Kelly *et al.*, 2001; Rucker and Creamer, 2002; Shi *et al.*, 2002). The hypothesis is that the backbone is highly exposed to solvent in PPII helices, allowing for a high degree of solvation, which in turn favors the conformation. Side chains modulate this, with bulky β-branched residues such as valine partially occluding backbone from solvent and thus being disfavored in PPII helices. Alanine and residues with long, flexible side chains do not occlude the backbone, or do so to a limited extent, and are therefore favored in this conformation.

If this hypothesis is true, one could expect the solvent-accessible surface area (ASA) of the polypeptide backbone in the PPII conformation to be correlated with measured PPII helix-forming propensities. In order to test this, Monte Carlo computer simulations of short peptides Ac-Ala-Xaa-Ala-NMe (Xaa = Ala, Asn, Gln, Gly, Ile, Leu, Met, Pro, Ser, Thr, and Val) were run. These particular residues were examined because their

PPII helix-forming propensities have been measured by Kelly *et al.* (2001) and A. L. Rucker, M. N. Campbell, and T. P. Creamer (unpublished results). In the simulations the peptide backbone was constrained to be in the PPII conformation, defined as $(\phi,\psi) = (-75 \pm 25°, +145 \pm 25°)$, using constraint potentials described previously (Yun and Hermans, 1991; Creamer and Rose, 1994). The AMBER/OPLS potential (Jorgensen and Tirado-Rives, 1988; Jorgensen and Severance, 1990) was employed at a temperature of 298°K, with solvent treated as a dielectric continuum of $\varepsilon = 78$. After an initial equilibration period of 1×10^4 cycles, simulations were run for 2×10^6 cycles. Each cycle consisted of a number of attempted rotations about dihedrals equal to the total number of rotatable bonds in the peptide. Conformations were saved for analysis every 100 cycles. Solvent-accessible surface areas were calculated using the method of Richmond (1984) and a probe of 1.40 Å radius.

The sum of the estimated average solvent-accessible surface areas, $\langle ASA \rangle$, for the peptide units ($-CO-NH-$) on either side of residue Xaa, plus the Cα of Xaa, in each peptide simulated are given in Table II. Also shown are the estimated PPII helix-forming propensities for each residue measured by Kelly *et al.* (2001) and A. L. Rucker, M. N. Campbell, and

TABLE II

Estimated PPII Helix-Forming Propensities and Average Sum of Backbone ASAs from Monte Carlo Computer Simulations of Peptides Ac-Ala-Xaa-Ala-NMe Restricted to the PPII Conformation[a]

Residue (Xaa)	%PPII Helix[b]	$\langle ASA_{Backbone} \rangle$[c] (Å²)	TS_{PPII} (kcal mol⁻¹)	$TS_{Disordered}$[d] (kcal mol⁻¹)	$T\Delta S$ (kcal mol⁻¹)
Ala	63 ± 2	63.0 ± 0.6	—	—	—
Asn	55 ± 2	52.9 ± 0.8	1.37	1.22	0.15
Gln	65 ± 2	54.6 ± 1.0	1.96	1.79	0.17
Gly	58 ± 2	106.3 ± 0.6	—	—	—
Ile	50 ± 1	48.2 ± 0.7	1.06	0.77	0.29
Leu	58 ± 2	54.8 ± 0.8	0.83	0.54	0.29
Met	57 ± 2	55.1 ± 1.1	1.79	1.60	0.19
Pro	66 ± 2	56.3 ± 0.5	—	—	—
Ser	57 ± 2	59.7 ± 0.6	0.65	0.60	0.05
Thr	52 ± 1	55.4 ± 0.6	0.64	0.48	0.16
Val	49 ± 1	50.7 ± 0.6	0.63	0.41	0.22

[a] Estimated side chain conformational entropy in the PPII conformation, TS_{PPII}, and in a model for disordered states, $TS_{Disordered}$, are given, along with the difference in entropy, $T\Delta S$.

[b] PPII contents of peptides Ac-(Pro)$_3$-Xaa-(Pro)$_3$-Gly-Tyr-NH$_2$ from Kelly *et al.* (2001) and A. L. Rucker, M. N. Campbell, and T. P. Creamer (unpublished results).

[c] Errors in ASA estimates are set equal to one standard deviation.

[d] From Creamer (2000).

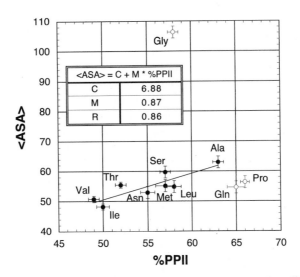

FIG. 5. Plot of estimated ⟨ASA⟩ from computer simulations against %PPII content (Kelly *et al.*, 2001; A. L. Rucker, M. N. Campbell, and T. P. Creamer, unpublished results). Line of best fit is computed only for filled circles.

T. P. Creamer (unpublished results). A plot of estimated ⟨ASA⟩ against %PPII content is given in Figure 5. At first glance, it would appear that there is little correlation between the two properties. However, three residues—proline, glycine, and glutamine—can be considered outliers, each for a specific reason. Proline has a high %PPII content in the polyproline-based host peptide used by Kelly *et al.* (2001) as a result of its unique properties as an imine. As discussed above, a proline that is followed in sequence by a second proline is restricted to the PPII conformation by steric interactions.

Glycine is also an outlier in Figure 5. Its high ⟨ASA⟩ in the simulated tripeptide is a direct consequence of its lack of a side chain. Such a high ⟨ASA⟩ would suggest that glycine has a very high PPII helix-forming propensity. Kelly *et al.* (2001) did measure a relatively high propensity for glycine, but not as high as the ⟨ASA⟩ would predict. This apparent disagreement is also due to a lack of a $C\beta$. Glycine possesses significantly more backbone conformational entropy than do residues that possess a side chain (D'Aquino *et al.*, 1996). Consequently, restricting glycine to the PPII region has a high entropic cost associated with it, leading to a reduction in its PPII helix-forming propensity.

Glutamine can also be considered an outlier if it forms a side chain-to-backbone hydrogen bond as hypothesized by Stapley and Creamer

(1999) and Kelly *et al.* (2001). Such a hydrogen bond, illustrated in Figure 3, will compensate for partial desolvation of the backbone, leading to a higher PPII helix-forming propensity than would be predicted from the estimated ⟨ASA⟩ (Fig. 5).

If proline, glycine, and glutamine are ignored, there is a relatively strong correlation between estimated ⟨ASA⟩ from the computer simulations and the measured %PPII contents (Fig. 5). The correlation coefficient is 0.86 for the remaining eight residues (0.67 if glutamine is included in the fit). Such a correlation suggests that backbone solvation does indeed play a major role in determining the PPII helix-forming propensities of the residues examined. However, since the correlation is far from perfect, it is clear that other factors must contribute, such as steric interactions between backbone atoms and contributions form the side chains other than partial desolvation of the backbone.

C. Side Chain Contributions

Other than an effect on backbone solvation, side chains could potentially modulate PPII helix-forming propensities in a number of ways. These include contributions due to side chain conformational entropy and, as discussed previously, side chain-to-backbone hydrogen bonds. Given the extended nature of the PPII conformation, one might expect the side chains to possess significant conformational entropy compared to more compact conformations. The side chain conformational entropy, TS_{PPII} (T = 298°K), available to each of the residues simulated in the Ac-Ala-Xaa-Ala-NMe peptides above was estimated using methods outlined in Creamer (2000). In essence, conformational entropy S can be derived from the distribution of side chain conformations using Boltzmann's equation

$$S = -R\sum_i p_i \ln p_i$$

where R is the gas constant, the summation is taken over all possible conformations of the side chain, and p_i is the probability that the side chain is in conformation i (Creamer, 2000).

Estimated side chain entropies TS_{PPII} for each residue simulated that possesses a mobile side chain are given in Table II along with entropy estimates, $TS_{Disordered}$, for side chains in a tripeptide model for disordered states taken from Creamer (2000), and the difference in entropy, $T\Delta S$. In all cases, TS_{PPII} in the peptides restricted to the PPII conformation is higher than that in disordered states, leading to positive values of $T\Delta S$. The gain in side chain entropy when restricted to the PPII conformation is, however, relatively small. Furthermore, there is no correlation between the estimated $T\Delta S$ and measured %PPII contents from Kelly *et al.* (2001), the correlation coefficient being -0.04. The gains in side chain conformational entropy will favor the occupation of the PPII conformation, but their magnitude and lack of correlation with PPII helix-forming propensities indicate that this is not a major determinant of the structure.

A side chain-to-backbone hydrogen bond between glutamine and the carbonyl oxygen of the next residue (Fig. 3) has been suggested to be the reason for the high PPII helix-forming propensity measured for glutamine (Kelly *et al.*, 2001). Although very few such hydrogen bonds were found when surveying known protein structures (see Table I), those glutamines involved in such hydrogen bonds were found to be predominantly in the β-region of (ϕ,ψ)-space, with many in the PPII region (Fig. 4A). Furthermore, almost all of the residues $i + 1$ to which the glutamine is hydrogen-bonded are also in the β-region, with many in the PPII region (Fig. 4B). This suggests that such a hydrogen bond stabilizes the PPII conformation. Fewer asparagines making this type of hydrogen bond were found, with even fewer of either residue involved found in the PPII region (Table I). This is in keeping with the relatively low PPII helix-forming propensity measured by Kelly *et al.* (2001). Glutamine and asparagine side chains differ by only a single $-CH_2-$ group. One is led then to ask whether the asparagine side chain is simply not long enough to form a PPII-stabilizing side chain-to-backbone hydrogen bond in this conformation.

To explore this issue, Monte Carlo computer simulations were run using the protocol outlined in the previous section. In these simulations, however, a peptide of sequence Ac-Ala-Xaa-Ala-Ala-NMe was employed (Xaa = Gln or Asn), the backbone was not constrained to the PPII conformation, and a side chain-to-backbone hydrogen bond was constrained using a potential function previously used to constrain α-helical backbone-to-backbone hydrogen bonds (Yun and Hermans, 1991; Creamer and Rose, 1994).

The results of the simulations are presented in Figure 6 in the form of Ramachandran plots. Panels A and B are plots of the (ϕ,ψ)-dihedral distributions for glutamine (residue i) and the next residue ($i + 1$), respectively. Panels C and D are the corresponding data for the peptide containing asparagine. It is clear from Figure 6A that glutamine, when constrained to hydrogen bond to the carbonyl oxygen of the next residue, is predominantly restricted to a region centered around $(\phi,\psi) = (-90°, +120°)$. This is very close to the PPII region. Notably, the next

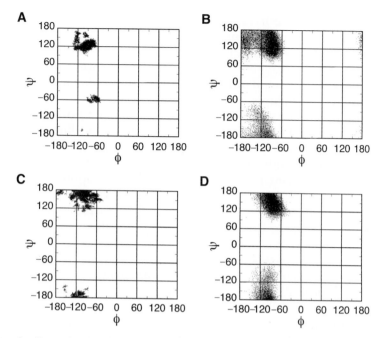

FIG. 6. Ramachandran plots for simulated Ac-Ala-Xaa-Ala-Ala-NMe peptides with Xaa = glutamine or asparagine, with a constrained side chain-to-backbone hydrogen bond. Conformational distribution for glutamine i (A) and for the residue $i + 1$ to which the glutamine is hydrogen-bonded (B). Conformational distribution for asparagine i (C) and for the residue $i + 1$ to which it is hydrogen-bonded (D).

residue in the simulated peptide, an alanine, is also very restricted (Fig. 6B). The majority of conformations are in a broad region clustered around $(\phi, \psi) = (-90°, +130°)$, again close to the PPII region.

Notably, when asparagine is constrained to form a side chain-to-backbone hydrogen bond, it is also restricted to the β-region of (ϕ, ψ)-space (Fig. 6C). The residue to which it is hydrogen-bonded is also restricted, mostly to the β-region (Fig. 6D). The asparagine, however, is mostly restricted to a region around $(\phi, \psi) = (-100°, +170°)$, farther from the PPII region than the glutamine. The next residue occupies a region clustered around $(\phi, \psi) = (-90°, +155°)$, close to the PPII region. These data appear to indicate that the asparagine can participate in such a hydrogen bond when not in the PPII helical conformation. On the other hand, glutamine can form a side chain-to-backbone hydrogen bond that results in both the glutamine and the next residue being restricted to conformations similar to the PPII conformation. This is in keeping with the experimental measurements of Kelly *et al.* (2001), who found that glutamine has a high PPII helix-forming propensity, while asparagine has a propensity similar to that of leucine, which is incapable of forming hydrogen bonds. From the simulations, it would appear that the side chain of asparagine is not long enough to form a side chain-to-backbone hydrogen bond with both the asparagine and the next residue in the PPII conformation.

Interestingly, asparagine was found to have a high Chou–Fasman frequency ($P_{PPII} = 1.3$) to be the residue immediately preceding a PPII helix in proteins of known structure (Stapley and Creamer, 1999). This observation, coupled with the simulation results (Fig. 6C,D), leads to the suggestion that asparagine might play a PPII helix-capping role analogous to that observed for α-helices (Presta and Rose, 1988; Aurora and Rose, 1998). Glutamine is common as the first residue in a PPII helix (Stapley and Creamer, 1999), suggesting that it is also involved in a PPII helix-capping role. It is not yet clear whether other polar residues can form side chain-to-backbone hydrogen bonds that can fulfill a capping role, although serine, threonine, arginine, and lysine are all common as the first residue in PPII helices and/or as the preceding residue (Stapley and Creamer, 1999). These residues also appear capable of forming side chain-to-backbone hydrogen bonds that might stabilize the PPII conformation (Table I). PPII helix-capping is worthy of further exploration.

VI. SUMMARY

Despite the fact that Tiffany and Krimm (1968a,b) formulated their hypothesis more than thirty years ago, it is only now that we are beginning to truly appreciate the importance of the PPII helical conformation. Recent experimental work has demonstrated that the polypeptide backbone possesses a significant propensity to adopt the PPII helical conformation (Kelly *et al.*, 2001; Rucker and Creamer, 2002; Shi *et al.*, 2002). The major determinant of this backbone propensity would appear to be backbone solvation, as was originally hinted at by Krimm and Tiffany (1974) and later suggested by a number of groups (Adzhubei and Sternberg, 1993; Sreerama and Woody, 1999; Kelly *et al.*, 2001; Rucker and Creamer, 2002; Shi *et al.*, 2002). The calculations and modeling described above provide data in support of this hypothesis.

Each residue has its own propensity to adopt the PPII conformation, with the backbone propensity being modulated by the side chain (Kelly *et al.*, 2001). Short, bulky side chains occlude backbone from solvent and thus disfavor the PPII conformation, while the lack of a side chain or long, flexible side chains tend to favor the conformation (Kelly *et al.*, 2001). Again, the described calculations support this. Steric interactions

(Pappu *et al.*, 2000) and side chain conformational entropy also contribute to the observed propensities. Furthermore, as suggested from surveys of PPII helices in proteins of known structure (Stapley and Creamer, 1999), side chain-to-backbone hydrogen bonds may well play a role in stabilizing PPII helices. The survey data, plus supporting calculations, also suggest that some polar residues may play a PPII helix-capping role analogous to that observed in α-helices (Presta and Rose, 1988; Aurora and Rose, 1998). Taken in sum, an atomic-level picture of the stabilization of PPII helices is beginning to emerge.

Once the determinants of PPII helix formation are known in more detail, it will become possible to apply them, along with the known determinants of the α-helical conformation, to the understanding of protein unfolded states. If, as suggested at the beginning of this article, protein unfolded states are dominated by residues in the PPII and α-conformations, these data will allow for modeling of the unfolded state ensembles of specific proteins with a level of realism that has not been previously anticipated.

ACKNOWLEDGMENTS

This work was supported in part by the National Science Foundation under Grant No. MCB-0110720. Acknowledgment is also made to the donors of The Petroleum Research Fund, administered by the American Chemical Society, for partial support of this research.

REFERENCES

Adzhubei, A. A., and Sternberg, M. J. E. (1993). *J. Mol. Biol.* **229,** 472–493.
Arunkumar, A. I., Kumar, T. K. S., and Yu, C. (1997). *Biochim. Biophys. Acta* **1338,** 69–76.
Aurora, R., Creamer, T. P., Srinivasan, R., and Rose, G. D. (1997). *J. Biol. Chem.* **272,** 1413–1416.
Aurora, R., and Rose, G. D. (1998). *Protein Sci.* **7,** 21–38.
Creamer, T. P. (1998). *Proteins* **33,** 218–226.
Creamer, T. P. (2000). *Proteins* **40,** 443–450.
Creamer, T. P., and Rose, G. D. (1994). *Proteins* **19,** 85–97.
D'Aquino, J. A., Gomez, J., Hilser, V. J., Lee, K. H., Amzel, L. M., and Freire, E. (1996). *Proteins* **25,** 143–156.
Drake, A. F., Siligardi, G., and Gibbons, W. A. (1988). *Biophys. Chem.* **31,** 143–146.
Dukor, R. K., and Keiderling, T. A. (1991). *Biopolymers* **31,** 1747–1761.
Helbecque, N., and Loucheux-Lefebvre, M. H. (1982). *Intl. J. Pept. Prot. Res.* **19,** 94–101.
Isemura, T., Okabayashi, H., and Sakakibara, S. (1968). *Biopolymers* **6,** 307–321.
Jorgensen, W. L., and Severance, D. L. (1990). *J. Am. Chem. Soc.* **112,** 4768–4774.
Jorgensen, W. L., and Tirado-Rives, J. (1988). *J. Am. Chem. Soc.* **110,** 1657–1666.
Kelly, M., Chellgren, B. W., Rucker, A. L., Troutman, J. M., Fried, M. G., Miller, A.-F., and Creamer, T. P. (2001). *Biochemistry* **40,** 14376–14383.
Kraulis, P. (1991). *J. Appl. Cryst.* **24,** 946–950.
Krimm, S., and Tiffany, M. L. (1974). *Isr. J. Chem.* **12,** 189–200.

Lattman, E. E., Fiebig, K., and Dill, K. A. (1994). *Biochemistry* **33,** 6158–6166.

Lumb, K. J., and Kim, P. S. (1994). *J. Mol. Biol.* **236,** 412–420.

MacArthur, M. W., and Thornton, J. M. (1991). *J. Mol. Biol.* **218,** 397–412.

Makarov, A. A., Adzhubei, I. A., Protasevich, I. I., Lobachov, V. M., and Fasman, G. D. (1994). *Biopolymers* **34,** 1123–1124.

Mattice, W. L., and Mandelkern, L. (1971). *J. Am. Chem. Soc.* **93,** 1769–1777.

McDonald, I. K., and Thornton, J. M. (1994). *J. Mol. Biol.* **238,** 777–793.

Mutter, M., Wohr, T., Gioria, S., and Keller, M. (1999). *Biopolymers* **51,** 121–128.

Pappu, R. V., Srinivasan, R., and Rose, G. D. (2000). *Proc. Natl. Acad. Sci. USA* **97,** 12565–12570.

Park, S. H., Shalongo, W., and Stellwagen, E. (1997). *Protein Sci.* **6,** 1694–1700.

Presta, L. G., and Rose, G. D. (1988). *Science* **240,** 1632–1641.

Richards, F. M., and Lim, W. A. (1994). *Quart. Rev. Biophys.* **26,** 423–498.

Richmond, T. J. (1984). *J. Mol. Biol.* **178,** 63–89.

Rucker, A. L., and Creamer, T. P. (2002). *Protein Sci.* **11,** 980–985.

Shi, Z., Olson, C. A., Rose, G. D., Baldwin, R. L., and Kallenbach, N. R. (2002). *Proc. Natl. Acad. Sci. USA* **99,** 9190–9195.

Shortle, D., and Ackerman, M. S. (2001). *Science* **293,** 487–489.

Sreerama, N., and Woody, R. W. (1994). *Biochemistry* **33,** 10022–10025.

Sreerama, N., and Woody, R. W. (1999). *Proteins* **36,** 400–406.

Stapley, B. J., and Creamer, T. P. (1999). *Protein Sci.* **8,** 587–595.

Stickle, D. F., Presta, L. G., Dill, K. A., and Rose, G. D. (1992). *J. Mol. Biol.* **226,** 1143–1159.

Tiffany, M. L., and Krimm, S. (1968a). *Biopolymers* **6,** 1767–1770.

Tiffany, M. L., and Krimm, S. (1968b). *Biopolymers* **6,** 1379–1382.

Wilson, G., Hecht, L., and Barron, L. D. (1996). *Biochemistry* **35,** 12518–12525.

Wimley, W. C., and White, S. H. (1996). *Nature Struct. Biol.* **3,** 842–848.

Woody, R. W. (1992). *Adv. Biophys. Chem.* **2,** 37–79.

Yun, R. H., and Hermans, J. (1991). *Protein Eng.* **4,** 761–766.

HYDRATION THEORY FOR MOLECULAR BIOPHYSICS

By MICHAEL E. PAULAITIS* and LAWRENCE R. PRATT†

*Department of Chemical Engineering, The Johns Hopkins University, Baltimore, Maryland 21218; and †Theoretical Division, Los Alamos National Laboratory, Los Alamos, New Mexico 87545

I. Introduction

The central role hydration plays in protein folding has long been recognized. That role has, however, several distinct and unique facets. Since folding requires packing nonpolar groups into the interior of the three-dimensional protein structure, the subject of hydrophobic interactions is essential in any consideration of the stability of the folded state of a protein relative to its unfolded state. The packing problem of sequestering nonpolar groups from water is directly related to a central problem in molecular theory of liquids of finding space for the solute in a dense liquid solvent. The structure of liquid water and the cooperative nature of water interactions are therefore expected to be important in a description of hydrophobic interactions. Because even polar groups must pack into the protein interior to some extent, this packing problem can be viewed more generally in terms of "context" hydrophobicity, i.e., the variation in hydrophobicity due to the local microenvironment.

Stabilizing the folded state also benefits from optimizing hydrogen-bonding interactions, with water providing donor and acceptor groups to polar side chains and main chain C=O and N—H groups left exposed on the surface of the folded structure. Direct evidence for such

ADVANCES IN
PROTEIN CHEMISTRY, Vol. 62

hydrogen-bonding interactions comes, for example, from protein X-ray structures, which show clusters of water molecules around solvent accessible C=O groups (Baker and Hubbard, 1984). Water molecules buried in the tertiary structure are likewise thought to stabilize the folded state by occupying cavities and forming hydrogen bonds with unsatisfied donors and acceptors in the protein interior (Williams *et al.*, 1994). The competition between main chain C=O and N—H groups forming intrachain hydrogen bonds or forming hydrogen bonds to solvent water molecules was noted by Edsall and McKenzie (1983) in their review of water and proteins as undoubtedly important in stabilizing folded protein structures. These ideas are discussed today, for example, in studies of the α-helix propensity of short polypeptide chains rich in alanine (Avgelj *et al.*, 2000; Levy *et al.*, 2001; Vila *et al.*, 2000; Thomas *et al.*, 2001). From this point of view, water molecules are distinct, chemically associating species that interact with proteins with stoichiometric specificity.

Hydration water will play an equally important role, or perhaps an even greater one, in stabilizing or destabilizing the unfolded state. Recent computational (Pappu *et al.*, 2000; Baldwin and Zimm, 2000) and experimental (Shortle and Ackerman, 2001; Plaxco and Gross, 2001) studies suggest that local excluded-volume interactions restrict substantially the conformational space accessible to unfolded polypeptides, creating structural biases that reduce the loss of conformational entropy on folding. Since polypeptide chains are expected to have greater solvent exposure in the unfolded state, solvent water molecules are also likely to participate in these local interactions, e.g., by providing hydrogen bond donor and acceptors and by occupying space to the exclusion of segments of the polypeptide chain. From this viewpoint, protein stabilization by the presence of water may in part be a manifestation of hydration water destabilizing the unfolded state (van Gunsteren *et al.*, 2001).

We present a molecular theory of hydration that now makes possible a unification of these diverse views of the role of water in protein stabilization. The central element in our development is the potential distribution theorem. We discuss both its physical basis and statistical thermodynamic framework with applications to protein solution thermodynamics and protein folding in mind. To this end, we also derive an extension of the potential distribution theorem, the quasi-chemical theory, and propose its implementation to the hydration of folded and unfolded proteins. Our perspective and current optimism are justified by the understanding we have gained from successful applications of the potential distribution theorem to the hydration of simple solutes. A few examples are given to illustrate this point.

The developments below sketch a thoroughgoing reconstruction of molecular statistical thermodynamic theory of solutions. There are several motivations for this effort, but particularly, the conventional molecular theories are not compelling for biophysical applications and do not make strong connections to the molecular intuitions involved in consideration of simulation and experiment. Some of the discussion below treats basic elements of statistical thermodynamics, known in specialized settings but weightier than is typical for this setting. We include examples to clarify and reinforce the basic concepts. Futhermore, the ends justify these means. These ideas have resulted in important recent progress on molecular issues of hydration related directly to biophysical applications, and the present development leads to a new theoretical suggestion [see Eq. (42)]. An important feature of these developments is their natural application to flexible, macromolecular solutes, which is clearly essential in any consideration of unfolded proteins or folded proteins that undergo large conformational changes, such as unfolding or conformational changes associated with their function.

Because this reconstruction has pedagogical aspects, we present a more developed preview of the ideas. As indicated above, we take the potential distribution theorem as a basis for our development. This is more than a convenient computational device; we emphasize that the potential distribution theorem should be viewed as a local partition function [see Eq. (6)] that serves as a generator of simple theories. Examples are given in Section III, which show the utility of defining a proximal region around the solute of interest and then paying close attention to the composition of that region. This is the basic idea of the quasi-chemical rules that are then derived. The quasi-chemical approximations appear as evaluations of a local partition function involving those composition variables, i.e., the occupancies of that defined proximal region.

A physical view that results from this theoretical structure is that solution components (water molecules!) may be viewed as "part" of the solute, playing an active role in the structure and function of proteins, for example. This notion is not foreign; it is common practice for protein crystallographers to identify conserved water locations in well-defined proximal regions surrounding the protein (Fig. 1, see color insert). This perspective suggests important ideas for some interesting features of protein structure and stability. For example, cold denaturation might be imagined to work by proximal water molecules binding to a globular protein more strongly as the temperature is lowered, thereby inducing a conformational change. Although this is a simple and reasonable view, such a molecular mechanism for cold denaturation has yet to be proved.

II. POTENTIAL DISTRIBUTION THEOREM AND PRELIMINARIES

The quantity of primary interest in our thermodynamic construction is the partial molar Gibbs free energy or chemical potential of the solute in solution. This chemical potential reflects the conformational degrees of freedom of the solute and the solution conditions (temperature, pressure, and solvent composition) and provides the driving force for solute conformational transitions in solution. For a simple solute with no internal structure (i.e., no intramolecular degrees of freedom), this chemical potential can be expressed as

$$\mu_i = kT \ln \rho_i \Lambda_i^3 + \mu_i^{\text{ex}} \tag{1}$$

where kT is the thermal energy, ρ_i is the solute number density, $\Lambda_i = \Lambda_i(T)$ is the thermal deBroglie wavelength of the solute (a known function of temperature), and μ_i^{ex} is the solute excess chemical potential. Since the concentration appears as a dimensionless quantity in this expression, the concentration dependence of the chemical potential comes with a choice of concentration units, and the first term in Eq. (1) expresses the colligative property of dilute solutions; i.e., the thermodynamic activity of the solute, $z_i \equiv e^{\mu_i/kT}$, is proportional to its concentration, ρ_i. The excess chemical potential accounts for intermolecular interactions between the solute and solvent molecules, and is given by the potential distribution theorem (Widom, 1963, 1982),

$$\mu_i^{\text{ex}} = -kT \ln \left\langle e^{-\Delta U/kT} \right\rangle_0 \tag{2}$$

where ΔU is the potential energy of solute–solvent interactions and the brackets indicate thermal averaging over all solution configurations of the enclosed Boltzmann factor. The subscript zero emphasizes that this average is performed in the absence of solute–solvent interactions. We note that the solvent can be a complex mixture of species. For dilute protein solutions, the solvent would typically include added salt and counterions in addition to water.

For a solute with internal degrees of freedom, the chemical potential is given by (Pratt, 1998)

$$\mu_i = kT \ln \left[\rho_i \Lambda_i^3 / q_i^{\text{int}} \right] - kT \ln \left\langle\!\left\langle e^{-\Delta U/kT} \right\rangle\!\right\rangle_0 \tag{3}$$

where q_i^{int} is the partition function for a single solute molecule, and the double brackets indicate averaging over the thermal motion of the solute and the solvent molecules under the condition of no solute–solvent interactions. The probability of finding a certain solute conformation

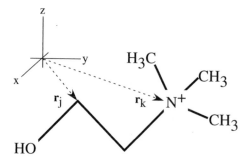

FIG. 2. Illustration of the definitions of conformational coordinate $\mathcal{R}^n$, e.g., $\mathcal{R}^n = \{\mathbf{r}_1, \mathbf{r}_2, \ldots, \mathbf{r}_n\}$. The conformational distribution $s_i^{(0)}(\mathcal{R}^n)$ is sampled for the single molecule in the absence of interactions with solvent by suitable simulation procedures using coordinates appropriate for those procedures. The normalization adopted in this development is $\int s_i^{(0)}(\mathcal{R}^n)\, d^n\mathcal{R} = V$, the volume of the system. Thus, the conformational average that corresponds to adding the second brackets in going from Eq. (4) to Eq. (3) is evaluated with the distribution function $s_i^{(0)}(\mathcal{R}^n) = V$.

in solution is related to this chemical potential. These probabilities are defined in terms of the number density of solute molecules, $\rho_i(\mathcal{R}^n)$, in conformation $\mathcal{R}^n$ (see Fig. 2),

$$\rho_i(\mathcal{R}^n) = s_i^{(0)}(\mathcal{R}^n)\left(e^{\mu_i/kT} q_i^{\text{int}}/\Lambda_i^3\right)\left\langle e^{-\Delta U/kT}\right\rangle_{0,\mathcal{R}^n} \qquad (4)$$

where $s_i^{(0)}(\mathcal{R}^n)$ is the normalized probability density for solute conformation $\mathcal{R}^n$ in the absence of interactions with the solvent, and the brackets indicate the thermodynamic average over all solvent configurations with the solute fixed in conformation $\mathcal{R}^n$. Equation (3) is obtained from Eq. (4) by averaging over all solute conformations, recognizing that $\rho_i(\mathcal{R}^n)$ is normalized to the total number of solute molecules.

A. Partition Function Perspective

Equations (2) and (3) relate intermolecular interactions to measurable solution thermodynamic properties. Several features of these two relations are worth noting. The first is the test-particle method, an implementation of the potential distribution theorem now widely used in molecular simulations (Frenkel and Smit, 1996). In the test-particle method, the excess chemical potential of a solute is evaluated by generating an ensemble of microscopic configurations for the solvent molecules alone. The solute is then superposed onto each configuration and the solute–solvent interaction potential energy calculated to give the probability distribution, $P_0(\Delta U/kT)$, illustrated in Figure 3. The excess

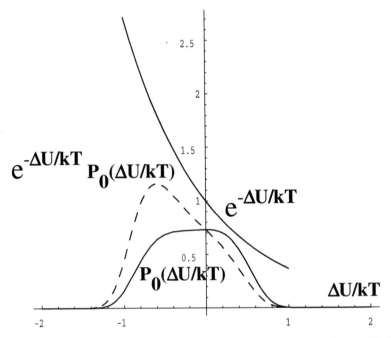

FIG. 3.   Functions in the integrand of the partition function formula Eq. (6). The lower solid curve labeled $P_0(\Delta U/kT)$ is the probability distribution of solute–solvent interaction energies sampled from the uncoupled ensemble of solvent configurations. The dashed curve is the product of this distribution with the exponential Boltzmann factor, $e^{-\Delta U/kT}$, the upper solid curve. See Eqs. (5) and (6).

chemical potential is obtained from

$$e^{-\mu_i^{ex}/kT} = \left\langle\!\left\langle e^{-\Delta U/kT}\right\rangle\!\right\rangle_0$$
$$= \int P_0(\Delta U/kT)e^{-\Delta U/kT}\,d(\Delta U/kT) \qquad (5)$$

The Boltzmann factor, $e^{-\Delta U/kT}$, preferentially weights the low-energy tail of the distribution, which amounts to reweighting $P_0(\Delta U/kT)$, giving even higher probabilities to those solvent configurations that are most favorable for solvation. Specifically, $P(\Delta U/kT) = e^{-(\Delta U - \mu_i^{ex})/kT}P_0(\Delta U/kT)$ as the potential distribution theorem requires. Equation (5) suggests the evaluation of a canonical partition function,

$$\int P_0(\Delta U/kT)e^{-\Delta U/kT}\,d(\Delta U/kT) \leftrightarrow \sum_i \Omega_i e^{-E_i/kT} \qquad (6)$$

The summation is over all energy levels and $\Omega_i(E_i)$ is the number of thermodynamic states with energy E_i (see Fig. 3).

We note that the calculation of ΔU will depend primarily on local information about solute–solvent interactions; i.e., the magnitude of ΔU is of molecular order. An accurate determination of this partition function is therefore possible based on the molecular details of the solution in the vicinity of the solute. The success of the test-particle method can be attributed to this property. A second feature of these relations, apparent in Eq. (4), is the evaluation of solute conformational stability in solution by separately calculating the equilibrium distribution of solute conformations for an isolated molecule and the solvent response to this distribution. This evaluation will likewise depend on primarily local interactions between the solute and solvent. For macromolecular solutes, simple physical approximations involving only partially hydrated solutes might be sufficient.

The test-particle description of solvation is often conceptualized by a molecular process of solute or test particle insertion, with the solute excess chemical potential calculated as a finite difference of the free-energy change for that process, e.g., in the canonical ensemble. This view is satisfactory, but not the only one, however. And the concept of literal insertion has limited practicality. It is important to recognize that the potential distribution theorem is particularly effective in crafting physical models as distinguished from a literal basis for computation. An alternative and equally valid view superposes the solution on the solute and suggests a theoretical development in which the solution is built up around the solute. In fact, we present below an alternative derivation of the potential distribution theorem based on the grand canonical ensemble of statistical thermodynamics. This alternative derivation more clearly suggests the applicability of the potential distribution theorem to protein solutions and leads to an implemention using quasi-chemical theory.

III. Applications of the Potential Distribution Theorem

A. Primitive Hydrophobic Effects: Information Theory

Our first example of an application of the potential distribution formula [Eq. (5)] is in the calculation of the excess chemical potential of a simple hydrophobic solute dissolved in water (Hummer et al., 1998a, 2000), which we consider to be a hard-core particle that excludes all water molecules from its molecular volume. We note that these purely repulsive solute–water interactions lead to values of either zero or one for the Boltzmann factor in Eq. (5), depending on whether any water molecules overlap the excluded volume of the solute for each water configuration sampled. Thermal averaging over all water configurations collects with a weight of unity all water configurations, leaving the solute excluded volume empty. This average is identically the probability, p_0, of finding an empty volume in pure water corresponding to the molecular volume of the solute—i.e., a cavity the size and shape of the hydrophobic solute. The potential distribution theorem is therefore given by

$$\mu_i^{\text{ex}} = -kT \ln p_0 \tag{7}$$

Thus, μ_i^{ex} for a hydrophobic solute is determined by quantifying the probability p_0 of successfully inserting a hard-core solute of the same size and shape into equilibrium configurations of water, as illustrated in Figure 4. A virtue of this approach is that the thermodynamics of hydrophobic hydration characterized by μ_i^{ex} is determined from the properties of pure water alone. The solute enters only through its molecular size and shape (see Fig. 4).

A recent breakthrough in molecular theory of hydrophobic effects was achieved by modeling the distribution of occupancy probabilities, the p_n depicted in Figure 4, rather than applying a more difficult, direct theory of p_0 for cavity statistics for liquid water (Pohorille and Pratt, 1990). This information theory (IT) approach (Hummer et al., 1996) focuses on the set of probabilities p_n of finding n water centers inside the observation volume, with p_0 being just one of the probabilities. Accurate estimates of the p_n, and p_0 in particular, are obtained using experimentally available information as constraints on the p_n. The moments of the fluctuations in the number of water centers within the observation volume provide such constraints.

For a given observation volume v in bulk water, the moments are determined from

$$\langle n^k \rangle = \sum_{n=0}^{\infty} p_n n^k \tag{8}$$

where the brackets indicate a canonical average. The zeroth, first, and second moments can be expressed in terms of experimentally accessible

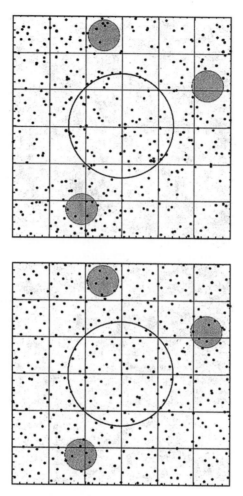

FIG. 4. A schematic two-dimensional illustration of the idea for an information theory model of hydrophobic hydration. Direct insertion of a solute of substantial size (the larger circle) will be impractical. For smaller solutes (the smaller circles) the situation is tractable; a successful insertion is found, for example, in the upper panel on the right. For either the small or the large solute, statistical information can be collected that leads to reasonable but approximate models of the hydration free energy, Eq. (7). An important issue is that the solvent configurations (here, the point sets) are supplied by simulation or X-ray or neutron scattering experiments. Therefore, solvent structural assumptions can be avoided to some degree. The point set for the upper panel is obtained by pseudo–random–number generation so the correct inference would be of a Poisson distribution of points and $\mu_i^{ex} = kT\rho v$ where v is the van der Waals volume of the solute. Quasi-random series were used for the bottom panel so those inferences should be different. See Pratt *et al.* (1999).

quantities as

$$\langle 1 \rangle = \sum_{n=0}^{\infty} p_n = 1$$

$$\langle n \rangle = \sum_{n=0}^{\infty} p_n n = \rho v$$

$$\langle n^2 \rangle = \sum_{n=0}^{\infty} p_n n^2 = \langle n \rangle + \rho^2 \int_v d\mathbf{r} \int_v d\mathbf{r'} g(|\mathbf{r} - \mathbf{r'}|) \qquad (9)$$

where ρ is the number density of bulk water and $g(r)$ is the radial distribution function between water-oxygen atoms in bulk water. The latter quantity can be derived from X-ray or neutron scattering measurements or from computer simulations. These moment conditions provide constraints on the p_n, guaranteeing that they are normalized and have the correct first and second moments.

The IT approach provides a systematic estimate of the p_n under the constraints of the available information, defined as the set $\{p_n\}$ that maximizes an information entropy η subject to the information constraints,

$$\max_{\{\text{constraints}\}} \eta(\{p_n\}) \qquad (10)$$

Generally, we adopt a relative or cross entropy,

$$\eta(\{p_n\}) = -\sum_{n=0}^{\infty} p_n \ln\left(\frac{p_n}{\hat{p}_n}\right) \qquad (11)$$

where $\hat{p}_n$ represents an empirically chosen "default model." We find that a flat distribution ($\hat{p}_n = 1$ for $n \le n_{\max}$ and $\hat{p}_n = 0$ otherwise), which results in a discrete Gaussian form of p_n with given mean and variance, is accurate for molecule-size cavities.

Maximizing the information entropy under the constraints of Eq. (9) leads to

$$p_n = \hat{p}_n e^{\lambda_0 + \lambda_1 n + \lambda_2 n^2} \qquad (12)$$

where λ_0, λ_1, and λ_2 are the Lagrange multipliers chosen to satisfy the moment conditions.

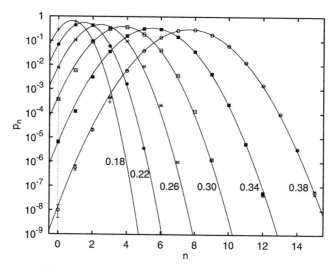

FIG. 5. Probabilities p_n of observing n water-oxygen atoms in spherical cavity volumes v. Results from Monte Carlo simulations of SPC water are shown as symbols. The parabolas are predictions using the flat default model in Eq. (11). The center-to-center exclusion distance d (in nanometers) is noted next to the curves. The solute exclusion volume is defined by the distance d of closest approach of water-oxygen atoms to the center of the sphere. (Hummer *et al.*, 1998a)

Figure 5 shows p_n distributions for spherical observation volumes calculated from computer simulations of SPC water. For the range of solute sizes studied, the ln p_n values are found to be closely parabolic in n. This result would be predicted from the flat default model, as shown in Figure 5 with the corresponding results. The corresponding excess chemical potentials of hydration of those solutes, calculated using Eq. (7), are shown in Figure 6. As expected, μ_i^{ex} increases with increasing cavity radius. The agreement between IT predictions and computer simulation results is excellent over the entire range $d \leq 0.36$ nm that is accessible to direct determinations of p_0 from simulation.

The simplicity and accuracy of such models for the hydration of small molecule solutes has been surprising, as well as extensively scrutinized (Pratt, 2002). In the context of biophysical applications, these models can be viewed as providing a basis for considering specific physical mechanisms that contribute to hydrophobicity in more complex systems. For example, a natural explanation of entropy convergence in the temperature dependence of hydrophobic hydration and the heat denaturation of proteins emerges from this model (Garde *et al.*, 1996), as well as a mechanistic description of the pressure dependence of hydrophobic

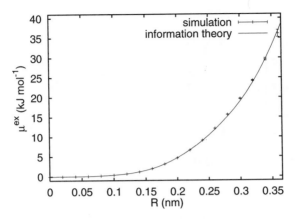

FIG. 6. Excess chemical potential of hard-sphere solutes in SPC water as a function of the exclusion radius d. The symbols are simulation results, compared with the IT prediction using the flat default model (solid line). (Hummer *et al.,* 1998a)

interactions, which suggests a mechanism for the observed pressure denaturation of proteins (Hummer *et al.,* 1998b).

B. *Electrostatic Solvation Free Energies*

A virtue of the potential distribution theorem approach is that it enables precise assessment of the differing consequences of intermolecular interactions of differing types. Here we use that feature to inquire into the role of electrostatic interactions in biomolecular hydration.

We note that if all electrostatic interactions between the solute and solvent are annulled, for example, by eliminating all solute partial charges in force field models, the potential distribution formula [Eq. (5)] sensibly describes the hydration of that hypothetical solute

$$e^{-\tilde{\mu}_i^{\text{ex}}/kT} = \left\langle\!\left\langle e^{-\Delta\tilde{U}/kT} \right\rangle\!\right\rangle_0 \tag{13}$$

The tilde indicates solute–solvent interactions without electrostatic interactions. The contribution of electrostatic interactions is then isolated as $\delta\mu_i^{\text{ex}} = \mu_i^{\text{ex}} - \tilde{\mu}_i^{\text{ex}}$, which we obtain from

$$e^{-\mu_i^{\text{ex}}/kT} = e^{-\tilde{\mu}_i^{\text{ex}}/kT}\frac{\left\langle\!\left\langle e^{-\Delta U/kT} \right\rangle\!\right\rangle_0}{\left\langle\!\left\langle e^{-\Delta\tilde{U}/kT} \right\rangle\!\right\rangle_0}$$

$$= e^{-\tilde{\mu}_i^{\text{ex}}/kT}\int \tilde{P}_0(\delta U/kT)e^{-\delta U/kT}\,d(\delta U/kT) \tag{14}$$

Here $\delta U = \Delta U - \Delta \tilde{U}$ is the electrostatic contribution to the solute–solvent interactions. The final expression is another partition function formula,

$$e^{-\delta \mu_i^{ex}/kT} = \int \tilde{P}_0(\delta U/kT) e^{-\delta U/kT} \, d(\delta U/kT) \qquad (15)$$

where $\tilde{P}_0(\delta U/kT)$ is the probability density of observing a value $\delta U/kT$ for solute–solvent electrostatic interactions when the sampling is carried out without them. We emphasize again that an important advantage of Eq. (15) is that the probability distributions, $\tilde{P}_0(\delta U/kT)$, can be modeled using available information about the system (Hummer *et al.*, 1997). Of course, those models should be reasonably accurate, but it is just as important that their information content be clear so that physical conclusions might be drawn from the observed accuracy.

1. Dielectric Models and Gaussian Theories

Equation (15) permits a straightforward analysis of dielectric continuum models of hydration that have become popular in recent decades. The dielectric model, also called the Born approximation, for the hydration free energy of a spherical ion of radius R with a charge q at its center is

$$\delta \mu_i^{ex} \approx -\frac{q^2}{2R}\left(\frac{\varepsilon - 1}{\varepsilon}\right) \qquad (16)$$

The dielectric constant of the external medium is ε. The most significant point of Eq. (16) is that $\delta \mu_i^{ex}$ is proportional to q^2. On this basis, we change variables, writing $\delta U = q\Phi$, and consider the Gaussian model (Hummer *et al.*, 1998c),

$$\tilde{P}_0(\Phi) \approx \frac{1}{\sqrt{2\pi \langle \delta \Phi^2 \rangle_0}} \exp\left[-\frac{1}{2}\frac{\Phi^2}{\langle \delta \Phi^2 \rangle_0}\right] \qquad (17)$$

Using this probability distribution in Eq. (15) produces

$$\delta \mu_i^{ex} \approx -\frac{q^2}{2kT}\langle \delta \Phi^2 \rangle_0 \qquad (18)$$

which we can compare with Eq. (16). The subscript zero on $\langle \cdots \rangle_0$ here indicates that averaging is performed with the solute molecule present, but in the absence solute–solvent electrostatic interactions. This formula avoids the issues of adjusting the Born radius R and of specifying the dielectric constant ε, both of which have been contentious. The most serious issue is the parameterization required for R in Eq. (16). It is clear from Eq. (18) that this parameter should depend on thermodynamic state—temperature, pressure, and composition of the system—and, in the more general case, on solute conformation as well. The dependences on temperature, pressure, composition, and solute conformation of the Gaussian model free energy have not received much attention. Where these variations have been checked, however, the results have been discouraging (Pratt and Rempe, 1999; Pratt et al., 1997; Tawa and Pratt, 1995, 1994). These issues are particularly relevant for unfolded proteins since variations in temperature, pressure, and composition are often used to induce unfolding, and we have less empirical evidence on what parameter values work for the unfolded state.

2. Multi-Gaussian Models

A number of directions can be taken to generalize Eq. (17). A general cumulant expansion is helpful, but typically not conclusive (Hummer et al., 1998c). A multi-Gaussian model is another alternative, and, in part, suggests the quasi-chemical theory which follows below. Here we assume that the distribution $\tilde{P}_0(\Phi)$ in Eq. (17) can be expressed as a linear combination of Gaussians corresponding to configurational substates of the system. As an example, the substates might be distinct configurations defined by different intra- and intermolecular hydrogen-bonding interactions for the solute and solvent molecules. Indexing those substates by s, we would then analyze the joint probability distribution $\tilde{P}_0(\Phi,s)$, assuming the conditional probability distributions $\tilde{P}_0(\Phi|s)$ are Gaussian and $p_s = \int \tilde{P}_0(\Phi, s) d\Phi$ are available from, for example, simulation. Then

$$\tilde{P}_0(\Phi) \approx \sum_s \tilde{P}_0(\Phi|s) \, p_s \qquad (19)$$

(see Fig. 7, color insert.) This approach effectively fixes the most immediate difficulties of dielectric models (Hummer et al., 1997). Two issues might be of concern. The first issue is the definition of "substate" and the second is the adoption of Gaussian models for $\tilde{P}_0(\Phi|s)$. The developments in Section VI suggest that the second of these issues is the greater concern. Furthermore, those developments emphasize that the composition law, Eq. (19), is generally valid, thus not a concern, and typically a helpful step.

Hydrogen bonds formed by the solute and solvent molecules were used as an example of a possible scheme for cataloging configurational substates. This example again hints that the partition function, Eq. (15), is local in nature; in this case, its evaluation relies on the local composition in the vicinity of the solute. That this local composition is an important general property in organizing configurational substates is the most basic concept for the quasi-chemical theory developed below.

IV. THE POTENTIAL DISTRIBUTION THEOREM REVISITED

With applications to protein solution thermodynamics in mind, we now present an alternative derivation of the potential distribution theorem. Consider a macroscopic solution consisting of the solute of interest and the solvent. We describe a macroscopic subsystem of this solution based on the grand canonical ensemble of statistical thermodynamics, accordingly specified by a temperature, a volume, and chemical potentials for all solution species including the solute of interest, which is identified with a subscript index 1. The average number of solute molecules in this subsystem is

$$\langle n_1 \rangle_{GC} = \frac{1}{\Xi(\mathbf{z}, V, T)} \sum_{\mathbf{n} \geq 0} n_1 \frac{\mathbf{z}^{\mathbf{n}}}{\mathbf{n}!} Q(\mathbf{n}, V, T) \tag{20}$$

where $\mathbf{n} \equiv \{n_1, n_2, \ldots\}$, $\mathbf{z} \equiv \{z_1, z_2, \ldots\}$, and $\mathbf{z}^{\mathbf{n}} \equiv \Pi_i z_i^{n_i}$ where n_i and z_i are the number of particles and the thermodynamic activity, respectively, of species i. $\Xi(\mathbf{z}, V, T)$ is the grand canonical partition function, which normalizes the sum in Eq. (20) over all possible non-negative occupancies of species in the subsystem, and $\langle \cdots \rangle_{GC}$ indicates grand canonical averaging over the population $\mathbf{n}$. $Q(\mathbf{n}, V, T)/\mathbf{n}!$ is the canonical partition function for $\mathbf{n}$ molecules with $\mathbf{n}! \equiv \Pi_i n_i!$.

We note that n_1 in the summand of Eq. (20) annuls terms with $n_1 = 0$ and permits the sum to start with $n_1 \geq 1$. Thus, Eq. (20) can be rearranged to have an explicit leading factor of z_1. This colligative property was noted above. Of course, a determination of z_1 establishes the solute chemical potential μ_1 we seek. We are motivated, therefore, to examine the coefficient multiplying z_1. To that end, we bring forward the explicit extra factor in z_1 and rearrange Eq. (20) to obtain

$$\langle n_1 \rangle_{GC} = \frac{z_1 Q_1(V, T)}{\Xi(\mathbf{z}, V, T)}$$

$$\times \sum_{\mathbf{n} \geq 0} \frac{\mathbf{z}^{\mathbf{n}}}{\mathbf{n}!} \frac{Q(\mathbf{n} \mid n_1 \to n_1 + 1, V, T)}{Q_1(V, T)} \tag{21}$$

where $(\mathbf{n} \mid n_1 \to n_1 + 1)$ means to add one molecule of species 1 to the population $\mathbf{n}$, and $Q_1(V, T) = Q(\{1, 0, \ldots\}, V, T)$ is the canonical partition function with exactly $n_1 = 1$ and no other molecules. In our notation above,

$$Q_1(V, T) \equiv q_1^{\text{int}} V / \Lambda_1^3 \qquad (22)$$

The additional factor of $Q_1(V, T)$ in Eq. (21) makes the leading term in the sum unity, as suggested by the usual expression for the cluster expansion in terms of the grand canonical partition function. Note that n_1 in the summand of Eq. (20) is not explicitly written in Eq. (21). It has been absorbed in the $\mathbf{n}!$, but its presense is reflected in the fact that the population is enhanced by one in the partition function numerator that appears in the summand. Equation (21) adopts precisely the form of a grand canonical average if we discover a factor of $Q(\mathbf{n}, V, T)$ in the summand for the population weight. Thus

$$\rho_1 = z_1 \left(q_1^{\text{int}} / \Lambda_1^3 \right) \left\langle \frac{Q(\mathbf{n} \mid n_1 \to n_1 + 1, V, T)}{Q_1(V, T)\, Q(\mathbf{n}, V, T)} \right\rangle_{\text{GC}} \qquad (23)$$

where $\rho_1 \equiv \langle n_1 \rangle_{\text{GC}} / V$. Finally, we observe that the ratio in this population average consists of configurational integrals involving the same coordinates in the numerator and the denominator. The one important difference between the configurational integrals in the numerator and denominator is that the integrand in the numerator is the Boltzmann factor for the system with one extra solute molecule. In contrast, the integrand in the denominator is a product of Boltzmann factors for that one extra solute molecule and for the rest of the solution, calculated separately and with no interactions between the extra solute molecule and the remaining molecules in solution. Thus,

$$\left\langle \frac{Q(\mathbf{n} \mid n_1 \to n_1 + 1, V, T)}{Q_1(V, T)\, Q(\mathbf{n}, V, T)} \right\rangle_{\text{GC}} = \left\langle\!\left\langle e^{-\Delta U / kT} \right\rangle\!\right\rangle_0 \qquad (24)$$

which combines population averaging with the normalized configurational integrations. This result leads directly to

$$\frac{\rho_1 \Lambda_1^3}{z_1 q_1^{\text{int}}} = \left\langle\!\left\langle e^{-\Delta U / kT} \right\rangle\!\right\rangle_0 \qquad (25)$$

where the brackets indicate a combination of configurational averaging and conformational averaging for the solute as Eq. (23) prescribes. This includes, of course, population fluctuations within the immediate locality of the solute. Equation (25) is the same as Eq. (3). Note that this derivation only depends on the structure of the grand partition function with the Boltzmann–Gibbs **n!**. Thus, quantum-mechanical treatments that retain these Boltzmann statistics will lead to a similar result.

The averaging of local populations will be developed as a central element of Eq. (25) and warrants closer inspection. The Boltzmann factor $e^{-\Delta U/kT}$ in this expression preferentially selects clusters of solvent molecules that have favorable energetic interactions with the solute from the probability distribution of $\Delta U/kT$ averaged over solvent configurations and solvent cluster sizes. For aqueous protein solutions, therefore, we might expect this population average to preferentially weight clusters of water molecules large enough to hydrate the hydrophilic groups of the folded or unfolded protein relative to smaller clusters or clusters of the same size having configurations that do not hydrate these groups as well.

V. QUASI-CHEMICAL THEORY OF SOLUTIONS

We noted above that the potential distribution theorem benefits from the fact that only local information on solute–solvent interactions is needed for ΔU in Eq. (5). Here we develop that idea into the quasi-chemical description for local populations involved in Eq. (25) (Pratt and LaViolette, 1998; Pratt and Rempe, 1999; Hummer *et al.*, 2000; Pratt *et al.*, 2001).

This quasi-chemical development is motivated by the observation that certain solute–solvent interactions can be characterized as chemical associations—i.e., strong interactions relative to the thermal energy, kT—although the majority of them are noncovalent interactions. Since these "chemical" interactions are also both short-range and specific in nature, we can identify an inner shell or proximal region around the solute that will accommodate strongly associating solvent molecules, and a second region, the outer shell, that corresponds to the remainder of the system volume. For protein solutions, we might define this inner shell to include regions the size of water molecules in close proximity to solvent-accessible, hydrophilic groups that are likely to bind water molecules. These sites could, for example, be identified from conserved crystallographic waters of hydration, as shown in Figure 1. These are customary considerations in describing hydrogen-bonding interactions in aqueous solutions; i.e., the definition of a hydrogen bond between pairs of water molecules, or between a water molecule and a particular amino acid of a protein. The inner shell can be specified by defining an indicator function b_j, such that $b_j = 1$ when solvent molecule j occupies this region and $b_j = 0$ when this solvent molecule is outside that region. With this definition, a natural outer shell contribution to the potential distribution theorem expression for the solute excess chemical potential is

$$\mu_i^{ex}/kT|_{outer} = -\ln \left\langle\!\!\left\langle e^{-\Delta U/kT} \prod_j (1 - b_j) \right\rangle\!\!\right\rangle_0 \qquad (26)$$

Note that the additional factor within the average, the $\prod_j (1 - b_j)$, would be zero for any solvent configuration in which a solvent molecule is found in the inner shell. Thus, this expression involves a potential distribution average under the constraint that no binding in the inner shell is permitted. We can formally write the full expression for the excess chemical potential as

$$\mu_i^{ex}/kT = -\ln \left[\frac{\langle\langle e^{-\Delta U/kT} \rangle\rangle_0}{\left\langle\!\!\left\langle e^{-\Delta U/kT} \prod_j (1 - b_j) \right\rangle\!\!\right\rangle_0} \right]$$
$$- \ln \left\langle\!\!\left\langle e^{-\Delta U/kT} \prod_j (1 - b_j) \right\rangle\!\!\right\rangle_0 \qquad (27)$$

This expression separates solute–solvent interactions into an outer shell contribution, which we submit can be described using simple physical approximations, and an inner shell contribution, which we treat using quasi-chemical theory.

A. Potential Distribution Formula for Averages

We now consider how averages associated with the system, including the solute of interest, are performed from the perspective of the potential distribution theorem. The average $\langle\langle \ldots \rangle\rangle_0$ does not involve the solute–solvent interactions, of course, but is an average over the thermal motion of the solute and solvent in the absence of energetic coupling between those subsystems. Consider then the average of some configurational function for the case in which interactions between the solute and the solvent are fully involved. We obtain this average by supplying the Boltzmann factor for the solute–solvent interactions with the proper normalization:

$$\langle F \rangle = \frac{\langle\langle e^{-\Delta U/kT}F\rangle\rangle_0}{\langle\langle e^{-\Delta U/kT}\rangle\rangle_0} \tag{28}$$

Note that just such a ratio of averages appears in Eq. (27), where the averaged quantity is just the indicator function $\Pi_j(1 - b_j)$. Since this function is equal to 1 for precisely those cases that the defined inner shell is empty, and is equal to 0 otherwise, we can recast Eq. (27) as

$$\mu_i^{\text{ex}}/kT = \ln x_0$$

$$- \ln \left\langle\left\langle e^{-\Delta U/kT} \prod_j (1 - b_j) \right\rangle\right\rangle_0 \tag{29}$$

where x_0 is the probability of those configurations for which the solute has no solvent molecules bound in the inner shell; i.e., it is the average of the indicator function.

B. Chemical Association Model for x_0

The final step in our quasi-chemical development is merely to recognize that a stoichiometric model for chemical association provides a correct description of x_0. We imagine following a specific solute molecule of interest through chemical conversions defined by changes in the inner shell populations,

$$\text{PW}_n + m\text{W} \rightleftharpoons \text{PW}_{m+n} \tag{30}$$

where P indicates the solute (e.g., a protein molecule) and W is the solvent species (water is the most common one). When these transformations are in equilibrium,

$$\frac{c_n}{c_0} = K_n \rho_W^n \tag{31}$$

where we have introduced the concentration, c_n, of a complex consisting of the protein and n water molecules to avoid confusion with the earlier different use of ρ_n. Equation (31) serves as a definition of K_n and is just the usual "products over reactants" chemical equilibrium ratio.

Supplying the appropriate normalization to evaluate x_0 explicitly gives

$$x_0 \equiv c_0 / \sum_n c_n = \frac{1}{1 + \sum_{n \geq 1} K_n \rho_W^n} \tag{32}$$

and substituting into Eq. (29) gives the desired final result:

$$\mu_i^{ex} / kT = - \ln \left[1 + \sum_{n \geq 1} K_n \rho_W^n \right]$$

$$- \ln \left\langle \left\langle e^{-\Delta U / kT} \prod_j (1 - b_j) \right\rangle \right\rangle_0 \tag{33}$$

Equation (33) offers a proper separation of inner and outer shell contributions so that different physical approximations might be used in these different regions and then properly matched. The description of inner shell interactions will depend on access to the equilibrium constants K_n. These are well defined and so might be the subject of either statistical thermodynamic computations or experiments. For simple solutes, such as the Li^+ ion, *ab initio* calculations can be carried out to obtain the K_n (Rempe *et al.*, 2000; Pratt and Rempe, 1999; Rempe and Pratt, 2001; Pratt *et al.*, 2000). For proteins, other methods must be found; e.g., by identifying inner shell waters of hydration from crystallographic waters, as suggested in Figure 1, and estimating the K_n from experimental data on the thermodynamics of transfer of small molecule analogs from the ideal gas phase into water (Makhatadze and Privalov, 1993). With definite quantitative values for these coefficients, the inner shell contribution in Eq. (33) appears just as a local grand canonical partition function involving the composition of the defined inner shell. We note that the net result of dividing the excess chemical potential in Eq. (33) into inner and outer shell contributions should not depend on the specifics of that division. This requirement can, in fact, provide a variational check that the accumulated approximations are well matched.

C. Packing Contributions

Folded and unfolded proteins in solution are dense materials characterized in large part by different degrees of conformational flexibility and solvent exposure. Thus, packing is a foundational issue for their solution thermodynamic properties. Although the developments above have been directed toward associative interactions, they can also be applied to treat excluded-volume interactions. We address the issue of packing here because the result will be used below in an approximate quasi-chemical formula (Pratt *et al.*, 2001).

Note that packing contributions are most obvious in the outer shell contribution to Eq. (33) because the indicator function $\Pi_j(1 - b_j)$ constrains that term to the case where no occupancy of the defined inner shell is permitted. These excluded volume interactions are the essense of packing contributions. To study those contributions, we consider a fictitious solute that does not interact with the solvent at all: $e^{-\Delta U/kT} = 1$. In that case, of course, μ_i^{ex} is zero and we write

$$0 = -\ln\left[1 + \sum_{n \geq 1} \tilde{K}_n \rho_W^n\right]$$
$$- \ln\left\langle\!\!\left\langle\prod_j (1 - b_j)\right\rangle\!\!\right\rangle_0 \tag{34}$$

where the tilde over the equilibrium constants indicates that these constants correspond to this specific fictitious case. The rightmost term of Eq. (34) gives the contribution to the chemical potential of a solute that perfectly excludes solvent from the region defined by $b_j = 1$ for all j, which we henceforth abbreviate as $\mathbf{b} = 1$. Thus, we have the formal result

$$\mu_i^{ex}|_{HC} = kT \ln\left[1 + \sum_{n \geq 1} \tilde{K}_n \rho_W^n\right] \tag{35}$$

where HC stands for hard-core. This result has a pleasing formal interpretation that develops from Eq. (31). Successive contributions to the partition function sum of Eq. (35) are evidently $\tilde{c}_n/\tilde{c}_0$. (The tilde, again, is a reminder that we are considering the fictitious species on which this argument is founded; $\tilde{c}_n$ is the concentration of those species with n ligands.) Noting also that $\tilde{c}_0/\tilde{c}_0 = 1$, the sum is then the total concentration of the fictitious species considered, divided by the concentration of those species with zero (0) ligands, $\tilde{c}_0$. This ratio is the inverse of the probability that a distinguished one of these fictitious species would have zero (0) ligands. Thus, Eq. (35) reduces formally to the well known $\mu_i^{ex}|_{HC} = -kT \ln p_0$ as in Eq. (7). In contrast to the more general quasi-chemical formula, Eq. (33), the hard-core case involves no outer shell term and a change in sign for the inner shell contribution.

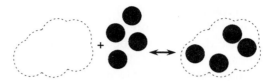

FIG. 8.   A quasi-chemical view of packing contributions. The dashed curve stencils the "fictitious" solute of interest, specifying its "inner shell" ($\mathbf{b} = 1$) exclusion region. To find space in a dense medium, we can consider chemical conversions that *extract* (right-to-left) solution elements from the space required.

This theory is fairly easy to implement in its most primitive form that uses the approximate value for the equilibrium constants obtained by neglecting external medium effects: $\tilde{K}_n \approx \tilde{K}_n^{(0)}$. The quantities $\mathbf{n}! \tilde{K}_n^{(0)}$ are then just n-molecule configurational integrals in which all n-molecules must be *inside* the inner region $\mathbf{b} = 1$ (see Fig. 8.) The geometric multipliers ρ_W in Eq. (35) serve to make that sum a grand canonical partition function for solvent molecules confined to the region $\mathbf{b} = 1$. Self-consistency with the specified solution density $\langle n_W \rangle_0 = v\rho_W$, where v is the volume of the $\mathbf{b} = 1$ region, can be achieved by augmenting this geometric weighting with a Lagrange multiplier γ that serves as a self-consistent mean field. But this might just as well be considered an activity coefficient that accounts for using the approximate values, $\tilde{K}_n^{(0)}$:

$$\frac{c_n}{c_0} = K_n \rho_W^n = \tilde{K}_n^{(0)} \gamma^n \rho_W^n \qquad (36)$$

VI.   PRIMITIVE QUASI-CHEMICAL APPROXIMATION

We can exploit the new results for packing contributions to reconsider the outer shell contribution in Eq. (33). For ionic solutes, the outer shell term would represent the Born contribution because it describes a hard ion stripped of any inner shell ligands. A Born model based on a picture of a dielectric continuum solvent is reasonable (see Section III,B, and Fig. 9, color insert). With that motivation, we first separate the outer shell term into an initial packing contribution and an approximate electrostatic contribution as

$$\ln \left\langle\!\!\left\langle e^{-\Delta U/kT} \prod_j (1 - b_j) \right\rangle\!\!\right\rangle_0 \approx \ln \left\langle\!\!\left\langle \prod_j (1 - b_j) \right\rangle\!\!\right\rangle_0 - \mu_i^{\mathrm{ex}}|_{\mathrm{BORN}}/kT$$

$$(37)$$

The electrostatic contributions, identified as $\mu_i^{\mathrm{ex}}|_{\mathrm{BORN}}$ and modeled on the basis of a dielectric continuum, are typically a substantial part of the numerical hydration free energy. But they are not subtle and are generally not accurate descriptors of variations of hydration free energies, with respect to either thermodynamic state or solute conformation. In this format, this Born contribution can be explicitly cast as

$$\mu_i^{\text{ex}}|_{\text{BORN}}/kT = -\ln\left[\frac{\left\langle\!\!\left\langle e^{-\Delta U/kT}\prod_j(1-b_j)\right\rangle\!\!\right\rangle_0}{\left\langle\!\!\left\langle \prod_j(1-b_j)\right\rangle\!\!\right\rangle_0}\right] \tag{38}$$

The other term on the right of Eq. (37) is a packing contribution of the type just analyzed. Adopting the simplest of those results directly we have

$$\ln\left\langle\!\!\left\langle e^{-\Delta U/kT}\prod_j(1-b_j)\right\rangle\!\!\right\rangle_0 \approx \ln\left[1+\sum_{n\geq 1}\tilde{K}_n^{(0)}\gamma^n\rho_W^n\right] - \mu_i^{\text{ex}}|_{\text{BORN}}/kT \tag{39}$$

Finally, we combine these simple approximations for the outer shell terms with the primitive quasi-chemical approximation for the inner shell term from Eq. (33) to obtain,

$$\mu_i^{\text{ex}} - \mu_i^{\text{ex}}|_{\text{BORN}} \approx -kT\ln\left[\frac{1+\sum\limits_{n\geq 1}K_n^{(0)}\gamma^n\rho_W^n}{1+\sum\limits_{n\geq 1}\tilde{K}_n^{(0)}\gamma^n\rho_W^n}\right] \tag{40}$$

The superscript zero again indicates that the equilibrium constants are obtained by neglecting external medium effects. We have noted above that the denominator of Eq. (40) is a grand canonical partition function for water molecules in the region $\mathbf{b}=1$. That region fluctuates, of course, as the solute conformations are sampled from $s_1^{(0)}(\mathcal{R}^n)$. The combination $\gamma\rho_W$ is the activity that governs evaluation of that denominator. We have chosen the same activity for the numerator sum so that the right side offers the evaluation of an average that looks familiar. In particular, substituting $K_n^{(0)} = (K_n^{(0)}/\tilde{K}_n^{(0)})\tilde{K}_n^{(0)}$ in the numerator gives

$$\frac{1+\sum\limits_{n\geq 1}K_n^{(0)}\gamma^n\rho_W^n}{1+\sum\limits_{n\geq 1}\tilde{K}_n^{(0)}\gamma^n\rho_W^n} = \left\langle\frac{K_n^{(0)}}{\tilde{K}_n^{(0)}}\right\rangle \tag{41}$$

utilizing again the notation of Eq. (23). The numerator and denominator of the ratio appearing on the right side of Eq. (41) are configurational integrals involving the same coordinates. The integrands differ only in the Boltzmann factor of solute–solution interactions. Thus, we can rewrite Eq. (40) as

$$\mu_i^{\mathrm{ex}} - \mu_i^{\mathrm{ex}}|_{\mathrm{BORN}} \approx -kT \ln \left\langle\!\left\langle e^{-\Delta U/kT} \right\rangle\!\right\rangle_{0,\mathbf{b}=1} \qquad (42)$$

This is a surprising result because, despite the strong approximations accumulated, it is remarkably similar to the general Eq. (3). Nevertheless, it is approximate and a number of additional points must be kept in mind in considering it. The right side of this expression indicates a grand canonical calculation for water in the proximal volume $\mathbf{b} = 1$, without the solute present, though the proximal volume fluctuates with solute conformational coordinates. The solute coordinates serve to establish a flexible "air-tight bag" for the water. The calculations require a specified activity. We have chosen the activity that makes sense for packing problems treated with these simple theories. It is amusing that the denominator for the averaging is acquired by considering packing questions. But after this form is appreciated, distinctions between attractive and repulsive interactions are not further required; all the solute–solution interactions are in $e^{-\Delta U/kT}$ and close-range configurations are properly sampled, though the theory is approximate.

Though carried further here, these theoretical ideas were first explored by Pratt and Rempe (1999). They argued that shape fluctuations were the most important concerns for simulations of biopolymers. This is particularly true for unfolded proteins. It deserves emphasis, therefore, that in this approach shape fluctuations are directly conditioned by $s_1^{(0)}(\mathcal{R}^n)$.

The direct calculation of the "reduced" partition function $\langle\!\langle e^{-\Delta U/kT} \rangle\!\rangle_{0,\mathbf{b}=1}$ is expected to be limited by the variance of the function $e^{-\Delta U/kT}$. However, representing the averaged quantity as a *ratio* opens possibilities for importance sampling, evaluating both numerator and denominator on the basis of a single sample designed to reduce the variance of $e^{-\Delta U/kT}$.

A.   Correspondence to Multi-Gaussian Models

The result, Eq. (41), is remarkable also because it takes the form anticipated intuitively by the multi-Gaussian theory, Eq. (19). The observation that $\tilde{p}_s = \tilde{K}_s^{(0)} \gamma^s \rho_W^s / (1 + \sum_{n\geq 1} \tilde{K}_n^{(0)} \gamma^n \rho_W^n)$ makes that clear. Here

FIG. 1. Top: Hydration waters surrounding T4 lysozyme. The 139 crystallographic waters for wild-type T4 lysozyme (4lzm) are displayed as green spheres. Also included are waters corresponding to the 40 best conserved water locations among 18 lysozyme mutants in 10 different crystal forms that have been superimposed on wild-type lysozyme. Water sites in the vicinity of the mutations are excluded. The conserved sites can be identified by clusters of green spheres. The molecular surface of the protein is colored light blue to indicate nonpolar residues and red to indicate polar and charged residues. Bottom: The same 139 crystallographic waters are shown without the protein. Data are from Zhang and Matthews (1994).

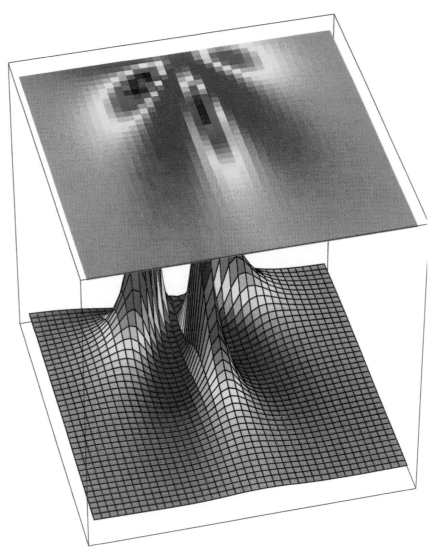

FIG. 7. Multi-Gaussian illustration: Three distinct peaks in probability corresponding to potential energy basins in the potential energy for solute–solvent interactions are depicted in a schematic two-dimensional configurational space. For this case of several configurationally distinct basins, $\tilde{P}_0(\delta U/kT)$ of Eq. (15) is a contraction onto the potential energy coordinate of the properties of those several basins; a single Gaussian function is likely to be problematic. Structural features identifying the distinct basins are then a first step in implementing a multi-Gaussian model, Eq. (19).

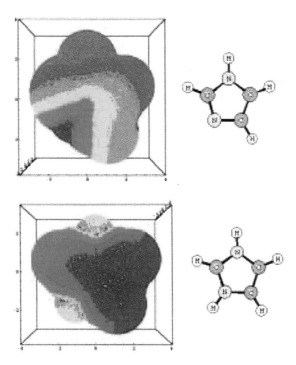

FIG. 9. Applicability of dielectric models to solute molecules of nonspherical shape. These maps show contours of the electrostatic potential on the dielectric surface of the molecules indicated on the right. More blue indicates a more negative value of the electrostatic potential, though the colors are not standardized to particular values of the electrostatic potential.

the $\tilde{p}_s$ are the occupancy probabilities for the $\mathbf{b} = 1$ region. For hard-core packing problems, this $\mathbf{b} = 1$ region is, of course, larger than the excluded volume; it might naturally include an anticipated first shell region. We can then generalize Eq. (41) beyond the thermodynamic potential distribution theorem to

$$P_0(\Delta U/kT) = \sum_{s \geq 0} P_0(\Delta U/kT|s)\,\tilde{p}_s \tag{43}$$

which can be compared to Eq. (19). For excluded volume (packing) problems, this expression would be written

$$p_n = \sum_{s \geq 0} P_0(n|s)\,\tilde{p}_s \tag{44}$$

This relation says that the probability of n solvent centers occupying the observation (excluded) volume is the product of two probabilities: The probability that the larger ($\mathbf{b} = 1$) region has s solvent centers *times* the conditional probability that the observation volume has n centers when the larger region has s centers. For packing problems, the approximation obtained above amounts to $P_0(n = 0|s) \approx K_s^{(0)}/\tilde{K}_s^{(0)}$ with a primitive quasi-chemical approximation for $\tilde{p}_s$. If a Gaussian model for $P_0(n|s)$ were satisfactory for Eq. (44), this would be a multi-Gaussian model. But the restriction to non-negative integers is conceptually and practically important for occupancy problems; $P_0(n|s)$ might be modeled with information theory approaches (Pratt *et al.*, 1999; Pratt, 2002). Because full solute–solution interactions are expressed in Eq. (42), "context" hydrophobicity should be correctly described. Being associated with the larger ($\mathbf{b} = 1$) volume, the probabilities $\tilde{p}_s$ might be estimated on the basis of macroscopic models (Pratt, 2002). But for idealized primitive hydrophobicity modeled by hard-core solutes, dewetting for mesoscale solutes (Pratt, 2002) should also be properly captured since incipient multiphase behavior of the solution should be expressed in $\tilde{p}_s$.

VII. CONCLUSIONS

The theories of hydration we have developed herein are built upon the potential distribution theorem viewed as a local partition function. We also show how the quasi-chemical approximations can be used to evaluate this local partition function. Our approach suggests that effective descriptions of hydration are derived by defining a proximal region around the solute of interest, carefully monitoring the composition of that region and, more primitively, conceptualizing the occupants of that region as integral parts of the solute. Progress over recent years in biophysical theories of hydration for small molecule solutes have been distinguished by precisely this perspective. The physical picture that develops for protein solutions is to take the solvent as an integral part of the protein, no less than as a determinant of protein structure and function. Applications of these approaches to proteins or other macromolecules will, however, be much more ambitious undertakings. These applications are, on the other hand, natural and essential in any consideration of unfolded proteins or large conformational changes in folded proteins. The new theoretical approximations derived here [Eq. (42)] appear to offer a practical scheme for those applications.

REFERENCES

Avgelj, F., Luo, P., and Baldwin, R. L. (2000). Energetics of the interaction between water and the helical peptide group and its role in determining helix propensities. *Proc. Natl. Acad. Sci. USA* **97,** 10786–10791.

Baker, E. N., and Hubbard, R. E. (1984). Hydrogen bonding in globular proteins. *Prog. Biophys. Mol. Biol.* **44,** 97–179.

Baldwin, R. L., and Zimm, B. H. (2000). Are denatured proteins ever random coils? *Proc. Natl. Acad. Sci. USA* **97,** 12391–12392.

Edsall, J. T., and McKenzie, H. A. (1983). Water and proteins II. The location and dynamics of water in protein systems and its relation to their stability and properties. *Adv. Biophys.* **16,** 53–184.

Frenkel, D., and Smit, B. (1996). *Understanding Molecular Simulation,* pp. 157–160. Academic Press, London.

Garde, S., Hummer, G., García, A. E., Paulaitis, M. E., and Pratt, L. R. (1996). Origin of entropy convergence in hydrophobic hydration and protein folding. *Phys. Rev. Lett.* **77,** 4966–4968.

Hummer, G., Garde, S., García, A. E., Paulaitis, M. E., and Pratt, L. R. (1998a). Hydrophobic effects on a molecular scale. *J. Phys. Chem. B* **102,** 10469–10482.

Hummer, G., Garde, S., García, A. E., Paulaitis, M. E., and Pratt, L. R. (1998b). The pressure dependence of hydrophobic interactions is consistent with the observed pressure denaturation of proteins. *Proc. Natl. Acad. Sci. USA* **95,** 1552–1555.

Hummer, G., Garde, S., García, A. E., Pohorille, A., and Pratt, L. R. (1996). An information theory model of hydrophobic interactions. *Proc. Natl. Acad. Sci. USA* **93,** 8951–8955.

Hummer, G., Garde, S., García, A. E., and Pratt, L. R. (2000). New perspectives on hydrophobic effects. *Chem. Phys.* **258,** 349–370.

Hummer, G., Pratt, L. R., and García, A. E. (1997). Multistate gaussian model for electrostatic solvation free energies. *J. Am. Chem. Soc.* **119,** 8523–8527.

Hummer, G., Pratt, L. R., and Garcia, A. E. (1998c). Molecular theories and simulation of ions and polar molecules in water. *J. Phys. Chem. A* **102,** 7885–7895.

Levy, Y., Jortner, J., and Becker, O. M. (2001). Solvent effects on the energy landscapes and folding kinetics of polyalanine. *Proc. Natl. Acad. Sci. USA* **98,** 2188–2193.

Makhatadze, G. I., and Privalov, P. L. (1993). Contribution of hydration to protein folding thermodynamics. *J. Mol. Biol.* **232**, 639–659.

Pappu, R. V., Srinivasan, R., and Rose, G. D. (2000). The Flory isolated-pair hypothesis is not valid for polypeptide chains: Implications for protein folding. *Proc. Natl. Acad. Sci. USA* **97**, 12565–12570.

Plaxco, K. W., and Gross, M. (2001). Unfolded, yes, but random? Never! *Nat. Struct. Biol.* **8**, 659–660.

Pohorille, A., and Pratt, L. R. (1990). Cavities in mol liq. and the theory of hydrophobic solubilities. *J. Am. Chem. Soc.* **112**, 5066–5074.

Pratt, L. R. (1998). Hydrophobic effects. *Encyclopedia of Computational Chemistry*, pp. 1286–1294.

Pratt, L. R. (2002). Theory of hydrophobic effects: "She's too mean to have her name repeated". *Annu. Rev. Phys. Chem.* **53**, 409–436.

Pratt, L. R., Hummer, G., and Garde, S. (1999). Theories of hydrophobic effects and the description of free volume in complex liquids. In *New Approaches to Problems in Liquid State Theory* (C. Caccamo, J.-P., Hansen, and G. Stell, eds.), vol. 529, pp. 407–420. Kluwer, Netherlands. NATO Science Series.

Pratt, L. R., and LaViolette, R. A. (1998). Quasi-chemical theories of associated liquids. *Mol. Phys.* **94**, 909–915.

Pratt, L. R., LaViolette, R. A., Gomez, M. A., and Gentile, M. E. (2001). Quasi-chemical theory for the statistical thermodynamics of the hardsphere fluid. *J. Phys. Chem. B* **105**, 11662–11668.

Pratt, L. R., and Rempe, S. B. (1999). Quasi-chemical theory and implicit solvent models for simulations. In *Simulation and Theory of Electrostatic Interactions in Solution. Computational Chemistry, Biophysics, and Aqueous Solutions* (L. R. Pratt and G. Hummer, eds.), vol. 492 of *AIP Conference Proceedings*, pp. 172–201. American Institute of Physics, Melville, NY.

Pratt, L. R., Rempe, S. B., Topol, I. A., and Burt, S. K. (2000). Alkali metal ion hydration and energetics of selectivity by ion-channels. *Biophys. J.* **78**, P2057–P2057.

Pratt, L. R., Tawa, G. J., Hummer, G., Garcia, A. E., and Corcelli, S. A. (1997). Boundary integral methods for the poisson equation of continuum dielectric solvation models. *Int. J. Quant. Chem.* **64**, 121–141.

Rempe, S. B., and Pratt, L. R. (2001). The hydration number of Na^+ in liquid water. *Fluid Phase Equilibria* **183**, 121–132.

Rempe, S. B., Pratt, L. R., Hummer, G., Kress, J. D., Martin, R. L., and Redondo, A. (2000). The hydration number of Li^+ in liquid water. *J. Am. Chem. Soc.* **122**, 966–967.

Shortle, D., and Ackerman, M. S. (2001). Persistence of native-like topology in a denatured protein in 8 M urea. *Science* **293**, 487–489.

Tawa, G. J., and Pratt, L. R. (1994). Tests of dielectric model descriptions of chemical charge displacements in water. *ACS Symposium Series* **568**, 60–70.

Tawa, G. J., and Pratt, L. R. (1995). Theoretical calculation of the water ion product k-w. *J. Am. Chem. Soc.* **117**, 1286–1294.

Thomas, S. T., Loladze, V. V., and Makhatadze, G. I. (2001). Hydration of the peptide backbone largely defines the thermodynamic propensity scale of residues at the c' position of the c-capping box of α helices. *Proc. Natl. Acad. Sci. USA* **98**, 10670–10675.

van Gunsteren, W. F., Bürgi, R., Peter, C., and Daura, X. (2001). The key to solving the protein-folding problem lies in an accurate description of the denatured state. *Angew. Chem. Int. Ed.* **40**, 351–355.

Vila, J. A., Ripoll, D. R., and Scheraga, H. A. (2000). Physical reasons for the unusual α helix stabilization afforded by charged or neutral polar residues in alanine-rich peptides. *Proc. Natl. Acad. Sci. USA* **97,** 13075–13079.

Widom, B. (1963). Some topics in the theory of fluids. *J. Chem. Phys.* **39,** 2808–2812.

Widom, B. (1982). Potential-distribution theory and the statistical mechanics of fluids. *J. Phys. Chem.* **86,** 869–872.

Williams, M. A., Goodfellow, J. M., and Thornton, J. M. (1994). Buried waters and internal cavities in monomeric proteins. *Protein Sci.* **3,** 1224–1235.

Zhang, X.-J., and Matthews, B. W. (1994). Conservation of solvent-binding sites in 10 crystal forms of t4 lysozyme. *Protein Sci.* **3,** 1031–1039.

INSIGHTS INTO THE STRUCTURE AND DYNAMICS OF UNFOLDED PROTEINS FROM NUCLEAR MAGNETIC RESONANCE

By H. JANE DYSON and PETER E. WRIGHT

Department of Molecular Biology, The Scripps Research Institute, La Jolla, California 92037

I. INTRODUCTION

Nuclear magnetic resonance (NMR) is unique in being able to provide detailed insights into the conformation of unfolded and partly folded proteins. Such studies are of importance not only in the context of protein folding research but also because of a growing awareness that many proteins are intrinsically unstructured in their biologically functional states. It is now recognized that many proteins and protein domains are only partially structured or are entirely unstructured under physiological conditions, folding to form stable structures only on binding to their biological receptors (Dyson and Wright, 2002; Wright and Dyson, 1999). Knowledge of the structural propensities of these domains, both free and bound to their targets, is essential for a proper understanding at the molecular level of their biological functions and interactions. Unfolded states of proteins also play a role in the transport of proteins across membranes and in the development of amyloid diseases.

ADVANCES IN
PROTEIN CHEMISTRY, Vol. 62

Structural characterization of non-native states of proteins and protein folding intermediates is of central importance for a detailed
understanding of protein folding mechanisms. The protein folding
problem remains one of the major unsolved problems in biology. Because of the speed and cooperativity of the folding process, detailed
structural and dynamic characterization of kinetic folding intermediates
is difficult and, in general, only low-resolution structural information
is available from protein engineering approaches or NMR hydrogen-
exchange pulse-labeling experiments. A valuable approach to characterize the structural and dynamic changes that accompany protein folding is to study partially folded states of proteins or peptide fragments
of proteins under equilibrium conditions. The power of modern NMR
and other spectroscopic techniques can then be applied directly to map
the protein folding energy landscape. NMR studies of fully denatured
states provide detailed insights (at the level of individual residues) into
the conformational ensemble at the starting point of folding, while
studies of partially folded states obtained under more mildly denaturing
conditions can reveal the intrinsic conformational propensities of the
polypeptide and identify potential folding initiation sites (Dyson and
Wright, 1996, 1998). These approaches can provide direct experimental
insights into the nature of the free-energy landscape for protein folding.

Owing to their intrinsically disordered nature, linear peptide fragments of proteins and unfolded full-length proteins are unsuitable for
structural study by X-ray crystallography. However, high-resolution NMR
experiments can provide detailed insights into the structure and dynamics of unfolded states, and studies of many proteins have been reported, including lysozyme (Buck *et al.*, 1994; Evans *et al.*, 1991; Schwalbe
et al., 1997), 434 repressor (Neri *et al.*, 1992a), FK506-binding protein
(Logan *et al.*, 1994), BPTI (Lumb and Kim, 1994), protein G (Frank *et al.*,
1995; Sari *et al.*, 2000), barnase (Arcus *et al.*, 1995; Wong *et al.*, 2000),
an SH3 domain (Zhang and Forman-Kay, 1995, 1997), a fibronectin
type III domain (Meekhof and Freund, 1999), fibronectin binding
protein (Penkett *et al.*, 1998), ubiquitin (Brutscher *et al.*, 1997), staphylococcal nuclease (Alexandrescu *et al.*, 1994), apoplastocyanin (Bai *et al.*,
2001), and apomyoglobin (Eliezer *et al.*, 1998; Yao *et al.*, 2001).

II. Conformational Propensities in Peptides

Because unidimensional ^1H NMR spectra of denatured proteins
lack resonance dispersion and resemble spectra of mixtures of free
amino acids, it was assumed that the denatured state behaves as a random coil (McDonald and Phillips, 1969). Calculations (Brant *et al.*,
1967; Zimm and Bragg, 1959) and spectroscopic measurements on

protein fragments (Epand and Scheraga, 1968; Hermans *et al.*, 1969; Taniuchi and Anfinsen, 1969) suggested that short linear peptides are also disordered in water solution. However, with advances in computational methods and development of improved experimental techniques, and particularly two-dimensional NMR, it was recognized that many peptides preferentially populate structured states in aqueous solution (for reviews, see Dyson and Wright, 1991; Wright *et al.*, 1988). For many years, NMR studies of intrinsic polypeptide conformational propensities were largely restricted to peptide fragments and small synthetic peptides because of the problems of resonance overlap in denatured protein spectra. Nevertheless, these studies formed the foundation for studies of full-length denatured proteins and provided fundamental insights into the local interactions that preferentially stabilize elements of secondary structure and hydrophobic clusters in nonglobular states.

In studies of linear peptides, the NMR parameters are averaged over all structures in the conformational ensemble. Fortunately, steric constraints impose restrictions on the allowed backbone conformations (Ramachandran *et al.*, 1963) so that only a limited range of backbone ϕ, ψ dihedral angles is significantly populated, predominantly in the α_R and, for glycine, α_L minima and in the broad β region (Fig. 1). This makes it possible to interpret NMR parameters such as nuclear Overhauser effects (NOEs), chemical shifts, and coupling constants as a population-weighted average over the minima in the conformational energy map. Although many structural analyses are performed at the purely qualitative level, methods have been developed for calculation of structural ensembles from experimental NMR constraints (Constantine *et al.*, 1995; Meirovitch and Meirovitch, 1996; Wang *et al.*, 1995; Yao *et al.*, 1994).

III.   NMR Studies of Unfolded and Partly Folded Proteins

With advances in multidimensional NMR technology and the development of methods for uniform labeling of proteins with ^{15}N and ^{13}C, direct NMR studies on unfolded states of full-length proteins, rather than protein fragments, became possible. These studies were aided by the advent of high-field NMR spectrometers, which provide the necessary dispersion and sensitivity. High sensitivity is a critical factor in the study of unfolded and partly folded proteins, since NMR experiments must often be performed at very low concentrations to prevent aggregation.

A.   *Resonance Dispersion and Assignment Methods*

Characterization of disordered states of proteins by NMR is particularly challenging because the polypeptide chain is inherently flexible

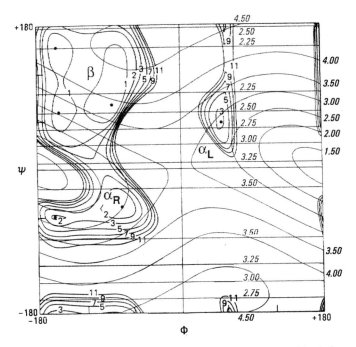

FIG. 1. Conformational energy diagram for the alanine dipeptide (adapted from Ramachandran *et al.*, 1963). Energy contours are drawn at intervals of 1 kcal mol^{-1}. The potential energy minima for β, α_R, and α_L are labeled. The dependence of the sequential $d_{NN}(i, i+1)$ distance (in Å) on the ϕ and ψ dihedral angles (Billeter *et al.*, 1982) is shown as a set of contours labeled according to interproton distance at the right of the figure. The $d_{\alpha N}(i, i+1)$ distance depends only on ψ for trans peptide bonds (Wright *et al.*, 1988) and is represented as a series of contours parallel to the ϕ axis. Reproduced from Dyson and Wright (1991). *Ann. Rev. Biophys. Chem.* **20**, 519–538, with permission from Annual Reviews.

and rapidly interconverts between multiple conformations. As a result, the chemical shift dispersion of most resonances, especially protons, is poor, and sequence-specific assignment of resonances is difficult. Isotope labeling with ^{13}C and ^{15}N and the use of multidimensional triple resonance NMR methods is essential. The chemical shifts of the backbone ^{15}N and ^{13}CO resonances are influenced both by residue type and by the local amino acid sequence; these resonances are well-dispersed, even in fully unfolded states (Braun *et al.*, 1994; Yao *et al.*, 1997; Zhang *et al.*, 1997a). The representative ^1H–^{15}N HSQC and H(N)CO spectra of acid-denatured apomyoglobin in Figure 2 demonstrate the superior dispersion available in the ^{15}N and ^{13}CO chemical shifts.

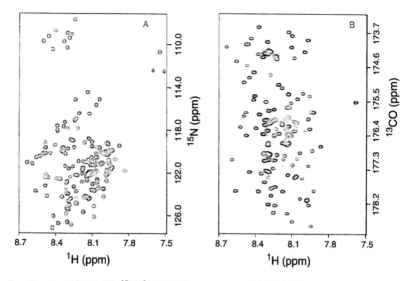

FIG. 2. (Left) 750-MHz ^{15}N–^1H HSQC spectrum and (right) 750-MHz HNCO spectrum of apomyoglobin at pH 2.3, 10 mM acetate-$d6$ and 5°C. Reproduced with permission from Yao *et al.* (1997).

The earliest approaches to sequence-specific resonance assignments utilized homonuclear (^1H) magnetization transfer methods to correlate resonances in the spectrum of the folded protein with the corresponding resonances in the denatured state (Evans *et al.*, 1991; Neri *et al.*, 1992b; Zhang *et al.*, 1994). This method lacks generality and is limited to smaller proteins for which exchange between folded and unfolded states occurs on an appropriate time scale. Three-dimensional triple resonance experiments to establish sequential connectivities via the well-resolved ^{15}N and ^{13}C resonances in uniformly $^{15}N,^{13}C$-labeled proteins provide a more general and robust method for obtaining unambiguous resonance assignments (Alexandrescu *et al.*, 1994; Eliezer *et al.*, 1998; Logan *et al.*, 1994). The pulse sequences that are commonly used have been reviewed elsewhere (Bax and Grzesiek, 1993; Kay, 1995). An assignment strategy that utilizes the superior chemical shift dispersion of the ^{13}CO resonance in unfolded proteins is especially valuable, given the excellent resolution and long T_2 relaxation time of these resonances (Yao *et al.*, 1997). A set of high-resolution constant-time triple resonance experiments that transfer magnetization sequentially along the amino acid sequence using carbonyl ^{13}C homonuclear isotropic mixing have been developed specifically for assignment of unfolded proteins (Liu *et al.*, 2000). These experiments have the advantage that the slow

[13]CO relaxation rates enable correlations to be established across proline residues.

An advantage of working with unfolded proteins is that the intrinsic flexibility of the polypeptide chain causes the resonances to be much narrower than they would be in a folded protein of comparable molecular weight. Unfortunately, the same is not true for partly folded proteins, or molten globule states, whose NMR spectra often contain a mixture of sharp (from unstructured regions) and broad resonances (from the structured subdomains). Indeed, many molten globules are extremely difficult to study by NMR because their lines are broadened by exchange processes occurring on intermediate time scales. For example, the molten globule state of α-lactalbumin gives rise to extremely broad resonances that largely preclude direct NMR analysis (Baum *et al.*, 1989; Kim *et al.*, 1999). Nevertheless, elegant NMR experiments have been brought to bear on the α-lactalbumin molten globule, with the result that it is now one of the best characterized molten globules (Wijesinha-Bettoni *et al.*, 2001).

B. NMR Parameters

Once backbone resonance assignments have been made, a number of NMR parameters can be used to characterize residual structure in unfolded and partly folded states. However, in interpreting the NMR data, it is important to bear in mind that, just as for linear peptides, unfolded and partly folded states of proteins are highly dynamic and that all parameters are a population-weighted average over all structures in the conformational ensemble. Conformational preferences are always identified, therefore, by comparison of experimental NMR parameters to those expected for a random coil state in which the polypeptide backbone dihedral angles adopt a Boltzmann distribution over the ϕ,ψ energy surface (Dyson and Wright, 1991; Smith *et al.*, 1996b). In addition, it is implicitly assumed that the dominant conformers in unfolded or partly folded polypeptides will have backbone dihedral angles that lie within the broad α and β minima on the ϕ,ψ conformational energy surface.

1. Chemical Shifts

The deviations of chemical shifts from random coil values, especially for $^{13}C^{\alpha}$, $^{13}C^{\beta}$, and $^{1}H^{\alpha}$, provide a valuable probe of secondary structural propensities (Spera and Bax, 1991; Wishart and Sykes, 1994). The ^{13}CO chemical shifts also provide information on secondary structure content and dihedral angle preferences, provided they are corrected for sequence dependence (Schwarzinger *et al.*, 2001). Theoretical calculations (de Dios *et al.*, 1993) show that the $^{13}C^\alpha$ and $^{13}C^\beta$ chemical shifts are determined primarily by the backbone ϕ, ψ dihedral angles. Empirical correlations confirm that these resonances, and especially $^{13}C^\alpha$, are highly sensitive indicators of secondary structure (Spera and Bax, 1991; Wishart and Nip, 1998; Wishart and Sykes, 1994). In α-helices, the $^{13}C^\alpha$ resonances are shifted downfield from their positions in random coil states by an average of 3.1 ± 1.0 ppm (Spera and Bax, 1991). The ^{13}CO resonances are also shifted downfield, while the $^{13}C^\beta$ and $^1H^\alpha$ resonances are shifted upfield. In β-sheets, the $^{13}C^\alpha$ resonance is shifted upfield [$\Delta\delta = -1.5 \pm 1.2$ ppm (Spera and Bax, 1991)] with respect to the random coil chemical shift, while the $^1H^\alpha$ and $^{13}C^\beta$ resonances are shifted downfield. In unfolded or partly folded proteins, rapid conformational averaging between the α and β regions of ϕ, ψ space reduces the secondary structure shifts below these extreme values. Indeed, the deviations of the chemical shifts from random coil values can be used to calculate the relative population of dihedral angles in the α or β regions or the population of helix in defined regions of the polypeptide (Eliezer *et al.*, 1998; Yao *et al.*, 2001). For example, the fractional helicity of a polypeptide between residues i to j can be determined from the $^{13}C^\alpha$ secondary shifts using the expression $\sum_{i,j} (\Delta\delta \, ^{13}C^\alpha_i)/3.1$. Similar expressions are used to calculate helicity from the secondary $^1H^\alpha$, $^{13}C^\beta$, or ^{13}CO chemical shifts.

An example of the use of chemical shifts to delineate residual secondary structure is given in Figure 3 for the molten globule state of apomyoglobin (Eliezer *et al.*, 1998; Eliezer *et al.*, 2000). Combined use of $^{13}C^\alpha$, $^1H^\alpha$, $^{13}C^\beta$, and ^{13}CO secondary shifts gives a more precise definition of secondary structure boundaries than use of $^{13}C^\alpha$ shifts alone (Eliezer *et al.*, 2000).

2. Coupling Constants

Coupling constants can also provide insights into the backbone conformational preferences in partly folded or unfolded proteins (Smith *et al.*, 1996a). The coupling constant $^3J_{HN\alpha}$ varies from about 8–10 Hz in β-strands to 4–5 Hz in helix. However, conformational averaging in unfolded states frequently results in intermediate values of $^3J_{HN\alpha}$ that are not particularly valuable as diagnostics of secondary structural propensities. Thus, for unfolded and partly folded proteins, as for peptides (Dyson and Wright, 1991), coupling constants are rarely as useful as chemical shifts for identification of residual structure.

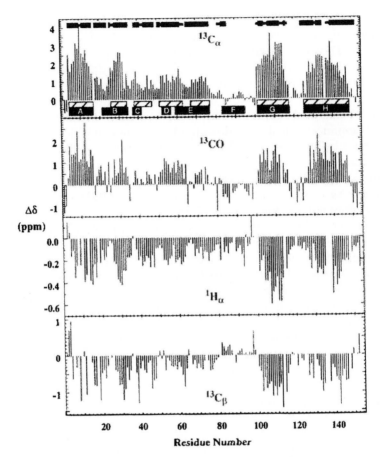

FIG. 3.   Secondary chemical shifts for $^{13}C^{\alpha}$, ^{13}CO, $^{1}H^{\alpha}$, and $^{13}C^{\beta}$ as a function of residue number in apomyoglobin at pH 4.1. Bars at the top of the figure indicate the presence of H_i^N–H_{i+1}^N NOEs; the smaller bars indicate that the NOE was ambiguous due to resonance overlap. Black rectangles at the base of the top panel indicate the locations of helices in the native holomyoglobin structure (Kuriyan *et al.*, 1986). Hashed rectangles indicate putative boundaries for helical regions in the pH 4 intermediate, based on the chemical shift and NOE data. Reproduced from Eliezer *et al.* (2000). *Biochemistry* **39**, 2894–2901, with permission from the American Chemical Society.

3. Temperature Coefficients

Amide proton temperature coefficients and hydrogen exchange rates can provide information about hydrogen-bonding interactions and solvent sequestration in unfolded or partly folded proteins (Dyson and Wright, 1991). Abnormally low temperature coefficients, relative to random coil values, are a clear indication of local structure and interactions.

Likewise, amide protons protected from exchange are a useful indication of structured regions in partly folded proteins and molten globules (Hughson *et al.*, 1990).

4. Nuclear Overhauser Effect

As in folded proteins, the NOE provides valuable information on secondary structure formation and long-range interactions in unfolded and partly folded proteins. However, in these disordered states, the NOE is difficult to interpret quantitatively because of the ubiquitous conformational averaging. Nevertheless, the $d_{\alpha N}(i, i+1)$, $d_{NN}(i, i+1)$, and $d_{\beta N}(i, i+1)$ NOEs between sequential amino acid residues do provide information on the local polypeptide backbone conformational preferences, i.e., on the relative population of dihedral angles in the α and β regions of ϕ,ψ space (Dyson and Wright, 1991). Such NOEs constitute a valuable supplement to chemical shifts in the analysis of backbone conformational preferences. The sequential $d_{\alpha N}(i, i+1)$ and $d_{NN}(i, i+1)$ NOEs provide information only on local ϕ and ψ dihedral angle preferences. They do not, by themselves, indicate the presence of folded conformations. Definitive identification of folded elements of secondary structure such as turns or helices requires additional information, either from medium-range NOEs (e.g., $d_{\alpha N}(i, i+2)$, $d_{\alpha N}(i, i+3)$, $d_{\alpha\beta}(i, i+3)$ NOE connectivities) or from other forms of spectroscopy such as circular dichroism (Dyson and Wright, 1991).

Because of the intrinsic flexibility and poor resonance dispersion of unfolded and partly folded proteins, long-range NOEs are generally very difficult to observe and assign. While observation of a long-range NOE between two protons provides a definitive indication that they are in close proximity in at least some structures in the conformational ensemble, determination of the nature of the folded structures is difficult unless an extensive network of NOEs can be observed. This has so far been achieved in only one case (Mok *et al.*, 1999).

While NOE connectivities in unfolded proteins can be observed in ^{15}N-edited 3D NOESY-HSQC spectra, the severe overlap of the ^{1}H resonances often reduces the utility of these experiments and seriously limits the number of NOE peaks that can be assigned. Improved resolution can be achieved for NOEs between NH groups by labeling both protons involved with their attached ^{15}N frequencies, using the 3D ^{15}N-HSQC-NOESY-HSQC experiment (Frenkiel *et al.*, 1990; Ikura *et al.*, 1990; Zhang *et al.*, 1997a). Although some long-range interactions can be observed in favorable cases (Mok *et al.*, 1999), this experiment primarily provides information about backbone conformational preferences. Detailed characterization of structured conformers requires identification and assignment of NOE connectivities involving side chain protons, which are, of course, poorly dispersed in disordered states. To overcome this problem, Kay and co-workers have developed a series of triple-resonance–based NOESY experiments that exploit the dispersion of the backbone ^{15}N and ^{13}CO resonances to report on NOEs involving side chain aliphatic protons, thereby removing problems associated with ambiguities in the aliphatic ^{1}H and ^{13}C chemical shifts (Zhang et al., 1997a,b).

5. Paramagnetic Relaxation Probes

Given the difficulty of detecting and interpreting long-range NOEs in unfolded and partly folded proteins, an excellent alternative is to utilize paramagnetic nitroxide spin labels, such as PROXYL, to probe the global structure. The method relies on introduction of spin labels at specific sites in the polypeptide. This is accomplished by site-directed mutagenesis, substituting amino acid residues one at a time by cysteine to provide a site for coupling of an iodoacetamide derivative of the spin label. If the protein contains natural cysteine residues in its sequence, it is necessary to mutate these to avoid multiple sites of paramagnetic labeling. By inserting spin labels at multiple sites, a sufficient number of long-range distance constraints can be obtained to allow determination of the global topology. The method has been applied successfully to characterize long-range structure in a denatured state of staphylococcal nuclease (Gillespie and Shortle, 1997a,b).

The problems inherent in the application of this method to a dynamic ensemble of conformations such as is found in disordered states of proteins have been previously described (Gillespie and Shortle, 1997a). Conformational averaging is likely to introduce a bias toward those conformations with the shortest contact distances, even though their populations within the ensemble may be quite small. Nevertheless, use of paramagnetic spin label probes can provide exceptionally important insights into the global topology and long-range interactions in unstructured states and partly folded proteins.

C. Measurement of Dynamics

NMR relaxation and diffusion experiments provide important insights into both the internal molecular dynamics and the overall hydrodynamic behavior of unfolded and partly folded states. Local variations in backbone dynamics are correlated with propensities for local compaction of the polypeptide chain that results in constriction of backbone motions (Eliezer et al., 1998, 2000). This can occur through formation of local hydrophobic clusters, formation of elements of secondary structure, or through long-range tertiary interactions in a compact folding intermediate. The measurements also provide insights into the effects of denaturants both on the overall hydrodynamic behavior of a polypeptide in solution and on its local structural elements.

Backbone dynamics are most commonly investigated by measurement of ^{15}N T_1 and T_2 relaxation times and the $\{^1H\}$–^{15}N NOE in uniformly ^{15}N-labeled protein. To circumvent problems associated with the limited dispersion of the NMR spectra of unfolded proteins, the relaxation and NOE data are generally measured using 2D HSQC-based methods (Farrow *et al.*, 1994; Palmer *et al.*, 1991).

For folded proteins, relaxation data are commonly interpreted within the framework of the model-free formalism, in which the dynamics are described by an overall rotational correlation time τ_m, an internal correlation time τ_e, and an order parameter S^2 describing the amplitude of the internal motions (Lipari and Szabo, 1982a,b). Model-free analysis is popular because it describes molecular motions in terms of a set of intuitive physical parameters. However, the underlying assumptions of model-free analysis—that the molecule tumbles with a single isotropic correlation time and that internal motions are very much faster than overall tumbling—are of questionable validity for unfolded or partly folded proteins. Nevertheless, qualitative insights into the dynamics of unfolded states can be obtained by model-free analysis (Alexandrescu and Shortle, 1994; Buck *et al.*, 1996; Farrow *et al.*, 1995a). An extension of the model-free analysis to incorporate a spectral density function that assumes a distribution of correlation times on the nanosecond time scale has recently been reported (Buevich *et al.*, 2001; Buevich and Baum, 1999) and better fits the experimental ^{15}N relaxation data for an unfolded protein than does the conventional model-free approach.

The assumptions inherent in model-free analysis can be avoided by direct mapping of the spectral density, either by measurement of additional relaxation parameters (Peng and Wagner, 1992) or by making simplifying approximations (Farrow *et al.*, 1995b; Ishima and Nagayama, 1995). The accuracy of the dynamics information obtained by spectral density mapping depends only on the accuracy of the experimental relaxation data, not on assumptions made about molecular motions. The disadvantage is that the method does not provide a physically intuitive description of the motions. In the reduced spectral density mapping approach, the spectral density is sampled at the frequencies 0, ω_N, and $0.87\omega_H$. $J(0)$ is sensitive both to fast internal motions on a picosecond–nanosecond time scale and to slow motions on the millisecond–microsecond time scale. Rapid internal motions tend to reduce the value of $J(0)$, while slow motions lead to anomalously large values of $J(0)$. In contrast, the high-frequency spectral density $J(0.87\omega_H)$ is sensitive only to fast internal motions, on a subnanosecond time scale; fast motions are reflected in relatively large values of $J(\omega_H)$. The $J(\omega_N)$ and $J(\omega_H)$ spectral densities are insensitive to the slow ms–μs time scale motions and so may be used to identify contributions of these motions to the $J(0)$ spectral densities.

As an example, we consider the application of reduced spectral density analysis to investigate the backbone dynamics in the pH 4 molten globule state of apomyoglobin (Eliezer *et al.*, 2000) as shown in Figure 4. Both $J(0)$ and $J(0.87\omega_H)$ are highly sensitive to variations in backbone motion, although $J(\omega_N)$ is much less informative. The largest values of $J(0)$ and the smallest values of $J(0.87\omega_H)$ are found in the A, B, G, and H helices, indicating restriction of motions on the subnanosecond time scale and packing to form a compact core in the molten globule state.

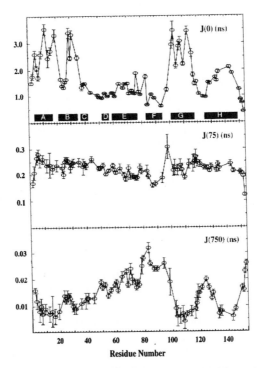

FIG. 4.   Spectral densities of backbone motions in the pH 4.1 apomyoglobin intermediate at three different frequencies. Reproduced from Eliezer *et al.* (2000). *Biochemistry* **39**, 2894–2901, with permission from the American Chemical Society.

Anomalously large values of $J(0)$ for certain residues in this core suggest the presence of additional motions on a slower μs–ms time scale. $J(0.87\omega_H)$ is exquisitely sensitive to motions on fast time scales and reveals large differences in motional behavior in different regions along the protein sequence. Reduced spectral density mapping suggests that motions are generally more uniform in urea- or guanidine-denatured states, although motional restriction [increased $J(0)$ values] have often been observed in regions of high hydrophobicity, probably due to local hydrophobic collapse of the polypeptide chain (Farrow et al., 1997; Schwalbe et al., 1997). ^1H–^{13}C dipole–dipole cross-correlated spin relaxation measurements are a useful probe of aromatic side chain mobility and are sensitive to side chain motional restrictions due to local hydrophobic interactions in the unfolded state (Yang et al., 1999).

D.   Translational Diffusion Measurements

The compactness of unfolded states can be determined from the translational diffusion coefficient, from which the Stokes radius can be calculated. Translational diffusion coefficients can be readily measured using the pulsed field gradient water-suppressed longitudinal encode–decode (water-sLED) experiment (Altieri et al., 1995). The method can be used qualitatively to monitor unfolding transitions (Jones et al., 1997) or to directly measure hydrodynamic radii (Jones et al., 1997; Wilkins et al., 1999). The effective dimensions of a polypeptide chain are strongly dependent on the extent of local compaction, through formation of local hydrophobic clusters or local elements of secondary structure. Recently, pulsed field gradient measurements have been utilized together with small-angle X-ray scattering techniques to model the size distribution of structures in an unfolded state of the drk N-terminal SH3 domain (Choy et al., 2002).

E.   Quantitative Approaches to Analysis of the Conformational Ensemble

As a result of the extensive conformational averaging that occurs in unfolded states, most investigators have been content to describe the conformational ensemble in qualitative or semiquantitative terms. When residual structures are highly populated, the structure of the dominant conformer can be calculated from NOE or other NMR restraints (Constantine et al., 1995; Neri et al., 1992a; Wang et al., 1995; Yao et al., 1994), provided conformational averaging is adequately accounted for. Distance restraints obtained from paramagnetic spin label broadening have been used to calculate a family of structures of a truncated fragment of staphylococcal nuclease (Gillespie and Shortle, 1997b). More recently, a novel computational approach has been developed for calculating ensembles of structures based on all available experimental constraints for unfolded states of proteins (Choy and Forman-Kay, 2001). The method has been used to describe the conformational ensemble of the unfolded drk SH3 domain, using constraints derived from NOEs, coupling constants, ^{13}C chemical shifts, diffusion coefficients, and fluorescence data. The results suggest that a small number of low-energy structures dominate the conformational ensemble, and indicate considerable chain compaction with significant native-like residual structure. Despite this impressive advance, much remains to be done to develop robust methods for quantitative analysis of the conformational ensembles of unfolded and partly folded proteins.

IV. INSIGHTS INTO STRUCTURE AND DYNAMICS OF UNFOLDED STATES

As a case study to illustrate the insights into the structure of unfolded and partly folded proteins that can be obtained from NMR, we review recent applications of the technology to various partly folded states of apomyoglobin. These experiments provide valuable insights into the changes in structure and dynamics, at the level of individual residues, which accompany compaction of the myoglobin polypeptide chain. We also briefly discuss characterization of intrinsically unstructured proteins under physiological conditions and the nature of coupled folding and binding events.

A. *Probing the Protein Folding Landscape: Equilibrium NMR Studies of Apomyoglobin*

Apomyoglobin has been the subject of extensive study in a number of laboratories and is a paradigm for studies of protein folding pathways. The apoprotein exhibits straightforward folding kinetics with well-defined folding intermediates and without the complications associated with proline *cis–trans* isomerism or disulfide bridge formation. Folding occurs by way of an obligatory on-pathway kinetic intermediate (Tsui *et al.*, 1999) that contains helical structure in the A, B, G, and H helix regions (Jamin and Baldwin, 1998; Jennings and Wright, 1993). Equally important, myoglobin forms a number of partially folded states under conditions suitable for detailed spectroscopic analysis. In particular, apomyoglobin forms an equilibrium molten globule intermediate (Griko *et al.*, 1988) that is amenable to direct NMR analysis and appears

$$U_{denaturant} \rightarrow U_{acid\ (pH\ 2)} \rightarrow E_{(pH\ 3,\ 20\ mM\ NaCl)} \rightarrow I_{MG\ (pH\ 4)} \rightarrow N_{apo\ (pH\ 6)}$$

$R_g(\text{Å})$	34-36	30	—	23	18
Fractional Helicity (θ_{222})	~0	0.14-0.21	~0.3	0.46-0.54	1.0

FIG. 5. Partly folded states of apoMb formed in acid solution. The radius of gyration for each state is shown (data from Eliezer et al., 1995; Gast et al., 1994; Nishii et al., 1994), as is the population of secondary structure that develops during compaction (Barrick and Baldwin, 1993; Gilmanshin et al., 2001; data from Griko and Privalov, 1994). The helicity is expressed relative to that of holoMb.

to be very similar in structure to the intermediate formed during kinetic refolding (Hughson et al., 1990). Studies of this molten globule are therefore directly relevant to the kinetic folding process.

By careful manipulation of the pH, salt concentration, and nature of the counterions, a number of partly folded states of apoMb of varying degrees of compaction can be formed. Studies of the structure and dynamics of these states provide insights into the energy landscape of the apoMb polypeptide and are therefore directly relevant for understanding the folding process. The major states of apoMb accessible at acid pH are summarized in Figure 5, together with radii of gyration and helicity. Strong denaturant is required to unfold apoMb completely and eliminate all detectable helical structure; the polypeptide is most extended in urea or guanidinium hydrochloride (R_g 34–36). This represents the fully denatured state and the starting point for kinetic refolding experiments.

The chain is slightly more compact (R_g 30) in the acid-unfolded state, formed at low salt concentrations near pH 2 (Nishii et al., 1995). This state retains significant helical structure (~10–15%) (Barrick and Baldwin, 1993; Griko and Privalov, 1994). As mentioned above, apoMb forms a well-characterized molten globule (often referred to as the I state) at pH ~4 and low salt concentrations (Griko et al., 1988; Griko and Privalov, 1994). This state is compact (R_g 23) (Eliezer et al., 1995; Gast et al., 1994; Nishii et al., 1995), ~35% helical (Griko and Privalov, 1994; Hughson et al., 1990), and stable to temperatures in excess of 50°C (Griko et al., 1988). Recently, an additional state of intermediate helicity, termed the E state, has been identified (Gilmanshin et al., 1997). The E state is ~25% helical, has a hydrophobic core composed of about 30–40 residues, and folds on a μs time scale from temperature-unfolded states (Gilmanshin et al., 2001; Gulotta et al., 2001). Finally, at pH ~6, apoMb folds into its native state with helical structure and a tertiary fold similar to that of the holo protein. Native apoMb is slightly less helical and slightly less compact (R_g 19 compared to 18 for holoMb) than the holoprotein; these differences reflect the stabilization of the F helix on binding heme (Eliezer and Wright, 1996).

It is clear from Figure 5 that apoMb can be coerced to form a number of states of varying helicity and compaction simply by manipulating the pH and salt concentration or by adding denaturant. In essence, this makes apoMb ideally suited for detailed mapping of the folding free-energy landscape using spectroscopic methods, especially NMR, which can in principle provide insights into structure and dynamics at every residue in the polypeptide chain. A schematic folding funnel for apoMb is shown in Figure 6, with the regions corresponding to the various partly folded states indicated. In the upper regions of the energy landscape, sampled in the pH 2 acid-unfolded state, NMR studies show that the polypeptide fluctuates between unfolded states and local hydrophobic clusters or transient elements of secondary structure (Eliezer *et al.*, 1998; Yao *et al.*, 2001). It is a reasonable assumption that these reflect the earliest events that occur after initiation of protein folding, while the chain is still relatively extended. The E state and the molten globule (I state)

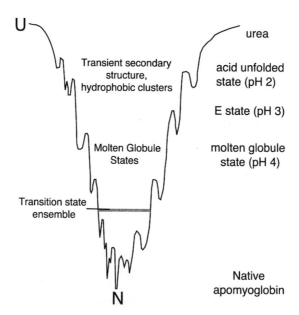

FIG. 6.   Schematic energy landscape for protein folding (folding funnel). The approximate regions of the energy landscape that correspond to the various partly folded states of apoMb are indicated on the right.

represent states further down the folding funnel. In the molten globule, the population of helix is increased substantially and the protein forms a compact core in which backbone motions are highly restricted (Eliezer *et al.*, 1998, 2000). Further helix stabilization occurs in the native state, at the bottom of the folding funnel, accompanied by specific side chain packing and further restriction of backbone motions. Thus, by taking advantage of the favorable NMR properties of apoMb in several partly folded states, remarkably detailed insights into the folding landscape can be obtained.

1. Denaturant-Unfolded Apomyoglobin

Apomyoglobin unfolded in 8 M urea at pH 2.3 displays distinct regions with different backbone mobility as monitored by NMR relaxation parameters (Schwarzinger *et al.*, 2002). These variations in backbone mobility can be correlated with intrinsic properties of the amino acids in the sequence. Clusters of small amino acids such as glycine and alanine show above average mobility. Increased backbone flexibility in glycine-rich regions has also been seen in the intrinsically unfolded fibronectin binding protein from *Staphylococcus aureus* (Penkett *et al.*, 1998). These observations are consistent with a model for the early events of apomyoglobin folding where regions of decreased flexibility interspersed with highly flexible regions are the key to the exploration of conformational space that occurs before the first steps in the folding process. The model derived from the behavior of apoMb in urea depends only on the most fundamental properties of the local amino acid sequence, and thus provides a feasible paradigm for the initiation of folding.

2. Acid-Unfolded Apomyoglobin

Apomyoglobin forms a denatured state under low-salt conditions at pH 2.3. The conformational propensities and polypeptide backbone dynamics of this state have been characterized in detail by NMR (Yao *et al.*, 2001). Nearly complete backbone and some side chain resonance assignments have been obtained using a triple-resonance assignment strategy tailored to low protein concentration (0.2 mM) and poor chemical shift dispersion. An estimate of the population and location of residual secondary structure has been made by examining deviations of $^{13}C^{\alpha}$, ^{13}CO, and $^{1}H^{\alpha}$ chemical shifts from random coil values (Fig. 7). Chemical shifts constitute a highly reliable indicator of secondary structural preferences, provided the appropriate random coil chemical shift references are used; but in the case of acid-unfolded apomyoglobin, $^{3}J_{HN\alpha}$ coupling constants are poor diagnostics of secondary structure formation (Yao *et al.*, 2001). The relative populations of α and β backbone dihedral

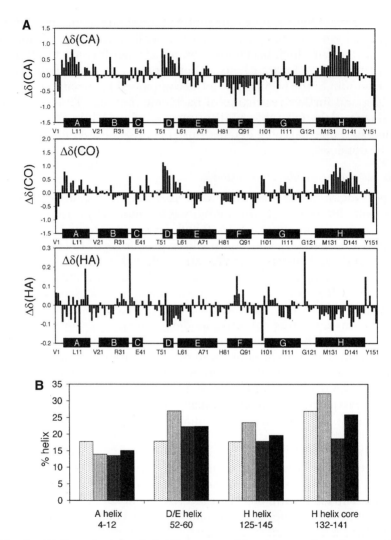

FIG. 7. (A) Secondary chemical shifts, corrected for sequence-dependent contributions, for $^{13}C^{\alpha}$ and ^{13}CO resonances of acid-unfolded apomyoglobin. Regions corresponding to the helices of the native protein are marked with black bars. (B) Populations of helix formed in acid-unfolded apomyoglobin, estimated from the deviations of $^{13}C^{\alpha}$, ^{13}CO, and $^{1}H^{\alpha}$ chemical shifts from random coil values. Sequence-dependent corrections were made (Schwarzinger *et al.*, 2001); reference random coil shifts were those determined for the peptides Ac-GGXGG-NH$_2$ in 8 M urea at pH 2.3 (Schwarzinger *et al.*, 2000). Adapted with permission from Yao *et al.* (2001). The percentage of helix was estimated from the secondary chemical shifts for each resonance, averaged over the regions indicated. The secondary shifts corresponding to fully formed helix were as follows: $^{13}C^{\alpha}$, 2.8 ppm; ^{13}CO,

angles can also be estimated from the relative intensity of the sequential $d_{\alpha N}(i, i + 1)$ and $d_{NN}(i, i + 1)$ NOEs (Waltho *et al.*, 1993). These NOE ratios show that the apoMb backbone fluctuates over both the α and β regions of ϕ, ψ space (Yao *et al.*, 2001). Nevertheless, substantial populations of helical structure, in dynamic equilibrium with unfolded states, are formed in regions corresponding to the A and H helices of the folded protein (Fig. 7B). In addition, the deviation of the chemical shifts from random coil values indicates the presence of helical structure encompassing the D helix and extending into the first turn of the E helix.

The polypeptide backbone dynamics of acid-unfolded apomyoglobin have been investigated using reduced spectral density function analysis of ^{15}N relaxation data (Yao *et al.*, 2001), and the results are shown in Figure 8. The spectral density $J(\omega_N)$ is particularly sensitive to variations in backbone fluctuations on the ps–ns time scale. The central region of the polypeptide spanning the C-terminal half of the E helix, the EF turn, and the F helix behaves as a free-flight random coil chain; but there is evidence from $J(\omega_N)$ of restricted motions on the ps–ns time scale in the A and H helix regions, where there is a propensity to populate helical secondary structure in the acid-unfolded state. Backbone fluctuations are also restricted in parts of the B and G helices owing to formation of local hydrophobic clusters. It is striking that the principal features of the plot of $J(\omega_N)$ versus residue number (central panel of Fig. 8) correlate remarkably well with the average buried surface area (Rose *et al.*, 1985), an intrinsic property of the amino acid sequence. Regions of restricted backbone flexibility are generally associated with large buried surface area, suggesting that the observed motional restrictions result from transient burial of side chains in local hydrophobic clusters. In other words, the $J(\omega_N)$ spectral density provides an exquisitely sensitive probe of the local hydrophobic collapse presumed to occur at the earliest stages of folding. Burial of hydrophobic surface appears to be of major importance in initiation of protein folding (Rose *et al.*, 1985); indeed, differences in the kinetic folding pathways of apomyoglobin and apoleghemoglobin can be rationalized on the basis of differences in buried surface area in the A and E helix regions (Nishimura *et al.*, 2000).

2.1 ppm (values averaged from those given in Wishart and Sykes (1994); $^1H^\alpha$, -0.3 ppm (Williamson, 1990). For each part of the apomyoglobin sequence (A helix, D/E helix, H helix and H helix core), the four columns represent percentages of helix derived from the secondary $^{13}C^\alpha$ (lightest shading), ^{13}CO (light gray), and $^1H^\alpha$ (dark gray) chemical shifts, and the mean of these three values (black). Adapted with permission from Yao *et al.* (2001).

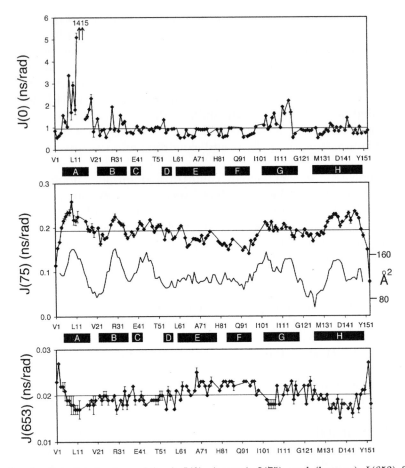

FIG. 8. Calculated values of (top) $J(0)$, (center) $J(75)$, and (bottom) $J(653)$ for acid-unfolded apoMb, pH 2.3, 25°C. The horizontal lines show the mean values of $J(75)$ (0.19 ns rad^{-1}) and $J(653)$ (0.020 ns rad^{-1}), and the 10% trimmed mean value of $J(0)$ (0.97 ns rad^{-1}). The average buried surface area, calculated using the values of Rose and co-workers (Rose *et al.*, 1985) and averaged over a seven-residue window, is also shown in the center figure (solid line, no data points, right-hand scale). Black bars indicate the positions of the helices in the folded structure of myoglobin.

A significant increase in $J(0)$ is observed for the NH resonances of some residues located in the A and G helices of the folded protein (upper panel of Fig. 8) and is associated with fluctuations on a μs–ms time scale that probably arise from transient, native-like contacts between these distant regions of the polypeptide chain. These results have been confirmed by spin labeling experiments, which reveal native-like long range inter-actions in the pH 2.3 acid-unfolded state (M. Lietzow, M. Jamin, H. J. Dyson, and P. E. Wright, unpublished data). Overall, our results indicate that the equilibrium unfolded state of apomyoglobin formed at pH 2.3 is an excellent model for the events expected to occur in the earliest stages of protein folding, providing insights into the regions of the polypep-tide that spontaneously undergo local hydrophobic collapse and sample native-like secondary structure.

3. Equilibrium Molten Globule Intermediate at pH 4

The partly folded state of apomyoglobin at pH 4 represents an excel-lent model for an obligatory kinetic folding intermediate. The structure and dynamics of this intermediate state have been extensively examined using NMR spectroscopy (Eliezer *et al.*, 1998, 2000). Secondary chemi-cal shifts, ^{1}H–^{1}H NOEs and amide proton temperature coefficients have been used to probe residual structure in the intermediate state, and NMR relaxation parameters T_1, T_2 and $\{^1$H$\}$–^{15}N NOE have been an-alyzed using spectral densities to correlate motion of the polypeptide chain with these structural observations. A significant amount of he-lical structure remains in the pH 4 state, indicated by the secondary chemical shifts of the ^{13}C$^\alpha$, ^{13}CO, ^{1}H$^\alpha$, and ^{13}C$^\beta$ nuclei (Fig. 3), and the boundaries of this helical structure are confirmed by the locations of ^{1}H–^{1}H NOEs. Hydrogen bonding in the structured regions is predom-inantly native-like according to the amide proton chemical shifts and their temperature dependence. The locations of the A, G, and H helix segments and the C-terminal part of the B helix are similar to those in native apomyoglobin, consistent with the early, complete protection of the amides of residues in these helices in quench-flow experiments (Jennings and Wright, 1993). These results confirm the similarity of the equilibrium form of apoMb at pH 4 and the kinetic intermediate ob-served at short times in the quench-flow experiment. Flexibility in this structured core is severely curtailed compared with the remainder of the protein, as indicated by the analysis of the NMR relaxation parame-ters. Regions with relatively high values of $J(0)$ and low values of $J(750)$ correspond well with the A, B, G, and H helices, an indication that ns time scale backbone fluctuations in these regions of the sequence are restricted (Fig. 4). Other parts of the protein show much greater flexibil-ity and greatly reduced secondary chemical shifts. Nevertheless, several regions show evidence of the beginnings of helical structure, including stretches encompassing the C-helix-CD loop, the boundary of the D and E helices, and the C-terminal half of the E helix. These regions are clearly not well-structured in the pH 4 state, unlike the A, B, G, and H helices, which form a native-like structured core. However, the proximity of this structured core most likely influences the region between the B and F helices, inducing at least transient helical structure.

4. Folded Apomyoglobin at pH 6

Multidimensional heteronuclear NMR spectroscopy has also been used to obtain structural information on isotopically labeled recombinant sperm whale apomyoglobin in the native state at pH 6.1 (Eliezer and Wright, 1996; Lecomte et al., 1999). Assignments for backbone resonances (^{1}HN, ^{15}N, and ^{13}C$^\alpha$) have been made for a large fraction of the residues in the protein; the secondary structure indicated by the observed chemical shifts is nearly identical to that found in the carbon monoxide complex of holomyoglobin (Eliezer and Wright, 1996). In addition, the chemical shifts themselves are highly similar in both proteins. This suggests that most of the apomyoglobin polypeptide chain adopts a well defined structure very similar to that of holomyoglobin. However, backbone resonances from a contiguous region of the apoprotein, corresponding to the EF loop, the F helix, the FG loop and the beginning of the G helix, are broadened beyond detection due to conformational fluctuations. It was proposed (Eliezer and Wright, 1996) that the polypeptide chain in this region exchanges between a holoprotein-like helical conformation and one or more unfolded or partly folded states. Such a model can explain the NMR data, the charge distributions observed by mass spectrometry, and the effects of mutagenesis.

5. Mapping the Apomyoglobin Energy Landscape

Taken together, the NMR studies of the urea-unfolded, pH 2.3, pH 4, and pH 6 states of apomyoglobin provide detailed insights into the changes in structure and dynamics which occur with compaction of the polypeptide chain during the folding process (Eliezer et al., 1998). Collapse of the apoMb polypeptide to form increasingly compact states leads to progressive accumulation of secondary structure (Fig. 9) and increasing restriction of polypeptide backbone fluctuations. Under weakly folding conditions (pH 2.3), the polypeptide fluctuates between unfolded states and local elements of structure (both secondary structure and hydrophobic clusters) that become extended and stabilized as the chain becomes more compact. Chain flexibility is greatest at the earliest stages of folding; as the folding protein becomes increasingly compact, backbone motions become more restricted, the hydrophobic core is formed, and nascent elements of secondary structure are progressively

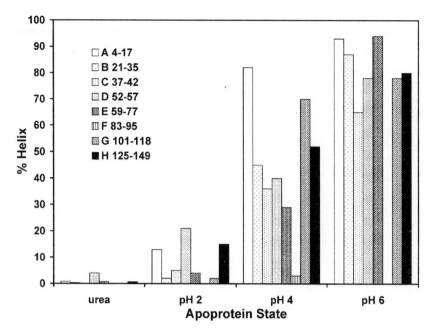

FIG. 9. Development of secondary structure concomitant with compaction of myoglobin.

stabilized. Ordered tertiary structure is formed only late in the folding process.

B. Characterization of Proteins That Are Unstructured under Nondenaturing Conditions

Most studies of unfolded proteins must of necessity be performed under denaturing conditions. Since denaturant appears to influence the conformational ensemble (Logan *et al.,* 1994; Zhang and Forman-Kay, 1995), characterization of the structural and dynamic behavior of proteins that are unfolded under nondenaturing conditions, i.e., at neutral pH and room temperature, are of particular interest. Several such studies have been reported. For a truncated staphylococcal nuclease fragment (Δ131Δ) (Alexandrescu *et al.,* 1994; Alexandrescu and Shortle, 1994), the N-terminal SH3 domain of *Drosophila* drk (Zhang and Forman-Kay, 1995), an intrinsically unfolded fibronectin binding protein from *Staphylococcus aureus* (Penkett *et al.,* 1998), and for apoplastocyanin under low salt conditions (Bai *et al.,* 2001). For all of these proteins, the observed NMR parameters deviate from the values expected for a statistical random coil, indicating local interactions determined by the amino acid sequence. In contrast, the unfolded states of the drk SH3 domain and the Δ131Δ fragment of staphylococcal nuclease are relatively compact and the conformational ensemble contains native-like structures (Choy *et al.*, 2002; Choy and Forman-Kay, 2001; Gillespie and Shortle, 1997a; Mok *et al.*, 1999). The unfolded states of the fibronectin binding protein and apoplastocyanin are much more expanded and persistent long-range interactions could not be detected (Bai *et al.*, 2001; Penkett *et al.*, 1998).

The low-salt denatured state of apoplastocyanin provides an excellent example of the sensitivity of NMR to transient local structure formation. The polypeptide behaves as a highly extended and extremely flexible chain; the protein displays motional characteristics similar to those of a random coil peptide, with no indication of global compaction (Bai *et al.*, 2001). The fact that $J(0)$ is similar in magnitude to $J(\omega_N)$ underscores the highly dynamic nature of the polypeptide chain (Fig. 10). Nevertheless, local maxima in the plot of $J(0)$ versus residue number show that the local amino acid sequence still influences chain flexibility. There is a small but detectable preference for the β-region of ϕ,ψ space in several contiguous regions of the polypeptide, including the regions that form β-strands in the folded plastocyanin (Fig. 11). Interestingly, a significant population (~30%) of non-native helical structure is formed between residues 59–64.

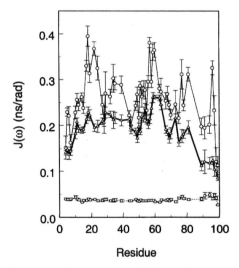

FIG. 10.   Calculated values of $J(0)$ (—○—), $J(50)$ (—△—), and $J(435)$ (···□···) for unfolded apoplastocyanin. Reproduced from Bai *et al.* (2001). *Protein Sci.* **10,** 1056–1066, with permission from Cold Spring Harbor Laboratory Press.

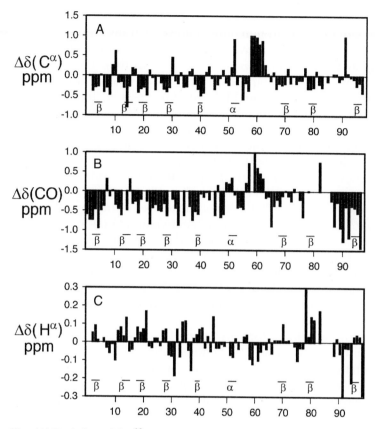

FIG. 11.   (A) Deviation of the $^{13}C^{\alpha}$ chemical shifts of unfolded apoplastocyanin from the sequence corrected random coil values of Wishart *et al.* (1995). (B) Deviation of the ^{13}CO chemical shifts of unfolded apoplastocyanin from the sequence corrected random coil values of Wishart *et al.* (1995). (C) Deviation of the H^{α} chemical shifts from the sequence-corrected random coil values of Merutka *et al.* (1995). The secondary structure in the folded apoplastocyanin is indicated. Reproduced from Bai *et al.* (2001). *Protein Sci.* **10**, 1056–1066, with permission from Cold Spring Harbor Laboratory Press.

C.   Intrinsically Unstructured Proteins: Coupled Folding and Binding Events

There are now numerous examples of proteins that are unstructured or only partially structured under physiological conditions yet are nevertheless functional (Dunker and Obradovic, 2001; Wright and Dyson, 1999). In many cases, such intrinsically disordered proteins adopt folded structures upon binding to their biological targets. As the proteins that constitute the transcriptional machinery have become better characterized in recent years, increased attention has been focused on protein–protein interactions in transcriptional regulation. Many transcriptional activation domains are either unstructured or partly structured and their interactions with their targets involve coupled folding and binding events. Intrinsically unstructured proteins also participate in various other biological processes, such as signal transduction, translation, and membrane fusion and transport.

One of the best characterized examples involves the kinase-inducible activation domain of the transcription factor CREB (cyclic AMP response element binding protein), which activates transcription of target genes, in part through direct interactions with the KIX domain of the coactivator CBP (CREB binding protein) in a phosphorylation-dependent manner. The solution structure of the complex formed by the phosphorylated kinase inducible domain (pKID) of CREB with KIX (Radhakrishnan et al., 1999) reveals that pKID undergoes a coil→helix folding transition on binding to KIX, forming two α-helices. The amphipathic helix αB of pKID interacts with a hydrophobic groove defined by helices α1 and α3 of KIX. The other pKID helix, αA, contacts a different face of the α3 helix. The phosphate group of the critical phosphoserine residue of pKID forms a hydrogen bond to the side chain of Tyr-658 of KIX. Although the phosphorylation of Tyr-658 is required before significant interaction is observed between KID and KIX, the presence of the phosphate group does not cause any appreciable change in the conformational ensemble of the KID in solution: Both KID and pKID are largely unstructured in the absence of KIX, except for a small propensity for helix formation toward the N-terminus (Hua et al., 1998; Radhakrishnan et al., 1998).

In the case of the interaction between the p160 coactivator ACTR and the binding domain of CBP, both of the isolated domains are intrinsically disordered. They combine with high affinity to form a cooperatively folded helical heterodimer of novel architecture (Demarest et al., 2002). Other examples of unfolded domains that fold on binding to the physiological partner include the transactivation domain of p53, which undergoes a coil-to-helix folding transition on binding to MDM2 (Kussie et al., 1996), and the acidic activation domain of VP16 which forms a helix on binding to hTAF$_{II}$31 (Uesugi et al., 1997). The intrinsically unstructured nature of these activation domains was originally established from studies of isolated protein fragments. Two recent papers confirm, using NMR and SAXS, respectively, that the p53 and VP16 activation domains remain unstructured in the context of the full length proteins (Ayed et al., 2001; Grossmann et al., 2001), providing strong supporting evidence for a physiological role of coupled folding and binding processes in transcriptional activation. The transcription elongation factor elongin C is also incompletely folded in its free state (Botuyan *et al.*, 1999, 2001; Buchberger *et al.*, 2000). Regions of stable secondary structure are present, but the molecule does not adopt a stable tertiary fold in the absence of ligand. There can be no doubt that many additional examples of functional unstructured proteins and coupled folding and binding events will be uncovered in the future. NMR will clearly play a major role in characterization of these proteins, as well as provide a detailed understanding of their biochemical mechanisms and biological function.

V. Conclusions

There are a large and increasing number of examples of unfolded and partly folded proteins and protein domains that are also functional. NMR is the method of choice to determine the conformational preferences inherent in these domains. NMR is also of the greatest utility in the elucidation of pathways of protein folding by allowing structural characterization of equilibrium and kinetic folding intermediates. Through the application of state-of-the-art heteronuclear NMR methods, new insights have been obtained into the nature of the conformational ensemble and the dynamics of denatured proteins. These studies show that the behavior of denatured proteins is typically far from that of a statistical random coil. This new view of the denatured state provides a basis for understanding the earliest events that occur during protein folding.

References

Alexandrescu, A. T., Abeygunawardana, C., and Shortle, D. (1994). *Biochemistry* **33,** 1063–1072.

Alexandrescu, A. T., and Shortle, D. (1994). *J. Mol. Biol.* **242,** 527–546.

Altieri, A. S., Hinton, D. P., and Byrd, R. A. (1995). *J. Am. Chem. Soc.* **117,** 7566–7567.

Arcus, V. L., Vuilleumier, S., Freund, S. M. V., Bycroft, M., and Fersht, A. R. (1995). *J. Mol. Biol.* **254,** 305–321.

Ayed, A., Mulder, F. A., Yi, G. S., Lu, Y., Kay, L. E., and Arrowsmith, C. H. (2001). *Nat. Struct. Biol.* **8,** 756–760.

Bai, Y., Chung, J., Dyson, H. J., and Wright, P. E. (2001). *Protein Sci.* **10,** 1056–1066.

Barrick, D., and Baldwin, R. L. (1993). *Biochemistry* **32,** 3790–3796.

Baum, J., Dobson, C. M., Evans, P. A., and Hanley, C. (1989). *Biochemistry* **28,** 7–13.

Bax, A., and Grzesiek, S. (1993). *Acc. Chem. Res.* **26,** 131–138.

Billeter, M., Braun, W., and Wüthrich, K. (1982). *J. Mol. Biol.* **155,** 321–346.

Botuyan, M. V., Koth, C. M., Mer, G., Chakrabartty, A., Conaway, J. W., Conaway, R. C., Edwards, A. M., Arrowsmith, C. H., and Chazin, W. J. (1999). *Proc. Natl. Acad. Sci. USA* **96,** 9033–9038.

Botuyan, M. V., Mer, G., Yi, G. S., Koth, C. M., Case, D. A., Edwards, A. M., Chazin, W. J., and Arrowsmith, C. H. (2001). *J. Mol. Biol.* **312,** 177–186.

Brant, D. A., Miller, W. G., and Flory, P. J. (1967). *J. Mol. Biol.* **23,** 47–65.

Braun, D., Wider, G., and Wüthrich, K. (1994). *J. Am. Chem. Soc.* **116,** 8466–8469.
Brutscher, B., Brüschweiler, R., and Ernst, R. R. (1997). *Biochemistry* **36,** 13043–13053.
Buchberger, A., Howard, M. J., Freund, S. M., Proctor, M., Butler, P. J., Fersht, A. R., and Bycroft, M. (2000). *Biochemistry* **39,** 12512.
Buck, M., Radford, S. E., and Dobson, C. M. (1994). *J. Mol. Biol.* **237,** 247–254.
Buck, M., Schwalbe, H., and Dobson, C. M. (1996). *J. Mol. Biol.* **257,** 669–683.
Buevich, A. V., and Baum, J. (1999). *J. Am. Chem. Soc.* **121,** 8671–8672.
Buevich, A. V., Shinde, U. P., Inouye, M., and Baum, J. (2001). *J. Biomol. NMR* **20,** 233–249.
Choy, W. Y., and Forman-Kay, J. D. (2001). *J. Mol. Biol.* **308,** 1011–1032.
Choy, W. Y., Mulder, F. A., Crowhurst, K. A., Muhandiram, D. R., Millett, I. S., Doniach, S., Forman-Kay, J. D., and Kay, L. E. (2002). *J. Mol. Biol.* **316,** 101–112.
Constantine, K. L., Mueller, L., Andersen, N. H., Tong, H., Wandler, C. F., Friedrichs, M. S., and Bruccoleri, R. E. (1995). *J. Am. Chem. Soc.* **117,** 10841–10854.
de Dios, A. C., Pearson, J. G., and Oldfield, E. (1993). *Science* **260,** 1491–1496.
Demarest, S. J., Martinez-Yamout, M., Chung, J., Chen, H., Xu, W., Dyson, H. J., Evans, R. M., and Wright, P. E. (2002). *Nature* **415,** 549–553.
Dunker, A. K., and Obradovic, Z. (2001). *Nat. Biotechnol.* **19,** 805–806.
Dyson, H. J., and Wright, P. E. (1991). *Annu. Rev. Biophys. Biophys. Chem.* **20,** 519–538.
Dyson, H. J., and Wright, P. E. (1996). *Annu. Rev. Phys. Chem.* **47,** 369–395.
Dyson, H. J., and Wright, P. E. (1998). *Nat. Struct. Biol.* **5,** 499–503.
Dyson, H. J., and Wright, P. E. (2002). *Curr. Opin. Struct. Biol.* **12,** 54–60.
Eliezer, D., Chung, J., Dyson, H. J., and Wright, P. E. (2000). *Biochemistry* **39,** 2894–2901.
Eliezer, D., Jennings, P. A., Wright, P. E., Doniach, S., Hodgson, K. O., and Tsuruta, H. (1995). *Science* **270,** 487–488.
Eliezer, D., and Wright, P. E. (1996). *J. Mol. Biol.* **263,** 531–538.
Eliezer, D., Yao, J., Dyson, H. J., and Wright, P. E. (1998). *Nat. Struct. Biol.* **5,** 148–155.
Epand, R. M., and Scheraga, H. A. (1968). *Biochemistry* **7,** 2864–2872.
Evans, P. A., Topping, K. D., Woolfson, D. N., and Dobson, C. M. (1991). *Proteins* **9,** 248–266.
Farrow, N. A., Muhandiram, R., Singer, A. U., Pascal, S. M., Kay, C. M., Gish, G., Shoelson, S. E., Pawson, T., Forman-Kay, J. D., and Kay, L. E. (1994). *Biochemistry* **33,** 5984–6003.
Farrow, N. A., Zhang, O., Forman-Kay, J. D., and Kay, L. E. (1995a). *Biochemistry* **34,** 868–878.
Farrow, N. A., Zhang, O., Forman-Kay, J. D., and Kay, L. E. (1997). *Biochemistry* **36,** 2390–2402.
Farrow, N. A., Zhang, O., Szabo, A., Torchia, D. A., and Kay, L. E. (1995b). *J. Biomol. NMR* **6,** 153–162.
Frank, M. K., Clore, G. M., and Gronenborn, A. M. (1995). *Protein Sci.* **4,** 2605–2615.
Frenkiel, T., Bauer, C., Carr, M. D., Birdsall, B., and Feeney, J. (1990). *J. Magn. Reson.* **90,** 420–425.
Gast, K., Damaschun, H., Misselwitz, R., Müller-Frohne, M., Zirwer, D., and Damaschun, G. (1994). *Eur. Biophys. J.* **23,** 297–305.
Gillespie, J. R., and Shortle, D. (1997a). *J. Mol. Biol.* **268,** 158–169.
Gillespie, J. R., and Shortle, D. (1997b). *J. Mol. Biol.* **268,** 170–184.
Gilmanshin, R., Gulotta, M., Dyer, R. B., and Callender, R. H. (2001). *Biochemistry* **40,** 5127–5136.
Gilmanshin, R., Williams, S., Callender, R. H., Woodruff, W. H., and Dyer, R. B. (1997). *Proc. Natl. Acad. Sci. USA* **94,** 3709–3713.
Griko, Y. V., and Privalov, P. L. (1994). *J. Mol. Biol.* **235,** 1318–1325.

Griko, Y. V., Privalov, P. L., Venyaminov, S. Y., and Kutyshenko, V. P. (1988). *J. Mol. Biol.* **202,** 127–138.

Grossmann, J. G., Sharff, A. J., O'Hare, P., and Luisi, B. (2001). *Biochemistry* **40,** 6267–6274.

Gulotta, M., Gilmanshin, R., Buscher, T. C., Callender, R. H., and Dyer, R. B. (2001). *Biochemistry* **40,** 5137–5143.

Hermans, J., Puett, D., and Acampora, G. (1969). *Biochemistry* **8,** 22–30.

Hua, Q. X., Jia, W. H., Bullock, B. P., Habener, J. F., and Weiss, M. A. (1998). *Biochemistry* **37,** 5858–5866.

Hughson, F. M., Wright, P. E., and Baldwin, R. L. (1990). *Science* **249,** 1544–1548.

Ikura, M., Bax, A., Clore, G. M., and Gronenborn, A. M. (1990). *J. Am. Chem. Soc.* **112,** 9020–9022.

Ishima, R., and Nagayama, K. (1995). *J. Magn. Reson. Ser. B* **108,** 73–76.

Jamin, M., and Baldwin, R. L. (1998). *J. Mol. Biol.* **276,** 491–504.

Jennings, P. A., and Wright, P. E. (1993). *Science* **262,** 892–896.

Jones, J. A., Wilkins, D. K., Smith, L. J., and Dobson, C. M. (1997). *J. Biomol. NMR* **10,** 199–203.

Kay, L. E. (1995). *Prog. Biophys. Mol. Biol.* **63,** 277–299.

Kim, S., Bracken, C., and Baum, J. (1999). *J. Mol. Biol.* **294,** 551–560.

Kuriyan, J., Wilz, S., Karplus, M., and Petsko, G. A. (1986). *J. Mol. Biol.* **192,** 133–154.

Kussie, P. H., Gorina, S., Marechal, V., Elenbaas, B., Moreau, J., Levine, A. J., and Pavletich, N. P. (1996). *Science* **274,** 948–953.

Lecomte, J. T., Sukits, S. F., Bhattacharjya, S., and Falzone, C. J. (1999). *Protein Sci.* **8,** 1484–1491.

Lipari, G., and Szabo, A. (1982a). *J. Am. Chem. Soc.* **104,** 4546–4559.

Lipari, G., and Szabo, A. (1982b). *J. Am. Chem. Soc.* **104,** 4559–4570.

Liu, A., Riek, R., Wider, G., Von Schroetter, C., Zahn, R., and Wüthrich, K. (2000). *J. Biomol. NMR* **16,** 127–138.

Logan, T. M., Thériault, Y., and Fesik, S. W. (1994). *J. Mol. Biol.* **236,** 637–648.

Lumb, K. J., and Kim, P. S. (1994). *J. Mol. Biol.* **236,** 412–420.

McDonald, C. A., and Phillips, D. C. (1969). *J. Am. Chem. Soc.* **91,** 1513.

Meekhof, A. E., and Freund, S. M. V. (1999). *J. Mol. Biol.* **286,** 579–592.

Meirovitch, E., and Meirovitch, H. (1996). *Biopolymers* **38,** 69–88.

Merutka, G., Dyson, H. J., and Wright, P. E. (1995). *J. Biomol. NMR* **5,** 14–24.

Mok, Y. K., Kay, C. M., Kay, L. E., and Forman-Kay, J. (1999). *J. Mol. Biol.* **289,** 619–638.

Neri, D., Billeter, M., Wider, G., and Wüthrich, K. (1992a). *Science* **257,** 1559–1563.

Neri, D., Wider, G., and Wüthrich, K. (1992b). *Proc. Natl. Acad. Sci. USA* **89,** 4397–4401.

Nishii, I., Kataoka, M., and Goto, Y. (1995). *J. Mol. Biol.* **250,** 223–238.

Nishii, I., Kataoka, M., Tokunaga, F., and Goto, Y. (1994). *Biochemistry* **33,** 4903–4909.

Nishimura, C., Prytulla, S., Dyson, H. J., and Wright, P. E. (2000). *Nat. Struct. Biol.* **7,** 679–686.

Palmer, A. G., Cavanagh, J., Wright, P. E., and Rance, M. (1991). *J. Magn. Reson.* **93,** 151–170.

Peng, J. W., and Wagner, G. (1992). *J. Magn. Reson.* **98,** 308–332.

Penkett, C. J., Redfield, C., Jones, J. A., Dodd, I., Hubbard, J., Smith, R. A. G., Smith, L. J., and Dobson, C. M. (1998). *Biochemistry* **37,** 17054–17067.

Radhakrishnan, I., Pérez-Alvarado, G. C., Dyson, H. J., and Wright, P. E. (1998). *FEBS Lett.* **430,** 317–322.

Radhakrishnan, I., Pérez-Alvarado, G. C., Parker, D., Dyson, H. J., Montminy, M. R., and Wright, P. E. (1999). *J. Mol. Biol.* **287,** 859–865.

Ramachandran, G. N., Ramakrishnan, C., and Sasisekharan, V. (1963). *J. Mol. Biol.* **7,** 95–99.

Rose, G. D., Geselowitz, A. R., Lesser, G. J., Lee, R. H., and Zehfus, M. H. (1985). *Science* **229,** 834–838.

Sari, N., Alexander, P., Bryan, P. N., and Orban, J. (2000). *Biochemistry* **39,** 965–977.

Schwalbe, H., Fiebig, K. M., Buck, M., Jones, J. A., Grimshaw, S. B., Spencer, A., Glaser, S. J., Smith, L. J., and Dobson, C. M. (1997). *Biochemistry* **36,** 8977–8991.

Schwarzinger, S., Kroon, G. J. A., Foss, T. R., Chung, J., Wright, P. E., and Dyson, H. J. (2001). *J. Am. Chem. Soc.* **123,** 2970–2978.

Schwarzinger, S., Wright, P. E., and Dyson, H. J. (2002). *Biochemistry* (in press).

Smith, L. J., Bolin, K. A., Schwalbe, H., MacArthur, M. W., Thornton, J. M., and Dobson, C. M. (1996a). *J. Mol. Biol.* **255,** 494–506.

Smith, L. J., Fiebig, K. M., Schwalbe, H., and Dobson, C. M. (1996b). *Fold. Design* **1,** R95–R106.

Spera, S., and Bax, A. (1991). *J. Am. Chem. Soc.* **113,** 5490–5492.

Taniuchi, H., and Anfinsen, C. B. (1969). *J. Biol. Chem.* **244,** 3864–3875.

Tsui, V., Garcia, C., Cavagnero, S., Siuzdak, G., Dyson, H. J., and Wright, P. E. (1999). *Protein Sci.* **8,** 45–49.

Uesugi, M., Nyanguile, O., Lu, H., Levine, A. J., and Verdine, G. L. (1997). *Science* **277,** 1310–1313.

Waltho, J. P., Feher, V. A., Merutka, G., Dyson, H. J., and Wright, P. E. (1993). *Biochemistry* **32,** 6337–6347.

Wang, J. J., Hodges, R. S., and Sykes, B. D. (1995). *J. Am. Chem. Soc.* **117,** 8627–8634.

Wijesinha-Bettoni, R., Dobson, C. M., and Redfield, C. (2001). *J. Mol. Biol.* **307,** 885–898.

Wilkins, D. K., Grimshaw, S. B., Receveur, V., Dobson, C. M., Jones, J. A., and Smith, L. J. (1999). *Biochemistry* **38,** 16424–16431.

Williamson, M. P. (1990). *Biopolymers* **29,** 1423–1431.

Wishart, D. S., Bigam, C. G., Holm, A., Hodges, R. S., and Sykes, B. D. (1995). *J. Biomol. NMR* **5,** 67–81.

Wishart, D. S., and Nip, A. M. (1998). *Biochem. Cell Biol.* **76,** 153–163.

Wishart, D. S., and Sykes, B. D. (1994). *Methods Enzymol.* **239,** 363–392.

Wong, K. B., Clarke, J., Bond, C. J., Neira, J. L., Freund, S. M. V., Fersht, A. R., and Daggett, V. (2000). *J. Mol. Biol.* **296,** 1257–1282.

Wright, P. E., and Dyson, H. J. (1999). *J. Mol. Biol.* **293,** 321–331.

Wright, P. E., Dyson, H. J., and Lerner, R. A. (1988). *Biochemistry* **27,** 7167–7175.

Yang, D. W., Mok, Y. K., Muhandiram, D. R., Forman-Kay, J. D., and Kay, L. E. (1999). *J. Am. Chem. Soc.* **121,** 3555–3556.

Yao, J., Chung, J., Eliezer, D., Wright, P. E., and Dyson, H. J. (2001). *Biochemistry* **40,** 3561–3571.

Yao, J., Dyson, H. J., and Wright, P. E. (1994). *J. Mol. Biol.* **243,** 754–766.

Yao, J., Dyson, H. J., and Wright, P. E. (1997). *FEBS Lett.* **419,** 285–289.

Zhang, O., and Forman-Kay, J. D. (1995). *Biochemistry* **34,** 6784–6794.

Zhang, O., and Forman-Kay, J. D. (1997). *Biochemistry* **36,** 3959–3970.

Zhang, O., Forman-Kay, J. D., Shortle, D., and Kay, L. E. (1997a). *J. Biomol. NMR* **9,** 181–200.

Zhang, O., Kay, L. E., Olivier, J. P., and Forman-Kay, J. D. (1994). *J. Biomol. NMR* **4,** 845–858.

Zhang, O., Kay, L. E., Shortle, D., and Forman-Kay, J. D. (1997b). *J. Mol. Biol.* **272,** 9–20.

Zimm, B. H., and Bragg, K. J. (1959). *J. Chem. Phys.* **31,** 526–535.

UNFOLDED STATE OF PEPTIDES

By XAVIER DAURA,[1] ALICE GLÄTTLI, PETER GEE, CHRISTINE PETER,
and WILFRED F. VAN GUNSTEREN

Laboratory of Physical Chemistry, Swiss Federal Institute of Technology Zurich,
ETH Hönggerberg, CH-8093 Zurich

I. INTRODUCTION

Polypeptide chains exist in an equilibrium between different conformations as a function of environment (solvent, other solutes, pH) and thermodynamic (temperature, pressure) conditions. If a polypeptide adopts a structurally ordered, stable conformation, one speaks of an equilibrium between a folded state, represented by the structured, densely populated conformer, and an unfolded state, represented by diverse, sparsely populated conformers. Although this equilibrium exists for polypeptide chains of any size, its thermodynamics and kinetics are typically different for oligopeptides and proteins. This can be broadly explained with reference to the different dimensionalities of the free-energy hypersurfaces of these two types of molecules.

Historically, research on polypeptide structure, whether by experimental or computational means, has focused on the folded state only. This is because an array of methods exist for detailed investigation of the folded state of a polypeptide, and because the folded state is commonly the functionally active one. There are, however, two scenarios in which a full characterization of the unfolded state becomes as essential as the determination of the folded conformation. The first is in the study of

---

[1] Current affiliation: Institució Catalana de Recerca i Estudis Avancats (ICREA) and Institute of Biotechnology and Biomedicine, Universitat Autònoma de Barcelona, E-08193 Bellaterra, Spain.

ADVANCES IN
PROTEIN CHEMISTRY, Vol. 62

the physical, chemical, or biological properties of peptides. The folded conformation of a peptide is, in general, only marginally more stable than the lowest-free-energy unfolded conformation. Moreover, the unfolded state of a peptide often has a higher probability than the folded state. As a result, any macroscopic observable of a peptide is weighted with both the folded and the unfolded states. Interpreting such observables in terms of the folded conformation only is therefore not correct. The second scenario is related to the study of peptide and protein folding (i.e., not structure prediction). The description of an equilibrium requires knowledge about each of the states involved. Therefore, it is fundamental to describe not only the folded state but also the unfolded state accurately in order to draw conclusions on the nature and mechanisms of peptide and protein folding. Only if the unfolded state is well represented can conclusions be made with respect to folding pathways, intermediate and transition states, time scales, and reaction coordinates.

It is only recently that some efforts have been made to characterize the unfolded state of specific polypeptides (Pappu *et al.*, 2000; Shortle and Ackerman, 2001; van Gunsteren *et al.*, 2001a; Choy *et al.*, 2002). Obtaining quantitative microscopic information from the unfolded state is difficult not only experimentally but also computationally. Whereas experiments suffer from the low population and transient nature of individual unfolded conformations, theoretical studies are bound by the time required to sample the conformations of the unfolded state with appropriate weights. In most cases, experimental studies report on low-resolution average properties of the unfolded state (Bai *et al.*, 2001; Shortle and Ackerman, 2001) or focus on particular substates such as molten globules (Kobayashi *et al.*, 2000; Chakraborty *et al.*, 2001), intermediates (Troullier *et al.*, 2000), or transition states (Matouschek *et al.*, 1989; Fersht *et al.*, 1992) (see also Evans and Radford, 1994; Plaxco and Dobson, 1996; Callender *et al.*, 1998; Dobson and Hore, 1998; Dyson and Wright, 1998). Computationally, sampling the unfolded state is bound to sampling the equilibrium ensemble of folding/unfolding pathways. Given the size of the problem, the challenge is to reduce the number of degrees of freedom of the model without perturbing its capacity to reproduce the behavior of the real system. The number of degrees of freedom can be kept to a computationally tractable number either by drastically simplifying the physical model for the protein and/or its environment (Onuchic *et al.*, 1997; Chan and Dill, 1998; Thirumalai and Klimov, 1999; Dinner *et al.*, 2000; Wang and Sung, 2000; Ferrara and Caflisch, 2001; Mirny and Shakhnovich, 2001) or by limiting the size of the molecular system to that of atomic-detail models of peptides in solution (Daura *et al.*, 1998; Duan and Kollman, 1998; Takano *et al.*,

1999; Hummer *et al.*, 2001; van Gunsteren *et al.*, 2001a). Alternatively, for an atomic-detail model of a protein in solution, if the native three-dimensional structure is known, one may construct a projection of the free energy on particular coordinates using a predefined set of unfolded conformations and biased sampling (Shea and Brooks, 2001), or one may perform unfolding simulations under denaturing conditions and assume that the sampled pathways are relevant to folding under native conditions (Daggett, 2002). Arguments in favor and against each of these simplifying approaches are abundant in the mentioned literature.

In this chapter we review the current knowledge on the unfolded state of peptides. Currently, the characterization of the populated microscopic states of a peptide is possible only by simulation methods. We put special emphasis on molecular dynamics simulations of spontaneous (i.e., lacking biasing potentials or directed-sampling algorithms), reversible folding of peptides in solution (i.e., with explicit solvent molecules), since the results from this type of studies are the least dependent on the methodology. Unfortunately, even in those studies where researchers have tried to sample the unfolded state of a peptide as thoroughly as possible, the description of specific folding pathways often eclipses that of the unfolded state. For this reason, most of the points discussed here are illustrated with original work from our group.

II. DEFINITIONS

For the sake of clarity, terms susceptible of different interpretations are unambiguously defined in this section. Note, however, that the following definitions are not necessarily applicable beyond the limits of this chapter.

When referring to a peptide, the term *structure* is used as an equivalent to the concept of *configuration* in physics, that is, a set of distinct atomic coordinates. A *conformation,* on the other hand, comprises a—potentially infinite—group of structures with high structural similarity and identical macroscopic properties. In this context, each *cluster of structures* from a (conformational) clustering analysis (see Section III,B) is considered to represent a distinct conformation of the peptide. Note that the clusters are determined on the exclusive basis of backbone conformation. The term *ensemble* is not, in general, used in a rigorous statistical mechanical sense, but rather to refer to a group of either structures or conformations.

From the definitions given above it follows that the *configurational space* (continuous) of a peptide is infinite, even if its *conformational space* (discrete) would not be. A concept often used to express different things is that of *accessible conformational space*. It has commonly been related to the *theoretical conformational space* of a peptide, that is, the space integrated by all the conformations that the peptide could adopt without incurring van der Waals clashes. This interpretation has its roots in the historical perception of the unfolded state of a peptide as a random state. Here, the concept of accessible or relevant conformational space is related to the space of probable (low free energy) conformations or, in the present context, the space of those conformations sampled in a molecular dynamics simulation of the folding/unfolding equilibrium in which the total number of visited conformers has converged. Any other conformation has no relevance. Note that with this definition the accessible conformational space of a peptide depends on the environment and thermodynamic conditions.

In Section III, a distinction is made between *experimental conformation* and *folded conformation*. Thus, the *most populated conformation* of a peptide is identified as the folded conformation, irrespective of experimental or secondary-structure considerations. The *folded state* is then associated to the folded or most populated conformation, while the *unfolded state* embodies all other conformations and includes any other substates.

Another concept that may cause confusion is that of stability. Here, the term *thermodynamic stability* is used in relation to populations, while the term *kinetic stability* is used in relation to average lifetimes. Thus, the thermodynamically most stable conformer is the most populated one, while the kinetically most stable conformer is the one with the longest average lifetime. Whether these are necessarily the same is discussed on the basis of the examples.

III.   A Sample of Unfolded States

The unfolded states of eight different peptides, sampled in a corresponding number of molecular dynamics simulations of reversible peptide folding in solution under equilibrium conditions, are analyzed in this section (see Table I).

A.   *Systems and Simulation Conditions*

All eight peptides are of new design and are built of different types of amino acids. Thus, peptides A to E are composed of β-amino acid residues, peptides F and G of α-amino acid residues, and peptide H of α-aminoxy acid residues. Their structural formulas are shown in Figure 1.

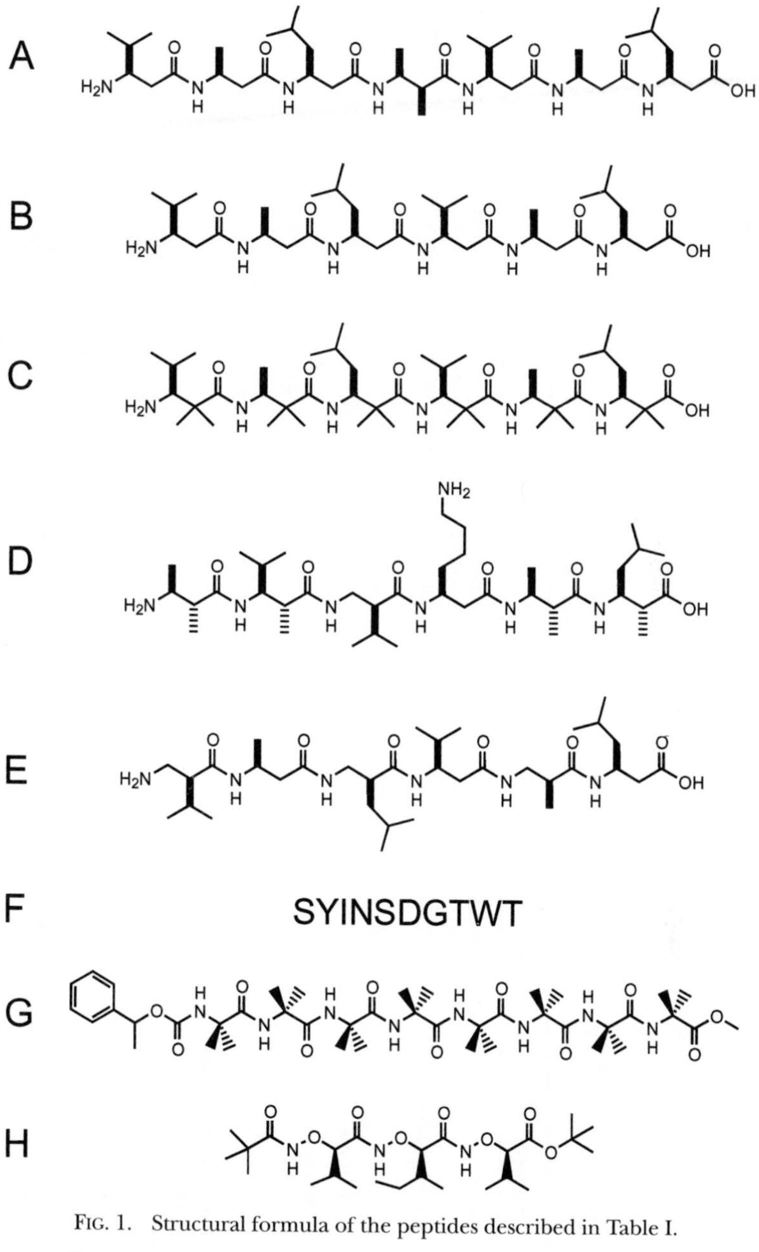

FIG. 1. Structural formula of the peptides described in Table I.

TABLE I
Reversible Folding of Eight Peptides[a]

System	A	B	C	D	E	F	G	H
Type of residues	β-amino acid	β-amino acid	β-amino acid	β-amino acid	β-amino acid	α-amino acid	α-amino acid	α-aminoxy acid
Experimental conformer	M-3_{14} helix	M-3_{14} helix	unknown	10-member turn (hairpin)	P-12/10 helix	β-hairpin 3:5	M,P,α, 3_{10} helices	P-1.8_8 helix
Type of solvent	CH_3OH	CH_3OH	CH_3OH	CH_3OH	CH_3OH	H_2O	DMSO	$CHCl_3$
Experimental temperature (K)	298	298	298	298	298	278	340	298
Simulation parameters								
No. of residues	7	6	6	6	6	10	8	3
No. of peptide atoms	64	56	68	64	56	110	75	39
No. of solvent atoms	2886	4389	4386	4359	4305	13755	4476	3360
No. of backbone torsional degrees of freedom	21	18	18	18	18	20	16	9
Temperature (K)	340	298	340	340	340	353	340	340
Simulation length (ns)	200	102	101	100	50	60	150	73
Simulated folding								
RMSD similarity cut-off (nm)	0.1	0.08	0.08	0.08	0.08	0.12	0.1	0.07
Sampling of EC	yes	yes		yes	yes	yes	yes	yes
Ranking of EC in cluster analysis	1	2		2	1	21	13	1

	1	2	3	4	5	6	7	8
Weight of EC (%)	30	15		10	16	1	2	19
Estimated free energy of folding to EC (kJ mol^{-1})	2	4		6	5	15	11	4
(Life time) of EC (ps)	463	750		90	205	113	116	74
No. of events of folding to EC	129	21		105	38	3	25	209
(Time) of folding to EC (ps)	1723	4133		616	694	7509	4000	220
(No. of conformers) visited during folding to EC	19	22		9	9	23	17	6
Weight of MPC (cluster 1) (%)	30	18	26	19	16	32	19	19
Estimated free energy of folding to MPC (kJ mol^{-1})	2	4	3	4	5	2	4	4
(Life time) of MPC (ps)	463	278	230	157	205	2079	367	74
No. of events of folding to MPC	129	68	123	131	38	9	81	209
(Time) of folding to MPC (ps)	1723	1245	204	473	694	2409	1465	220
(No. of conformers) visited during folding to MPC	19	10	3	7	9	13	9	6
Total number of conformers	360	200	76	286	129	111	179[b]	148
No. of unfolded conformers (to 99% weight)	234	131	39	208	88	78	108	100
No. of unfolded conformers (to 75% weight)	28	19	7	36	13	14	23	15
No. of unfolded conformers (to 50% weight)	6	6	3	9	5	5	9	6

[a] EC, experimental conformer; MPC, most populated conformer; see Sections III, A and B for a description of this table.
[b] The total number of clusters reported for this peptide in van Gunsteren et al. (2001a) was incorrect.

The structural properties of these peptides at room temperature (278 K for peptide F and 340 K for peptide G) have been studied by NMR. In both experiment and simulation, peptides A to E were solvated with methanol, peptide F was solvated with water, peptide G was solvated with dimethyl sulfoxide, and peptide H was solvated with chloroform. The restraints derived from the NMR data are indicative of an M-3_{14} helix for peptides A and B (Seebach *et al.*, 1996), a hairpin with a 10-membered hydrogen-bonded turn for peptide D (Daura *et al.*, 2001), a P-12/10 helix for peptide E (Seebach *et al.*, 1997, 1998), a β-hairpin 3:5 with an I + G1 β-bulge turn for peptide F (de Alba *et al.*, 1997), and a P-1.8_8 helix for peptide H (Yang *et al.*, 1999). The lowest-energy NMR model structure is shown in Figure 2 (see color insert) for each of these peptides. There was no ordered conformation detected for peptide C (Seebach *et al.*, 2000), and the NMR data from peptide G were compatible with a mixture of M- and P-, 3_{10}- and α-helices (Bellanda *et al.*, 2001). Models of these helices are also shown in Figure 2 (G1 to G4). The number of residues, number of backbone torsional degrees of freedom, and number of peptide and solvent atoms are given in Table I for each of the systems. The examples have been selected from a bigger pool of simulations at different temperatures. Thus, whenever possible, a long molecular dynamics simulation at a temperature close to 340 K has been chosen. This choice follows from two considerations. First, because of the higher kinetic energy per degree of freedom, at higher temperatures there is a larger conformational space accessible to the peptide. This implies that by performing the simulation at around 340 K, a temperature that can be reached experimentally, we obtain a true upper limit for the number of conformations of the peptide within the range of temperatures of interest, that is, somewhere between 293 and 313 K. Second, at higher temperatures the kinetic stability of the most populated conformation decreases. This means that more events of folding and unfolding can be sampled in the same simulation time, and the calculated equilibrium properties become statistically more significant. In addition, it is important to note that the weight of a particular conformer in the temperature-dependent ensemble of accessible conformers typically varies with temperature. Furthermore, the relative weights of particular conformers may also change. Therefore, with the exception of peptides B and G, the most populated conformations in the simulations reported in Table I do not need to coincide with the experimental conformation derived at a lower temperature. The simulation time ranges from 50 ns for system E to 200 ns for system A. Further details on the simulations have been given by Daura *et al.* (1998) for peptide A, Glättli *et al.* (2002) for peptides B and C, Daura *et al.* (2001) for peptide D, Daura *et al.*

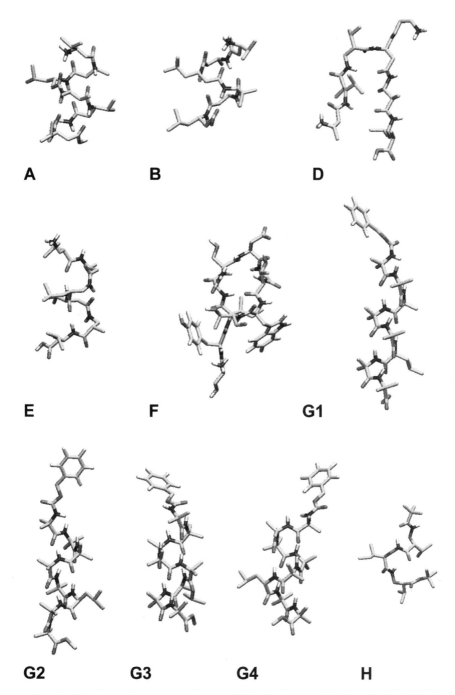

FIG. 2. Experimental structure, when available, for the peptides described in Table I Models A, B, D, E, F, and H: lowest-energy NMR model structures for the corresponding peptides. The NMR data from peptide G was compatible with a mixture of M-3_{10}-helix (model G1), P-3_{10}-helix (model G2), M-α-helix (model G3), and P-α-helix (model G4). There was no ordered conformation detected for peptide C.

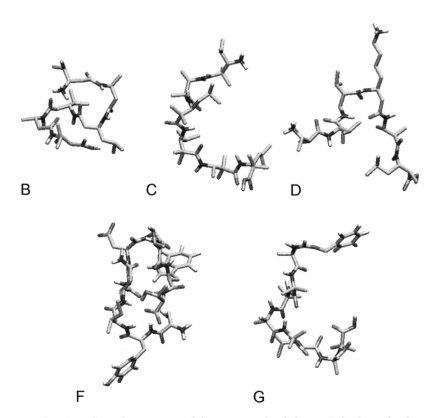

FIG. 3. Central member structure of the most populated cluster. Only shown for those peptides whose most populated conformer is different from the experimental conformer under the simulation conditions (see Table I).

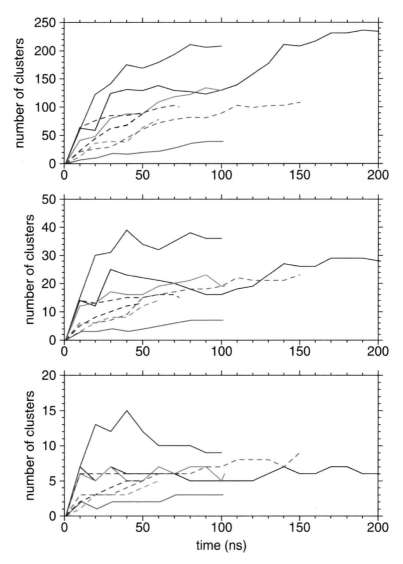

FIG. 4. Number of clusters as a function of time. Upper panel: Number of clusters with a 99% joined weight in the unfolded state. Middle panel: Number of clusters with a 75% joined weight in the unfolded state. Lower panel: Number of clusters with a 50% joined weight in the unfolded state. Black solid lines, peptide A; red solid lines, peptide B; green solid lines, peptide C; blue solid lines, peptide D; black dashed lines, peptide E; red dashed lines, peptide F; green dashed lines, peptide G; blue dashed lines, peptide H.

(1999a) for peptide E, Santiveri *et al.* (2002) for peptide F, Bürgi *et al.* (2001) for peptide G, and Peter *et al.* (2000) for peptide H. Whether the conclusions about the unfolded states of this heterogeneous set of non-natural peptides in a diversity of solvents can be extended to natural α-peptides (peptides of α-amino acids) in an aqueous medium is a question that cannot be readily answered. The physical principles governing the behavior of all these systems are, however, the same.

B. Clustering Analysis

The analysis of the simulations is summarized in four blocks of data in Table I. For each simulation, a conformational clustering analysis was performed on a set of peptide structures taken at 10-ps intervals from the trajectory, using the backbone atom-positional root-mean-square difference (RMSD) as similarity criterion. Structures separated in time by less than 10 ps are in general correlated. Except for the tripeptide H, the first and last residues of the peptide chains were not considered when performing the translational and rotational superposition of the structures and when calculating the RMSD values. The first nanosecond of each trajectory was excluded from the clustering analysis to avoid any bias from the initial structure, which corresponded to the experimental conformer for peptide A and to the fully extended conformer (all backbone torsional dihedral angles in trans) for peptides B to H. The RMSD similarity cutoff or maximum cluster radius used in the clustering analysis is given in Table I for each peptide. They were chosen proportional to the number of residues and are approximately equal to the maximum atom-positional RMSD between the various NMR model structures derived for each peptide. The clustering algorithm has been described elsewhere (Daura *et al.*, 1999b). Important features of this algorithm are that it favors the most populated cluster and ensures a minimum atom-positional RMSD between centers of clusters equal to the RMSD similarity cutoff. It also results in many clusters having only one member. In general, these single-member clusters are not isolated in the RMSD space but lie at the border of other clusters. They are a result of the hard constraint that no member of a cluster may be more than a specified distance from the cluster's central member. To identify the set of structures from the simulation that would be representative of the experimental conformer, the NMR model structure with lowest energy was imposed during the clustering process as central member of a cluster. This arbitrary choice of the center of one of the clusters avoids the distribution of the experimental conformer over more than one cluster, a question of practical interest that does not alter the properties of the unfolded state. For peptide G, the cluster representing the experimental conformer was generated with the model structures G1 to G4 (see Fig. 2).

As already mentioned, no ordered conformation could be detected by NMR for peptide C. In all other cases, the experimental conformer was sampled also in the simulation. The following properties in Table I were evaluated on the basis of the clustering results. The ranking of the experimental conformer, its percentage weight in the ensemble, and the free energy of folding to the experimental conformer were calculated from the relative population of the cluster centered at the NMR model structure. Kinetic properties were evaluated from the time sequence of sampling of clusters, with a resolution of 10 ps. The average lifetime of the experimental conformer and the number of events of (re)folding to it were calculated by recording the times at which the cluster centered at the NMR model structure was accessed and left. If the cluster was left but accessed again within 20 ps, the sampling of the experimental conformer was considered continuous and the refolding event was not counted. Note that this criterion differs from that used by van Gunsteren *et al.* (2001a) to evaluate the number of folding events. The average time of (re)folding to the experimental conformer and the average number of conformers visited during the (re)folding process were calculated from the subset of (re)folding events involving more than one unfolded conformer. In those cases where the experimental conformer was not the most populated one (cluster 1), the same analysis was performed for the latter (see Table I and Fig. 3, see color insert). As already mentioned, if no other indication is given, folding refers to the process that leads to the most populated conformer. The total number of conformers determined by clustering analysis is indicated. The number of unfolded conformers (i.e., other than cluster 1) is given as a function of their joined weight in the unfolded state. For example, in the simulation of peptide A, clusters 2 to 235, 2 to 29, and 2 to 7 have a 99%, 75%, and 50% joined weight in the sampled unfolded space, respectively. Note that by giving the numbers for a 99% weight the single-member clusters are partially or totally excluded. As already mentioned, single-member clusters are an artifact of the clustering algorithm.

Before commenting on the values that appear in Table I, the level of convergence of the number of clusters with time needs to be assessed. This question is addressed in Figure 4 (see color insert). The number of clusters with a 99%, 75%, and 50% joined weight in the unfolded state is plotted as a function of time in the upper, middle, and lower panels, respectively. Even for the longest simulation (i.e., 200 ns), it is not possible to assess whether the number of clusters with a 99% weight in the unfolded state has converged. However, from the evolution of the systems, it seems unlikely that this number could be higher than 500 clusters for any of the peptides studied if the simulations were extended to convergence. This implies that the unfolded state of these peptides is not random (Plaxco and Gross, 2001; van Gunsteren *et al.*, 2001a). The number of clusters with a 75% weight in the unfolded state shows a faster progression to convergence, and it seems reasonable to state that the number of clusters with a 50% weight has converged in most simulations. A correlation analysis (see Section III,C) shows that the simulation time correlates with the total number of clusters but not with the number of clusters having a 99%, 75%, and 50% weight in the unfolded state.

One set of results in Table I deserves special mention: Even at the relatively high temperatures applied, a small number of conformers dominate the unfolded state. This has clear consequences for folding to the thermodynamically most stable conformer. For example, the average number of conformers visited during (re)folding is consequently small, that is, in the interval between the number of conformers with a 50% and a 75% weight in the unfolded state. That these peptides have a small number of relevant conformations is the basis for the observed fast folding (van Gunsteren *et al.*, 2001a,b).

C. *Correlation Analysis*

To find possible correlations between pairs of variables in Table I, a systematic correlation analysis has been performed. Pearson's correlation coefficient, ρ, has been used to measure the strength of the linear relationship between two variables. Coefficient ρ can take values from -1 (perfect negative correlation) to 1 (perfect positive correlation). The statistical significance of ρ has been tested with a t-test. A low p-value for this test (typically <0.05) means that there is a statistically significant relationship between the two variables. In principle, the results of this analysis cannot be extrapolated to other peptides, since the sample is not statistically representative. The number of peptides studied is small, the distribution of peptide lengths is narrow, the nature of the peptides is diverse, and the solvation and temperature conditions are also inhomogeneous. Nevertheless, this correlation analysis opens up a number of thought-provoking questions that call for further investigation. Note, in addition, that the correlation between any two independent variables in Table I cannot be perfect and the variance must be necessarily large. For example, it is clear that two peptides of identical length may have very different folding properties under identical conditions. Furthermore, the somewhat arbitrary choice of criteria used for the definition of clusters, lifetimes, etc., may also have an influence on the degree of correlation between two variables.

The correlation analysis was performed for every pair of independent variables in Table I—pairs of dependent variables such as weight of the most populated conformer vs. free energy of folding were excluded— using the data for the most populated conformer. Correlations with a p-value smaller than 0.05 were found only for the following pairs: (1) number of residues vs. average lifetime of the most populated conformer ($\rho = 0.80$, $p = 0.017$), (2) number of residues vs. number of events of folding to the most populated conformer ($\rho = -0.79$, $p = 0.020$), (3) number of residues vs. average time of folding to the most populated conformer ($\rho = 0.85$, $p = 0.007$), (4) number of peptide atoms vs. average lifetime of the most populated conformer ($\rho = 0.90$, $p = 0.002$), (5) number of peptide atoms vs. average time of folding to the most populated conformer ($\rho = 0.75$, $p = 0.031$), (6) simulation length vs. total number of conformers ($\rho = 0.72$, $p = 0.042$), (7) weight of the most populated conformer vs. average lifetime of the most populated conformer ($\rho = 0.71$, $p = 0.048$), (8) average lifetime of the most populated conformer vs. average time of folding to the most populated conformer ($\rho = 0.80$, $p = 0.017$), (9) average time of folding to the most populated conformer vs. average number of conformers visited during folding to the most populated conformer ($\rho = 0.79$, $p = 0.021$), (10) total number of conformers vs. number of conformers with a 99% weight in the unfolded state ($\rho = 0.99$, $p < 0.001$), (11) total number of conformers vs. number of conformers with a 75% weight in the unfolded state ($\rho = 0.88$, $p = 0.004$), (12) number of conformers with a 99% weight in the unfolded state vs. number of conformers with a 75% weight in the unfolded state ($\rho = 0.91$, $p = 0.002$), (13) number of conformers with a 75% weight in the unfolded state vs. number of conformers with a 50% weight in the unfolded state ($\rho = 0.83$, $p = 0.011$).

The conclusions of this analysis are in general nontrivial. The number of residues of the peptide correlates with the average folding time and the average lifetime of the thermodynamically most stable conformer (Fig. 5). As a result, it anticorrelates with the number of folding events observed in the simulation. It is tempting to suggest that while the average folding time grows linearly with the number of residues (Fig. 5, middle panel), the average lifetime grows with an exponential law (Fig. 5, upper panel). A relation of this type would mean that as the length of the peptide increases, the gain in kinetic stability overcasts the loss in folding speed, making folding a more efficient process for longer

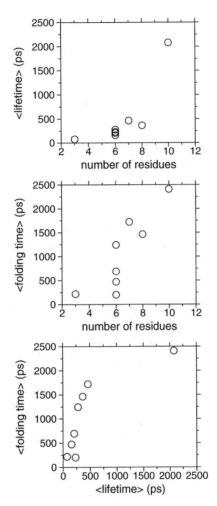

FIG. 5.   Correlations between variables: Number of residues, average lifetime of the most populated conformer, and average time of folding to the most populated conformer (see Table I).

chains. The correlation found between the folding time and the average lifetime of the thermodynamically most stable conformer points in the same direction (Fig. 5, lower panel). However, although such a relation between these two critical variables would make much sense, the data are clearly insufficient to draw a conclusion. Unexpectedly, it is the number of residues and not the number of backbone torsional degrees of freedom that correlates with the average folding time and the average lifetime of the thermodynamically most stable conformer. This implies that the extra torsional dihedral per residue in the β-peptides and the peptide of α-aminoxy acid residues does not confer additional flexibility to the backbone. Less surprisingly, the analysis shows a correlation between thermodynamic stability (weight of the most populated conformer) and kinetic stability (average lifetime of the most populated conformer) and a correlation between the average folding time and the average number of conformers visited during folding. There is a high correlation between the total number of conformers and the number of conformers having a 99% weight in the unfolded state, even if the two quantities are remarkably different. On the other hand, the number of conformers with a 50% weight does not correlate with any of the previous two. Indeed, the number of conformers adding up to a 50% weight is very similar for all the peptides studied. To confirm our hypothesis that the paradoxically—in the terms put forward by Levinthal—fast folding of peptides is made possible by the existence of only a small number of accessible conformers, a correlation between the number of unfolded conformers and the average number of conformers visited during (re)folding may be expected. This correlation is not apparent in the data set presented. That for a given peptide the average number of conformers visited during (re)folding increases with an increasing number of accessible unfolded conformers should be evident, since all the accessible (sampled) conformers are by definition part of a (re)folding pathway. When comparing different peptides, however, the ratio between the two quantities does not need to be constant, but depends on the exact thermodynamic and kinetic properties of the peptide. Thus, the correlation may not be apparent unless a large number of cases is analyzed. This point serves to stress again that the correlation analysis presented here necessarily suffers from the extreme limitations of the data set. Nevertheless, it offers a number of ideas that may be focus of future research.

D. Thermodynamic vs. Kinetic Stability

Another question of interest is whether the thermodynamically most stable conformer of a peptide is necessarily the kinetically most stable one, too. This information can be easily extracted from the simulations. Thus, for peptides A, E, F, G, and H, the thermodynamically most stable conformer is also the kinetically most stable one. For peptides B, C, and D, however, these two properties are associated with different conformers. Interestingly, the kinetically most stable conformers of

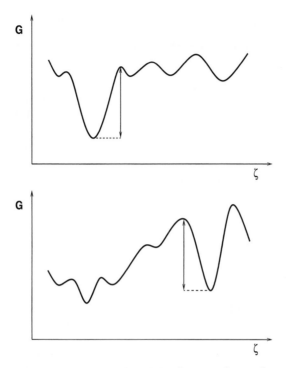

FIG. 6.   Schematic representation of the relation between thermodynamic and kinetic stability in the studied equilibria.

peptides B and D are the respective experimental conformers, which in this case are not the most populated ones. These two scenarios can be rationalized using the schematic representations shown in Figure 6. In the first case (upper panel), the highest free-energy barriers would involve the transition from the folded state to the unfolded state, while in the second case (lower panel) they would involve the transition from a low-free-energy unfolded conformer to other conformers. One determinant for the second situation may be the presence of a highly structured accessible conformer that thermodynamically is not the most stable one under the given conditions. This is the case, for example, for peptides B and D.

E.   Folding Funnels

How do the results reported in this chapter integrate in the so-called landscape perspective of polypeptide folding? The essence of this perspective is that folding occurs through multiple routes on a continuous funnel-like energy landscape (Bryngelson *et al.*, 1995; Dill and Chan, 1997). The funnel shape is meant to imply a bias toward the lowest free-energy state. The simulations reported here do not particularly support or oppose this perspective. First, it should be noted that one can explain almost any result, even seemingly contradictory ones, with this model; it all depends on how wide, how deep, and how rugged the funnel is. A walk through the vast literature on the topic clearly illustrates this point. Second, the folding funnel picture has emerged from the projection of entire systems (usually represented by minimalist models) on one or two coordinates, typically functions of the number of native contacts or other global quantities like the radius of gyration. That one can properly characterize a complex molecular system with such projections is at the very least questionable. After all, what is a funnel in a $3N + 1$ (N atoms + 1 free-energy coordinate) dimensional space?

F. Conclusions from the Literature

The literature on peptide folding by simulation techniques is less extensive than one might expect, especially when compared to that on the more complex problem of protein folding. Some representative examples that did not fall within the scope of this chapter can be found in Schaefer *et al.* (1998), Bursulaya and Brooks (1999), Wang and Sung (2000), Hansmann and Onuchic (2001), Higo *et al.* (2001), Colombo *et al.* (2002), García and Sanbonmatsu (2002). In but a handful of reports on peptide folding by molecular dynamics simulation methods has a reasonable sampling of the unfolded state been performed and an effort to describe it made. The most relevant examples come from the groups of van Gunsteren (Daura *et al.*, 1999b), Karplus (Dinner *et al.*, 1999), and Caflisch (Ferrara *et al.*, 2000), the latter two using an atomic-detail model with an implicit representation of the solvent. There is qualitative agreement between these two sets of studies in that the unfolded state is not random but dominated by a relatively small number of conformers. Nevertheless, open discussions exist regarding the validity of implicit solvent models in the context of peptide-folding simulations (Bursulaya and Brooks, 2000; Schäfer *et al.*, 2001) and the scaling of the number of conformers with chain length (Dinner and Karplus, 2001; van Gunsteren *et al.*, 2001b). An interesting study probing the conformational space of peptides was reported by Pappu *et al.* (2000), who enumerated all sterically allowed conformations for short polyalanyl chains and found that systematic local steric effects can extend beyond nearest-chain neighbors and restrict the size of the accessible conformational space significantly.

A different aspect of the unfolded state emphasized by Daura *et al.* (1999b) and Hummer *et al.* (2001) is its dependence, and that of folding pathways, on temperature. As already mentioned, the free-energy hypersurface of a system depends on molecular environment and thermodynamic conditions. The high temperatures used in some studies of protein folding result in an enlargement of the accessible conformational space and a change of the relative weights of specific conformers. As a consequence, the folding pathways populated at the relevant temperature are not necessarily maintained.

IV. Epilogue

The independent investigation of folding pathways of single peptides in the greatest detail is unlikely to shed light on the general mechanisms of peptide folding. This is because the folding properties of a peptide cannot be readily extrapolated even to similar peptides. Only when a statistically relevant sample of peptides has been analyzed—as a whole—will it be possible to extract general rules about folding. A first attempt in this direction has been presented here. Although the set of peptides studied is still far from constituting a statistically relevant sample, there is one conclusion that strongly comes out: One key factor for the observed fast folding of these peptides is the surprisingly small number of conformers accessible. And this is the case even at a high temperature.

References

Bai, Y. W., Chung, J., Dyson, H. J., and Wright, P. E. (2001). Structural and dynamic characterization of an unfolded state of poplar apo-plastocyanin formed under nondenaturing conditions. *Protein Sci.* **10**, 1056–1066.

Bellanda, M., Peggion, E., Bürgi, R., van Gunsteren, W. F., and Mammi, S. (2001). Conformational study of an Aib-rich peptide in DMSO by NMR. *J. Pept. Res.* **57**, 97–106.

Bryngelson, J. D., Onuchic, J. N., Socci, N. D., and Wolynes, P. G. (1995). Funnels, pathways, and the energy landscape of protein folding: A synthesis. *Proteins: Struct. Funct. Genet.* **21**, 167–195.

Bürgi, R., Daura, X., Mark, A., Bellanda, M., Mammi, S., Peggion, E., and van Gunsteren, W. F. (2001). Folding study of an Aib-rich peptide in DMSO by molecular dynamics simulations. *J. Pept. Res.* **57**, 107–118.

Bursulaya, B. D., and Brooks, C. L. (1999). Folding free energy surface of a three-stranded β-sheet protein. *J. Am. Chem. Soc.* **121**, 9947–9951.

Bursulaya, B. D., and Brooks, C. L. (2000). Comparative study of the folding free energy landscape of a three-stranded β-sheet protein with explicit and implicit solvent models. *J. Phys. Chem. B* **104**, 12378–12383.

Callender, R. H., Dyer, R. B., Gilmanshin, R., and Woodruff, W. H. (1998). Fast events in protein folding: The time evolution of primary processes. *Annu. Rev. Phys. Chem.* **49**, 173–202.

Chakraborty, S., Ittah, V., Bai, P., Luo, L., Haas, E., and Peng, Z. Y. (2001). Structure and dynamics of the alpha-lactalbumin molten globule: Fluorescence studies using proteins containing a single tryptophan residue. *Biochemistry* **40,** 7228–7238.

Chan, H. S., and Dill, K. A. (1998). Protein folding in the landscape perspective: Chevron plots and non-Arrhenius kinetics. *Proteins: Struct. Funct. Genet.* **30,** 2–33.

Choy, W. Y., Mulder, F. A. A., Crowhurst, K. A., Muhandiram, D. R., Millett, I. S., Doniach, S., Forman-Kay, J. D., and Kay, L. E. (2002). Distribution of molecular size within an unfolded state ensemble using small-angle X-ray scattering and pulse field gradient NMR techniques. *J. Mol. Biol.* **316,** 101–112.

Colombo, G., Roccatano, D., and Mark, A. E. (2002). Folding and stability of the three-stranded beta-sheet peptide betanova: Insights from molecular dynamics simulations. *Proteins: Struct. Funct. Genet.* **46,** 380–392.

Daggett, V. (2002). Molecular dynamics simulations of the protein unfolding/folding reaction. *Acc. Chem. Res.* **35,** 422–429. Available at http://pubs.acs.org.

Daura, X., Gademann, K., Jaun, B., Seebach, D., van Gunsteren, W. F., and Mark, A. E. (1999a). Peptide folding: When simulation meets experiment. *Angew. Chem. Int. Ed.* **38,** 236–240.

Daura, X., Gademann, K., Schäfer, H., Jaun, B., Seebach, D., and van Gunsteren, W. F. (2001). The β-peptide hairpin in solution: Conformational study of a β-hexapeptide in methanol by NMR spectroscopy and MD simulation. *J. Am. Chem. Soc.* **123,** 2393–2404.

Daura, X., Jaun, B., Seebach, D., van Gunsteren, W. F., and Mark, A. E. (1998). Reversible peptide folding in solution by molecular dynamics simulation. *J. Mol. Biol.* **280,** 925–932.

Daura, X., van Gunsteren, W. F., and Mark, A. E. (1999b). Folding-unfolding thermo-dynamics of a β-heptapeptide from equilibrium simulations. *Proteins: Struct. Funct. Genet.* **34,** 269–280.

de Alba, E., Rico, M., and Jiménez, M. A. (1997). Cross-strand side-chain interactions versus turn conformation in β-hairpins. *Protein Sci.* **6,** 2548–2560.

Dill, K. A., and Chan, H. S. (1997). From Levinthal to pathways to funnels. *Nat. Struct. Biol.* **4,** 10–19.

Dinner, A. R., and Karplus, M. (2001). Comment on the communication "The key to solving the protein-folding problem lies in an accurate description of the denatured state" by van Gunsteren *et al. Angew. Chem. Int. Ed.* **40,** 4615–4616.

Dinner, A. R., Lazaridis, T., and Karplus, M. (1999). Understanding β-hairpin formation. *Proc. Natl. Acad. Sci. USA* **96,** 9068–9073.

Dinner, A. R., Sali, A., Smith, L. J., Dobson, C. M., and Karplus, M. (2000). Understanding protein folding via free-energy surfaces from theory and experiment. *Trends Biochem. Sci.* **25,** 331–339.

Dobson, C. M., and Hore, P. J. (1998). Kinetic studies of protein folding using NMR spectroscopy. *Nat. Struct. Biol.* **5,** 504–507.

Duan, Y., and Kollman, P. A. (1998). Pathways to a protein folding intermediate observed in a 1-microsecond simulation in aqueous solution. *Science* **282,** 740–744.

Dyson, H. J., and Wright, P. E. (1998). Equilibrium NMR studies of unfolded and partially folded proteins. *Nat. Struct. Biol.* **5,** 499–503.

Evans, P. A., and Radford, S. E. (1994). Probing the structure of folding intermediates. *Curr. Opin. Struct. Biol.* **4,** 100–106.

Ferrara, P., Apostolakis, J., and Caflisch, A. (2000). Thermodynamics and kinetics of folding of two model peptides investigated by molecular dynamics simulations. *J. Phys. Chem. B* **104,** 5000–5010.

Ferrara, P., and Caflisch, A. (2001). Native topology or specific interactions: What is more important for protein folding? *J. Mol. Biol.* **306,** 837–850.

Fersht, A. R., Matouschek, A., and Serrano, L. (1992). The folding of an enzyme. 1. Theory of protein engineering analysis of stability and pathway of protein folding. *J. Mol. Biol.* **224,** 771–782.

García, A. E., and Sanbonmatsu, K. Y. (2002). α-Helical stabilization by side chain shielding of backbone hydrogen bonds. *Proc. Natl. Acad. Sci. USA* **99,** 2782–2787.

Glättli, A., Daura, X., Seebach, D., and van Gunsteren, W. F. (2002). Can one derive the conformational preference of a peptide from its CD spectrum. Submitted.

Hansmann, U. H. E., and Onuchic, J. N. (2001). Thermodynamics and kinetics of folding of a small peptide. *J. Chem. Phys.* **115,** 1601–1606.

Higo, J., Galzitskaya, O. V., Ono, S., and Nakamura, H. (2001). Energy landscape of a β-hairpin peptide in explicit water studied by multicanonical molecular dynamics. *Chem. Phys. Lett.* **337,** 169–175.

Hummer, G., Garcia, A. E., and Garde, S. (2001). Helix nucleation kinetics from molecular simulations in explicit solvent. *Proteins: Struct. Funct. Genet.* **42,** 77–84.

Kobayashi, T., Ikeguchi, M., and Sugai, S. (2000). Molten globule structure of equine beta-lactoglobulin probed by hydrogen exchange. *J. Mol. Biol.* **299,** 757–770.

Matouschek, A., Kellis, J. T., Serrano, L., and Fersht, A. R. (1989). Mapping the transition-state and pathway of protein folding by protein engineering. *Nature* **340,** 122–126.

Mirny, L., and Shakhnovich, E. (2001). Protein folding theory: From lattice to all-atom models. *Annu. Rev. Biophys. Biomolec. Struct.* **30,** 361–396.

Onuchic, J. N., LutheySchulten, Z., and Wolynes, P. G. (1997). Theory of protein folding: The energy landscape perspective. *Annu. Rev. Phys. Chem.* **48,** 545–600.

Pappu, R. V., Srinivasan, R., and Rose, G. D. (2000). The Flory isolated-pair hypothesis is not valid for polypeptide chains: Implications for protein folding. *Proc. Natl. Acad. Sci. USA* **97,** 12565–12570.

Peter, C., Daura, X., and van Gunsteren, W. F. (2000). Peptides of aminoxy acids: A molecular dynamics simulation study of conformational equilibria under various conditions. *J. Am. Chem. Soc.* **122,** 7461–7466.

Plaxco, K. W., and Dobson, C. M. (1996). Time-resolved biophysical methods in the study of protein folding. *Curr. Opin. Struct. Biol.* **6,** 630–636.

Plaxco, K. W., and Gross, M. (2001). Unfolded, yes, but random? Never! *Nat. Struct. Biol.* **8,** 659–660.

Santiveri, C. M., Jiménez, M. A., Rico, M., van Gunsteren, W. F., and Daura, X. (2002). β-Hairpin folding and stability: Molecular dynamics simulations of designed peptides in aqueous solution. Submitted.

Schaefer, M., Bartels, C., and Karplus, M. (1998). Solution conformations and thermodynamics of structured peptides: Molecular dynamics simulation with an implicit solvation model. *J. Mol. Biol.* **284,** 835–848.

Schäfer, H., Daura, X., Mark, A. E., and van Gunsteren, W. F. (2001). Entropy calculations on a reversibly folding peptide: Changes in solute free energy cannot explain folding behavior. *Proteins: Struct. Funct. Genet.* **43,** 45–56.

Seebach, D., Abele, S., Gademann, K., Guichard, G., Hintermann, T., Jaun, B., Matthews, J. L., and Schreiber, J. V. (1998). β^2- and β^3-peptides with proteinaceous side chains: Synthesis and solution structures of constitutional isomers, a novel helical secondary structure and the influence of solvation and hydrophobic interactions on folding. *Helv. Chim. Acta* **81,** 932–982.

Seebach, D., Ciceri, P. E., Overhand, M., Jaun, B., Rigo, D., Oberer, L., Hommel, U., Amstutz, R., and Widmer, H. (1996). Probing the helical secondary structure of short-chain β-peptides. *Helv. Chim. Acta* **79,** 2043–2066.

Seebach, D., Gademann, K., Schreiber, J. V., Matthews, J. L., Hintermann, T., Jaun, B., Oberer, L., Hommel, U., and Widmer, H. (1997). 'Mixed' β-peptides: A unique helical secondary structure in solution. *Helv. Chim. Acta* **80,** 2033–2038.

Seebach, D., Sifferlen, T., Mathieu, P. A., Hane, A. M., Krell, C. M., Bierbaum, D. J., and Abele, S. (2000). CD spectra in methanol of β-oligopeptides consisting of β-amino acids with functionalized side chains, with alternating configuration, and with geminal backbone substituents. Fingerprints of new secondary structures? *Helv. Chim. Acta* **83,** 2849–2864.

Shea, J. E., and Brooks, C. L. (2001). From folding theories to folding proteins: A review and assessment of simulation studies of protein folding and unfolding. *Annu. Rev. Phys. Chem.* **52,** 499–535.

Shortle, D., and Ackerman, M. S. (2001). Persistence of native-like topology in a denatured protein in 8 M urea. *Science* **293,** 487–489.

Takano, M., Yamato, T., Higo, J., Suyama, A., and Nagayama, K. (1999). Molecular dynamics of a 15-residue poly(L-alanine) in water: Helix formation and energetics. *J. Am. Chem. Soc.* **121,** 605–612.

Thirumalai, D., and Klimov, D. K. (1999). Deciphering the timescales and mechanisms of protein folding using minimal off-lattice models. *Curr. Opin. Struct. Biol.* **9,** 197–207.

Troullier, A., Reinstädler, D., Dupont, Y., Naumann, D., and Forge, V. (2000). Transient non-native secondary structures during the refolding of alpha-lactalbumin detected by infrared spectroscopy. *Nat. Struct. Biol.* **7,** 78–86.

van Gunsteren, W. F., Bürgi, P., Peter, C., and Daura, X. (2001a). The key to solving the protein-folding problem lies in an accurate description of the denatured state. *Angew. Chem. Int. Ed.* **40,** 351–355.

van Gunsteren, W. F., Bürgi, R., Peter, C., and Daura, X. (2001b). Comment on the communication "The key to solving the protein-folding problem lies in an accurate description of the denatured state" by van Gunsteren *et al.*—Reply. *Angew. Chem. Int. Ed.* **40,** 4616–4618.

Wang, H. W., and Sung, S.-S. (2000). Molecular dynamics simulations of three-strand β-sheet folding. *J. Am. Chem. Soc.* **122,** 1999–2009.

Yang, D., Qu, J., Li, B., Ng, F. F., Wang, X. C., Cheung, K. K., Wang, D. P., and Wu, Y. D. (1999). Novel turns and helices in peptides of chiral α-aminoxy acids. *J. Am. Chem. Soc.* **121,** 589–590.

A NEW PERSPECTIVE ON UNFOLDED PROTEINS

By ROBERT L. BALDWIN

Department of Biochemistry, Stanford University, Stanford, California 94305

I. INTRODUCTION

In 1968, thanks to the groundbreaking work of Charles Tanford and his co-workers, unfolded proteins were understood to be structureless random chains (Tanford, 1968), a safe starting point for investigating how the folding process begins when a denatured protein is diluted out of 6 M guanidinium chloride (GdmCl). Today there are strong suspicions that the seeds of the folding process are already present in the denatured protein. What has changed?

Tanford (1968) found that all residual structures measurable by chiral activity are eliminated in 6 M GdmCl (he used optical rotatory dispersion, the precursor of circular dichroism used today). He and co-workers also found, on studying denatured proteins in 5–7.5 M GdmCl, that the intrinsic viscosities ($[\eta]$) of 16 proteins, of chain lengths (n) varying from 26 to 1739 residues, fall on a straight-line plot of $\log[\eta]$ versus $\log n$ with slope 0.666, which is in the range expected for random chain molecules. Although the slope is larger than expected for an unperturbed random chain (0.5) with a Gaussian distribution of chain conformations, 0.666 is a reasonable value for a random chain with some excluded volume in a good solvent. The presence of excluded volume was confirmed for several proteins by using osmotic pressure measurements to determine their second virial coefficients. The existence of a common plot over this large range of chain lengths reinforced the premise that these proteins are truly unfolded in 5–7.5 M GdmCl. Tanford and co-workers found some residual structure in heat-denatured proteins, using optical rotatory dispersion, and advised denaturing a protein in 6 M GdmCl in order to make kinetic studies of its refolding pathway starting from a structureless chain molecule. Tanford's measurements stand unchallenged today, and his interpretation of the properties of denatured proteins in 6 M

*ADVANCES IN
PROTEIN CHEMISTRY, Vol. 62*

GdmCl likewise stands; however, the study of the folding process has taken a new turn, one not foreseen in 1968.

II.   Hydrophobic Clusters in Urea-Denatured Proteins

It is now known that one class of residual structures, hydrophobic clusters of nonpolar side chains, withstands denaturation in concentrated urea solutions, and possibly also in 6 M GdmCl, where NMR studies are more difficult to make because of a loss of signal strength. In urea solutions, hydrophobic clusters can be observed in denatured proteins by various NMR probes, such as the nuclear Overhauser effect (Neri *et al.*, 1992), chemical shifts of aromatic side chain protons (Saab-Rincon *et al.*, 1996), or transverse relaxation rates of amide groups in the polypeptide backbone (Schwalbe *et al.*, 1997). In reduced, unfolded hen lysozyme, six hydrophobic clusters are detected by measurements of transverse relaxation rates, and a mutagenesis study reveals that the six clusters form a network connected by cooperative interactions (Klein-Seetharaman *et al.*, 2002). However, hydrophobic clusters are not usually believed to guide the folding process to the native structure unless the nonpolar cluster is so arranged that it stabilizes a secondary structure, an α-helix or a β-hairpin, found in the native protein. So where does the renewed interest in denatured proteins come from? A few main factors contribute to the change in perspective.

III.   Rationale for Studying Denatured Proteins in Water

The commonly held view of the folding process has changed, especially in the last few years. In 1968, a small protein folding by the two-state mechanism ($D \Leftrightarrow N$) was supposed to exist detectably at equilibrium in just two forms: as the denatured protein D and the native protein N. This restriction applied both to proteins in the presence of denaturant (6 M GdmCl or 8 M urea) and those in the presence of water. Today it is known that secondary structures, α-helices and β-hairpins, are often stable in water. And, for reasons discussed below, it has become desirable to investigate the structures of denatured proteins in water. A few years ago, the transition state for folding was thought to contain only a small nucleus of native-like structure. The existence of a correlation between folding rate and contact order (Plaxco *et al.*, 1998) has changed that view, because it implies that the topology of the native state is present in the transition state, which is now regarded as a distorted native structure. The nucleation–condensation model for folding has become the extended nucleus model (Fersht, 2000). Likewise, the denatured state in water is now commonly thought to be an embryonic folding intermediate.

Why should protein folders now want to study the denatured state in water? There are two main reasons. First, simulations of the folding process are providing increasingly valuable insights into the folding process, and simulations are made in water, not in 6 M GdmCl. The starting point for a simulation is usually an extended β-strand, rather than an attempt at simulating a structureless, random chain. The local structures formed rapidly at the start of folding are accessible to simulation. The second reason is that the kinetics of α-helices and β-hairpins can now be measured by fast-reaction methods (Eaton *et al.*, 2000; Dyer *et al.*, 1998). Thus, theory and experiment are converging on early folding events in this fast time range (10–100 ns).

IV. Stiffness of the Random Chain

A continuing puzzle has been just how stiff the unfolded polypeptide chain is. Attention was focused on the problem by the early measurements (Brant and Flory, 1965a) of Flory's characteristic ratio for some synthetic polypeptides, measured both in aqueous solution (polyglutamate, polylysine) and in an organic solvent (poly-β-benzyl aspartate). Their measured values were nearly twice as large as the values predicted for free rotation about backbone bonds. For synthetic flexible-chain polymers, the values predicted for free rotation are generally in good agreement with measured values (Flory, 1969). Brant and Flory (1965b) proposed that the answer to the puzzle lies in strong dipole–dipole interactions among the peptide NH and CO groups. In a single peptide unit, the peptide NH and CO dipoles are antiparallel in the extended β-strand conformation but parallel in the more compact α-conformation (see Avbelj and Moult, 1995). Consequently, the dipole–dipole interactions strongly favor the β- over the α-conformation for short runs of residues in a single conformation, too short to form an α-helix stabilized by peptide H-bonds.

The measurements of chain stiffness of denatured proteins are made in the presence of a strong denaturant, such as 8 M urea or 6 M GdmCl, in which peptide H-bonds are weak and peptide helices unfold (Scholtz *et al.*, 1995; Smith and Scholtz, 1996), and the possible presence of α-helices or β-hairpins is not an issue in these denaturants. The careful and thorough measurements of intrinsic viscosities made by Tanford and co-workers (1968), discussed above, yield a substantially lower estimate for chain stiffness than the work of Flory and co-workers. A comparison is made by Tanford (1968) between the proportionality coefficient relating the mean-square end-to-end distance of the unperturbed chain to the number of residues. The proportionality coefficient given by Flory is twice as large as the one given by Tanford. One possible explanation is simply that synthetic polypeptides in aqueous solution have different stiffness from denatured proteins in 6 M GdmCl or 8 M urea. [Tanford (1968) gives limited data showing that viscosities in 8 M urea are closely similar to those in 6 M GdmCl.] Another possible explanation lies in the glycine contents of natural proteins: The glycyl peptide bond is predicted by Flory and co-workers to be substantially more flexible than the peptide bonds of amino acids with Cβ carbon atoms, and limited data with glycine–glutamate copolymers support this proposal (Miller *et al.*, 1967).

Modern work on the stiffness problem has been limited. Zhou (2002) has produced a simple phenomenological model of the excluded-volume effect as a function of chain length that reproduces both Tanford's (1968) viscosity data and more recent measurements of the Stokes radius, obtained chiefly from translational diffusion coefficients. The latter can now be measured conveniently for denatured proteins in concentrated urea solutions by pulsed field gradient NMR (Wilkins *et al.*, 1999). Zhou's model was generated with sequences of backbone ϕ,ψ angles chosen at random from a library of loop regions of protein structures, with provision for excluded volume modeled by hard spheres centered at the Cα atoms. The distances between residues were found to have approximately Gaussian distributions. The actual dimensions (i.e., radius of gyration) of unfolded proteins can now be measured directly by low-angle X-ray or neutron scattering (for review, see Millett *et al.*, 2002), but scattering measurements made in concentrated urea or GdmCl solutions suffer from low contrast (see Wilkins *et al.*, 1999).

V. Preferred Backbone Conformations

Tanford's (1968) work on the chain dimensions of denatured proteins fixed the view that they are structureless random chain molecules. Moreover, Flory's "isolated-pair" hypothesis (Flory, 1969) about denatured protein chains was widely accepted until recently. According to this hypothesis, the ϕ,ψ backbone angles of one residue are independent of those of the residues preceding and following it. Recently, an exhaustive enumeration study of the backbone conformations in short alanine peptides, given by random sampling of ϕ,ψ space, found that the isolated-pair hypothesis is invalid in a major region of ϕ,ψ space encompassing particularly the right-hand α-conformation (Pappu *et al.*, 2000).

The authors found that the isolated-pair hypothesis holds fairly well in the region around the β-strand conformation.

The concept of preferred backbone conformations in denatured proteins began to be taken seriously when the ϕ,ψ angles of individual residues in loop regions of protein structures, outside the regular secondary structures, were found to be dominated by the same β- and α-conformations found in β-sheets and α-helices, although regular secondary structures were excluded (Kleywegt and Jones, 1996; Smith *et al.*, 1996). The left-hand polyproline II conformation was also found to be a major backbone conformation in loop regions (Adzhubei and Sternberg, 1993; Kleywegt and Jones, 1996).

These observations provided an impetus for finding methods of determining backbone conformations in "unstructured" peptides. An additional impetus was finding that the CD spectra of denatured proteins and unstructured peptides suggest some form of regular backbone structure related to polyproline II [see Tiffany and Krimm (1968), and Woody (1992), and references therein]. Recently, studies using sophisticated methods of optical spectroscopy found that trialanine has the polyproline II backbone conformation (Wouterson and Hamm, 2000, 2001; Schweitzer-Stenner *et al.*, 2001), and so does a blocked dialanine peptide analyzed by NMR with the aid of residual dipolar couplings (Poon *et al.*, 2000). Moreover, NMR analysis of the backbone conformations of a seven-residue alanine peptide gives the dominant conformation at $0°C$ as polyproline II but also demonstrates a gradual thermal transition to β-strand with increasing temperature (N. Kallenbach, personal communication, 2002). Rucker and Creamer (2002), using circular dichroism, find that a seven-residue ionized lysine peptide likewise has predominantly the polyproline II structure.

These results strongly suggest that "unstructured" peptides have definite backbone conformations and that the concept of a denatured protein as a structureless random chain breaks down when backbone conformations of individual residues are described, although the random chain concept may still be useful when describing the overall chain conformation.

What is the origin of the energy difference between the polyproline II and β-strand backbone conformations? Brant and Flory (1965b) emphasize the important roles of steric clash, dipole–dipole interactions (see also Avbelj and Moult, 1995), and the torsional potentials governing rotation about the backbone ϕ,ψ angles (see also Flory, 1969). An *ab initio* quantum mechanics study (Han *et al.*, 1998; see also references therein to earlier work) finds that solvation by water is important. The authors examine the predicted stabilities of eight conformers of

N-acetylalanine-N-methylamide with four water molecules H-bonded to the peptide NH and CO groups and compare the results with no water present. They find that, with water molecules present, the polyproline II backbone has the lowest energy, followed first by β-strand and then by the right-hand α-conformation. Without water, neither polyproline II nor right-hand α is a stable conformation, and β-strand becomes the most stable backbone. Molecular dynamics simulations of an eight-residue alanine peptide in explicit water indicate that bridging water molecules are important in stabilizing the polyproline II structure (Sreerama and Woody, 1999).

Using only the repulsive term of a van der Waals potential, Pappu and Rose (2002) examine the effect of steric clash, or excluded volume, on preferred backbone conformations of blocked alanine peptides of varying lengths, including a seven-residue peptide. They find that excluded volume is sufficient to give preferred backbone conformations, and they also find that polyproline II is the most preferred. Their work has interesting connections both to the simulation results of Srinivasan and Rose (1999), who find native-like patterns of backbone conformations in some denatured proteins in water, and to the statistical analysis by Shortle (2002) of the relation between sequence and backbone conformation in loop regions of proteins. Shortle suggests that steric clashes between side chains and backbone lead to the long-range order, observed by means of residual dipolar couplings, in denatured staphylococcal nuclease in 8 M urea (Shortle and Ackerman, 2001).

Experimental and theoretical approaches are now converging on the polyproline II backbone conformation as the most stable structure for short alanine peptides in water. It becomes of urgent importance to determine the energy differences between polyproline II and other possible backbone conformations, as well as to determine how amino acid composition and sequence affect backbone conformation.

ACKNOWLEDGMENTS

I thank Franc Avbelj, George Rose and Huan-Xiang Zhou for discussion.

REFERENCES

Adzhubei, A. A., and Sternberg, M. J. (1993). *J. Mol. Biol.* **229,** 472–493.
Avbelj, F., and Moult, J. (1995). *Biochemistry* **34,** 755–764.
Brant, D. A., and Flory, P. J. (1965a). *J. Am. Chem. Soc.* **87,** 2788–2791.
Brant, D. A., and Flory, P. J. (1965b). *J. Am. Chem. Soc.* **87,** 2791–2800.
Dyer, R. B., Gai, F., Woodruff, W. H., Gilmanshin, R., and Callender, R. H. (1998). *Acc. Chem. Res.* **31,** 709–716.

Eaton, W. A., Muñoz, V., Hagen, S. J., Gouri, S. J., Lapidus, L. J., Henry, E. R., and Hofrichter, J. (2000). *Annu. Rev. Biophys. Biomol. Struct.* **29,** 327–359.

Fersht, A. R. (2000). *Proc. Natl. Acad. Sci. USA* **97,** 1525–1529.

Flory, P. J. (1969). *Statistical Mechanics of Chain Molecules.* Wiley, New York.

Han, W. G., Jalkanen, K. J., Elstner, M., and Suhai, S. (1998). *J. Phys. Chem. B* **102,** 2587–2602.

Klein-Seetharaman, J., Oikawa, M., Grimshaw, S. B., Wirmer, J., Duchardt, E., Ueda, T., Imoto, T., Smith, L. J., Dobson, C. M., and Schwalbe, H. (2002). *Science* **295,** 1719–1722.

Kleywegt, G. J., and Jones, T. A. (1996). *Structure* **4,** 1395–1400.

Miller, W. G., Brant, D. A., and Flory, P. J. (1967). *J. Mol. Biol.* **23,** 67–80.

Millett, I. S., Doniach, S., and Plaxco, K. W. (2002). Submitted.

Neri, D., Billeter, M., Wider, G., and Wüthrich, K. (1992). *Science* **257,** 1559–1563.

Pappu, R. V., Srinivasan, R., and Rose, G. D. (2000). *Proc. Natl. Acad. Sci. USA* **97,** 12565–12570.

Pappu, R. V., and Rose, G. D. (2002). *Protein Sci.,* In press.

Plaxco, K. W., Simons, K. T., and Baker, D. (1998). *J. Mol. Biol.* **277,** 985–994.

Poon, C. D., Samulski, E. T., Weise, C. F., and Weisshaar, J. C. (2000). *J. Am. Chem. Soc.* **122,** 5642–5643.

Rucker, A. L., and Creamer, T. P. (2002). *Protein Sci.* **11,** 980–985.

Saab-Rincon, G., Gualfetti, P. J., and Matthews, C. R. (1996). *Biochemistry* **35,** 1988–1994.

Scholtz, J. M., Barrick, D., York, E. J., Stewart, J. M., and Baldwin, R. L. (1995). *Proc. Natl. Acad. Sci. USA* **92,** 185–189.

Schwalbe, H., Fiebig, K. M., Buck, M., Jones, J. A., Grimshaw, S. B., Spencer, A., Glaser, S. J., Smith, L. J., and Dobson, C. M. (1997). *Biochemistry* **36,** 8977–8991.

Schweitzer-Stenner, R., Eker, F., Huang, Q., and Griebenow, K. (2001). *J. Am. Chem. Soc.* **123,** 9628–9633.

Shortle, D. (2002). *Protein Sci.* **11,** 18–26.

Shortle, D., and Ackerman, M. S. (2001). *Science* **293,** 487–489.

Smith, J. S., and Scholtz, J. M. (1996). *Biochemistry* **35,** 7292–7297.

Smith, L. J., Bolin, K. A., Schwalbe, H., MacArthur, M. W., Thornton, J. M., and Dobson, C. M. (1996). *J. Mol. Biol.* **255,** 494–506.

Sreerama, N., and Woody, R. W. (1999). *Proteins Struct. Funct. Genet.* **36,** 400–406.

Srinivasan, R., and Rose, G. D. (1999). *Proc. Natl. Acad. Sci. USA* **96,** 14258–14263.

Tanford, C. (1968). *Adv. Protein Chem.* **23,** 121–282.

Tiffany, M. L., and Krimm, S. (1968). *Biopolymers* **6,** 1767–1770.

Wilkins, D. K., Grimshaw, S. B., Receveur, V., Dobson, C. M., Jones, J. A., and Smith, L. J. (1999). *Biochemistry* **38,** 16424–16431.

Woody, R. W. (1992). *Adv. Biophys. Chem.* **2,** 37–79.

Wouterson, S., and Hamm, P. (2000). *J. Phys. Chem. B* **104,** 11316–11320.

Wouterson, S., and Hamm, P. (2001). *J. Chem. Phys.* **114,** 2727–2737.

Zhou, H.-X. (2002). *J. Phys. Chem. B.* **106,** 5769–5775.

AUTHOR INDEX

A

Aamouche, A., 116
Abaturov, L. V., 208, 216, 218
Abele, S., 348
Abeygunawardana, C., 5, 6, 8, 253, 312, 315, 333
Abkevich, V. I., 41, 46
Abramov, V. M., 34, 245, 253
Acampora, G., 313
Acharya, K. R., 220
Ackerman, E. J., 42
Ackerman, M. S., 7, 11, 13, 108, 163, 233, 257, 258, 265, 284, 342, 366
Acquaviva, R., 38
Adler, A. J., 168
Adzhubei, A. A., 65, 66, 70, 77, 231, 264, 268, 269, 274, 280, 365
Adzhubei, I. A., 77
Agard, D. A., 67
Agashe, V. R., 258
Agbaje, I., 126
Ahearn, J. M., Jr., 214
Aizawa, S., 38
Akasaka, K., 46, 243, 244, 245, 247, 248, 249, 250, 251, 252
Albrecht, K., 210
Alexander, P., 312
Alexandrescu, A. T., 5, 6, 8, 253, 312, 315, 321, 333
Al-Hashimi, H. M., 10, 13
Allison, L. A., 214
Almlof, T., 212
Alonso, D. O., 243, 259
Altieri, A. S., 323
Altmann, H., 38, 212
Altschul, S. F., 35
Ambrose, E. J., 125
Amemiya, Y., 243, 244, 247, 254
Amouche, A., 115, 116, 124

Amzel, L. M., 276
Amzel, M., 241
Andersen, N. H., 313, 323
Anderson, A. G., 173, 175
Anderson, D. J., 64
Anderson, J. M., 169
Andreu, D., 38
Andreu, J. M., 38
Andries, J. C., 169
Anfinsen, C. B., 4, 313
Antonic, J., 135
Apostolakis, J., 356
Arai, M., 51, 80, 216, 226, 227, 228
Arai, S., 247, 250, 251
Aranda, C., 38
Archer, D. B., 73
Arcus, V. L., 163, 312
Arico-Muendel, C. C., 71
Arnold, G. E., 32, 33
Arnsdorf, M. F., 83
Arrowsmith, C. H., 336, 337
Arunkumar, A. I., 264
Asadourian, A., 125
Asher, S. A., 119, 126, 135, 141
Ashvar, C. S., 115, 116, 124
Atkins, P. W., 54
Aune, K. C., 163
Aurora, R., 264, 265, 273, 280, 281
Ausio, J., 44, 46, 210, 253
Austin, R. H., 108, 242, 258
Avbelj, F., 284, 363, 365
Aviles, F. J., 38
Axhausen, P., 102
Ayed, A., 336

B

Backmann, J., 115
Bacquet, R., 38

M

SUBJECT INDEX

A

AAMA, *see* N-Acetyl-ʟ-alanine
  N'-methylamide
AAT, *see* Atomic axial tensors
Ab initio techniques, peptides, 131, 137–138
N-Acetyl-ʟ-alanine *N'*-methylamide,
  172–175
Acid denaturation
  apomyoglobin, 327–331
  nuclease Δ131Δ, 13
Alexa488, unfolded protein FCS studies,
  104–105
α-helix, Raman optical activity, 62–64
Alzheimer's disease, protein misfolding, 82
Amino acid sequence
  disordered protein encoding, database
    comparisons
    composition comparisons, 28–32
    evolutionary characteristics, 34–37
    ordered/disordered proteins, 28
    overview, 27
    sequence attributes, 32–34
  disordered protein encoding predictions,
    26–27
  ordered/disordered protein prediction
    overview, 37
    PONDR accuracies, 40–42
    PONDRs, 39
Amyloid fibril formation, 82–84
ApoMb, *see* Apomyoglobin
Apomyoglobin
  circular dichroism, 223–225
  equilibrium NMR studies
    acid-unfolded apoMb, 327–331
    denaturant-unfolded apoMb, 327
    energy landscape mapping, 332–333
    folded apoMb at pH 6, 332
    molten globule at pH 4, 331–332
    overview, 324–327

Apoplastocyanin, NMR studies, 334
APT, *see* Atomic polar tensors
Atomic axial tensors
  IR and VCD simulation, 124
  unfolded peptides, 137
Atomic polar tensors
  IR and VCD simulation, 124
  unfolded peptides, 137

B

Baboon α-lactalbumin, 220
Backbone solvation, polyproline II helix,
  274–277
Beta meander, nuclease Δ131Δ, molten
  structure, 7–8
β-sheet, Raman optical activity, 64–65
Biomolecules, chiral, Raman optical activity,
  57–58
Biophysics, *see* Molecular biophysics
BLA, *see* Bovine α-lactalbumin
Bovine α-lactalbumin
  CD, 216, 218, 226–227
  IR and VCD, 150–151
  Raman optical activity, 71–73
BSE, protein misfolding, 82

C

Carbonic anyhydrase, circular dichroism,
  221–223
α_s-Casein, circular dichroism, 209
β-Casein, Raman optical activity, 78–79, 85
CD, *see* Circular dichroism
Cereal seed proteins, circular dichroism,
  212–214
Chemical association model, x_0, hydration,
  301–302

389

W

X

V